TABLET TO JOSEPH ASPDIN
(Accredited Inventor of Portland Cement)

PORTLAND CEMENT TECHNOLOGY

by

J. C. WITT, M.E., Ph.D.

Consulting Engineer
Chicago

SECOND EDITION

CHEMICAL PUBLISHING COMPANY, INC.

New York

1966

Portland Cement Technology

ISBN: 978-0-8206-0014-7

Chemical Publishing Company:
www.chemical-publishing.com
www.chemicalpublishing.net

First edition:

Chemical Publishing Company, Inc. - New York 1966
Second Impression:
Chemical Publishing Company, Inc. - 2011

Printed in the United States of America

TO
THE EVER-ACCELERATING ADVANCEMENT
OF THE
PORTLAND CEMENT INDUSTRY

Technology is not the only factor
in the advancement of portland cement
but it is the most important one

Preface to Second Edition

"What's past is prologue"—Shakespeare

The first edition of *Portland Cement Technology* summarized the development of the cement industry for well over a century, preceded by history from prehistoric times. This edition represents an additional period of approximately twenty years.

In such a short time, many changes are to be anticipated in an active industry. But in portland cement these changes came about so gradually that they may not become apparent until before-and-after data are compared.

In 1947, the production was 186,519,347 barrels. The rising annual production crossed the 300-million mark in 1956. There have been no major changes in raw materials, nor in manufacturing procedures, affecting the industry as a whole.

In common with industry in general, the manufacture of cement encounters the so-called lag between the results of research and commercial application. This regrettable lag has long been recognized, but it continues. Many factors are involved—such as inertia, competition of existing equipment and practices, and over-conservatism. Frequently, the improvement of a product requires an increase in costs of production, but occasionally a decrease can be anticipated. One wonders how rapidly the industry would progress if the lengths of these lags could be reduced to zero.

Two points stand out. It is evident that the importance of the cement industry is not only increasing, but also accelerating, and this is gratifying. The other point is that the versatility of concrete not only suggests new departures in construction, but also permits more economy than would otherwise be obtainable.

To restrict this edition to one volume without making it unwieldy, some material has been omitted. Unless otherwise mentioned, statements refer to the manufacturing of cement in the United States, and to one type of cement only. Blended cements are not included. If a material other than cement, water, and aggregate is to be a component of concrete, this material can be handled separately. If anything of the kind is mixed with cement before shipment, some such marking as Cement Plus . . . [a given item] would be desirable.

Five types of chemical admixtures are under consideration: (a) water-reducing; (b) retarding; (c) accelerating; (d) water-reducing and retarding; (e) water-reducing and accelerating. In fact, the subject of chemical admixtures is becoming so extensive and complex, and there are so many differences of opinion, that a separate volume could be written on the subject.

Cement is a low-priced, heavy product that can be shipped only within a short radius. Someone has said ". . . while it sells cheaply, it moves expensively." The detail of control cannot approach that of pharmaceuticals, nor that of many other expensive products. As Benjamin Franklin Affleck, one of the immortals in the cement industry, used to say: "Some purchasers expect tailor-made suits at the price of ready-made."

Manufacturers of cement must understand the basic principles of concrete in order to make satisfactory products. Information on the manufacturing of cement can be very helpful to producers of concrete. In this volume, an effort has been made to present information useful to everyone interested in the manufacture and use of cement.

Unless otherwise noted, the specifications are from the American Society for Testing and Materials. An example is "ASTM Designation C 183–61T," in which C signifies cementitious material; *183*, the number of the Specification; *61*, the date of the adoption, or of the latest revision. (*T*, if present, indicates that the item is tentative, rather than standard.)

Formerly, a complete set of ASTM Standards was published every three years. Now a complete set appears annually. Some standards, either in full or in part, are presented here in the interest of general information, or to serve as ready references.

No specifications can be considered permanent. Committees, subcommittees, and sub-subcommittees are always active in review and revision. For example. *Standard C 150*, first published in 1941, was revised twelve times in twenty years. Any user of an ASTM Specification should be careful to obtain the latest revision.

As the preparation of this second edition of *Portland Cement Technology* has taken considerable time, and as the annual reports of various organizations are frequently delayed a year or more, many figures mentioned here are not necessarily those of the date of publication.

May 1965   J. C. Witt

Preface to First Edition

With my first responsible affiliation in the cement industry came the suggestion that I undertake the serious study of cement. At the technology department of the Carnegie library in Pittsburgh, a request for books on cement manufacture brought five or six. Scanning all of them, I selected *Portland Cement*, by Richard K. Meade, as the best suited to my purpose, and it has been one of my favorite reference books ever since.

Portland Cement was the only general book on the manufacture of cement that had been written and published in the United States. When Meade passed on in 1930, the cement industry suffered a great loss. No more of his valuable books were to appear. Although his text remains a book of value, it does not cover the developments of the industry during the most active years of its history. Four years ago, at the request of the Chemical Publishing Company, I undertook the revision of Meade's book, then revision with some additions, and finally the writing of a new book on the subject. As the work progressed, a change in title suggested itself.

Portland Cement Technology is based on years of experience in many phases of the technology of portland cement and some other materials—prospecting for raw materials, plant operation, engineering, research, development, consultation, administration. It is a pleasure to acknowledge my indebtedness to the following persons, with nearly all of whom I have been associated, for their kindness and helpfulness. Each is a specialist in some phase of the subject matter of *Portland Cement Technology*, and each has reviewed one or more of its chapters:

A. C. Fieldner, Chief
Fuels and Explosives Service, Bureau of Mines
United States Department of the Interior

A. T. Goldbeck, Engineering Director
National Crushed Stone Association

H. J. Love, Managing Director
National Slag Association

J. L. McConnell, Consulting Engineer
J. L. McConnell and Associates

M. C. Miller, Vice President
Hawkeye Portland Cement Company

W. C. Morse, Director
Mississippi Geological Survey

C. W. Nash, Associate Professor of Electrical Engineering
Illinois Institute of Technology

J. C. Peebles, Dean of Engineering
Illinois Institute of Technology

C. M. Price, Consulting Engineer
(Formerly Chemical Engineer, Calaveras Cement Company)

S. J. Robison, Engineer
Western Precipitation Corporation
(Formerly, Chief Engineer, Universal Atlas Cement
Company)

N. M. Stineman, Engineer
Veterans' Administration
(Formerly, Editor of *Concrete*)

T. J. Sullivan, Western Service Manager
Babcock and Wilcox Company

C. M. Tuttle, Traffic Manager
Marquette Cement Manufacturing Company

Stanton Walker, Director of Engineering
National Sand and Gravel Association, and
National Ready Mixed Concrete Association

G. W. Ward, Chairman
Section of Inorganic Chemistry
Midwest Research Institute

November 1946   *J. C. Witt*

Contents

Chapter 1

INTRODUCTION

Select some raw materials, proportion them, grind, heat, cool, and grind again. Reduced to simplest terms, this is the process for manufacturing portland cement. Mixing the cement, thus produced, with aggregates and water results in mortar or concrete, to obtain which cement is manufactured. What raw materials should be selected? How are they proportioned? How finely are they ground? To what temperature are they heated? The attempt is made to answer these and many other somewhat similar questions, on the basis of the information now available in the industry. The acquisition of this information has resulted from the manufacture of portland cement for well over a century, from research, and from numerous trial and error procedures.

This is a practical text and reference book. It is written for persons, technical and non-technical, inside and outside the cement and concrete industries, and in fact for everyone sufficiently interested in cement to reach for a book on the subject. Theoretical considerations have been kept at the minimum, consistent with a clear understanding of the subject.

In conformity with the same policy, the number of formulas and tables has been restricted in the interest of continuity of the text. To some extent, references have been substituted for them. There would have been no difficulty in writing a book that is nearly all formulas and tables, with only sufficient other material to hold them together. There is available such a large mass of data of all kinds that any type of book could be prepared.

1

The treatment of the various chapters is based primarily on materials, rather than on plant equipment. Equipment is designed to carry materials through various manufacturing processes. Materials are not selected to be handled by equipment units. There is some tendency to confuse cause and effect.

Most statements containing numerical data are intended as examples only—not means, averages, maxima, nor minima. One million British thermal units may be mentioned as the heat required to produce a barrel of clinker. This is simply an example, and it does not in any way imply the average for the industry. Incidentally, English, rather than metric, units are employed, unless otherwise stated. Temperatures are expressed in Fahrenheit degrees. For some conversion factors, see Chapter 24.

Most of the information of any industry, as it develops, remains unwritten. A portion finds its way into the literature, which serves as the depository. Any technical book, no matter how extensive, is largely a résumé of the literature, although it may contain much material that has not appeared previously. In this book, I have drawn from my published papers, unpublished manuscripts, notes, and calculations. It has not been necessary to consult so many references as otherwise would have been required. In no sense can it be claimed that *Portland Cement Technology* is complete. It is said that Izaak Walton (1593–1683), who wrote the famous treatise *The Compleat Angler*, continued to add to its completeness for a quarter of a century.

When the book *I Write as I Please* * was announced, something about the title appealed to me because the statement is applicable to my own writing. Many points about the manufacture and use of cement are highly controversial, and on some points I find myself with the minority. However, I assume individual responsibility for all my statements. Some may be challenged, and some may be found incorrect.

There are many bases on which the manufacture of cement may be discussed. It is a chemical industry because a mix-

---

* Walter Durante, *I Write as I Please*, The World Publishing Co., New York, 1942.

ture of chemical compounds is synthesized. Every step in the process involves chemical reactions or preparation for chemical reactions. It is a process-engineering industry, because every step may be considered a unit operation. (Conversely there are examples of almost every unit operation in the industry, and in some individual plants.) It is also a mechanical-engineering industry, because mechanical equipment is essential. Without engineering, cement would be so expensive that it could be found only in museum exhibits. The annual production of hundreds of millions of barrels would be impossible.

As a matter of fact, practically every profession and every trade are involved, directly or indirectly, in the industries of cement and concrete.

Manufacturing industries may be divided into two groups: chemical and non-chemical. The first group includes all industries in which the conversion of raw materials into finished products requires chemical changes. Portland cement is such a product. There are relatively few industries in the second group. Some examples are such mechanical processes as woodworking, sheet metal working, and weaving, in which the chemical compositions of the raw materials and of the finished products are identical. The increased value of the product over that of the raw materials is due to such operations as cutting, mixing, and fabricating. The non-chemical industries are closely related to the chemical industries, and in many respects are dependent upon them. For examples, most items made of wood receive protective coatings, which are made by chemical processes; sheet metal plants depend on chemical industries for their raw materials; and most textiles are dyed before they are marketed.

The chemical industries may be placed in two divisions. The members of the first had appeared before science had been developed to such an extent that it could be used for directing and controlling the processes. Further developments were retarded until scientific methods could be applied. The members of the second division did not appear until much progress had been made in science. Raw materials, resulting from other highly developed processes, and technique of a high order were also necessary. Portland cement may be said to occupy an intermediate position, but in most respects it belongs to the first division. However, very crude analytical methods were employed early in the development of portland cement.

Cement is manufactured by either the dry or the wet process. The principal difference is in the condition in which the raw materials are ground, but most departments of a plant are affected to some extent. Statements in which the process is not mentioned apply to either process.

The industry has a rather elaborate terminology. Some terms are common to industry in general, and others have been coined as occasions have arisen. In some cases, more than one term is used for the same item. The names for materials, departments, processes, and personnel vary in the industry and may vary from plant to plant of the same company. In this book, multiple terms have been avoided. In order that the following chapters may be understood with the least effort, the meaning of some of the most important terms is explained in advance.

RAW MATERIALS DEPOSITS: Most raw materials are natural substances, such as limestone and clay. Deposits of these have resulted from geological processes. The term, *raw materials deposits*, is applied to raw materials that have not been moved from their original locations.

RAW MATERIALS: After raw materials have been quarried or mined, they are placed into process. It is considered that they retain their identity as separate materials until the raw mix is produced.

RAW MIX: This is the product obtained when the raw materials have been proportioned and the mixture has been subjected to the final grinding operation. Each material retains its original chemical composition.

SLURRY: In the wet process, the raw mix carries water, averaging 30 to 40 per cent of the total weight. This product is commonly called *slurry*.

KILN FEED: In both processes, the raw mix may undergo some changes on its way to the kiln. Adjustments in composition and blending are examples. The term, *kiln feed*, applies to the mixture as it enters the kiln.

Clinker: Clinker is the product discharged from the kiln. It results from chemical changes in the kiln feed. There are also some physical changes.

Cement: This is the product obtained when clinker is ground with a relatively small quantity of a retarder, usually gypsum.

In the United States, the unit weight of cement is the *barrel*, 376 pounds net. This is equivalent to 4 sacks of 94 pounds each. The volume of a barrel is approximately 4 cubic feet. Operating and accounting figures are based on this unit. For example, the fuel requirements are expressed as pounds of coal, gallons of oil, or cubic feet of gas per barrel; the power, as kilowatt hours per barrel; and the labor, as man-hours per barrel. The unit originated when cement was shipped in barrels, a practice that is now obsolete. In some foreign countries, quantities of cement and fuel are expressed in tons, or in metric tons. For this reason, the comparison of foreign practice with our own involves tedious calculations. Chapter 24 contains some factors to facilitate conversion of values.

In comparison with manufacturing processes in general, cement raw materials have some advantages and some disadvantages. Among the advantages may be mentioned the wide distribution of suitable materials, and the facility by which they may be quarried, or mined, and transported to the plants. At this point, however, some of the disadvantages appear. In most chemical reactions, the reacting particles are submicroscopic in size. Although kiln feed is ground fine, the size is relatively very large. The object of crushing and grinding is to increase the contact surface of the particles. To produce a barrel of cement, it is necessary to grind approximately 1000 pounds of solids to a 200-mesh fineness. This requires a great deal of power; in fact, most of the power of the whole industry.

The relation of the cement industry to the manufacturing industry, in general, could be discussed at great length. Equipment and technique for mining, quarrying, materials handling, crushing, grinding, dust collecting, power generating, and power transmitting have been adopted from other industries. On the other hand, equipment and technique developed by the cement industry, or developed primarily for it, have been widely adopted outside the industry. Pulverized coal and a system for transporting pulverized materials may be mentioned as important examples. Still greater contributions of the cement industry to industry in general are concrete, and various concrete products, for the construction of plants, plant equipment, and unlimited transportation facilities to and from the plants. In this book, operations are explained as they take place under normal conditions of supply, labor, and transportation.

Chapter 2

HISTORY OF THE CEMENT INDUSTRY

The development of the cement industry, in common with that of many other large industries, has come about from a series of incidents that started many centuries ago. The history is by no means continuous, and it has been customary to fill in the breaches with conjecture. This has been the practice so long that some of the boundaries between fact and conjecture are not well defined. It is known, however, that there has been a successive development of active materials, used for cementing other materials that are essentially inactive.

Of the various important factors that have been directly or indirectly responsible for this development, five may be mentioned:

1. Recognition of some of the imperfections of cementitious materials
2. Advancements in science, which have suggested explanations of these imperfections and procedures for developing better cementitious materials
3. The development of mechanical equipment, which has made it possible to manufacture products in large quantities without prohibitive hand labor
4. Developments in other industries
5. The advancement of technology in general, which has suggested, and in some cases required, better and better cementitious materials.

7

Cementitious Materials

A plastic is a material which may acquire plasticity temporarily. During the period of plasticity, the material may be pressed, cast, molded, or in some other manner formed into objects of definite shape. When the plasticity disappears, the material hardens or sets, and the shape is retained. There are two classes of plastics, the organic and the inorganic. However, the term plastic is usually restricted to the organic group.

A plastic that may be employed to cement one or more other materials, known as fillers or aggregates, is a *cementitious material*. Five cementitious materials, which are closely related to the development of portland cement, will be discussed. These are clays, common limes, hydraulic limes, natural cements, and pozzolana cements. These materials have an important characteristic in common. Plasticity results when they are mixed with water. When the water is removed by any means, such as evaporation, sedimentation, chemical reaction, or by any combination of these, the plasticity disappears. The order in which these materials have been listed corresponds, in general, to the increasing degree of processing that is required before the addition of water. This book is restricted to a few inorganic water-plastics, and nearly all the subject matter is concerned with the manufacture and use of one of them, portland cement.

Clays

As clays may be used alone or mixed with other substances, they are considered here as cementitious materials. Clays are finely-divided materials formed by the disintegration of minerals containing aluminum. Generally, it is considered that the base of clays is the kaolin minerals, of which there are three varieties, kaolinite, nacrite, and dickite. The calculated oxide composition is 46.5 per cent of silica, 39.5 per cent of alumina, and 14.0 per cent of water. Starting with kaolin, the composition of clays varies widely because of the different proportions of silicon and aluminum, and because of the presence of other metals, including magnesium, sodium, and potassium. Varying composition causes differences in plasticity, fusing point, and other characteristics.

It is probable that clay was the first cementitious material, because it was widely distributed and required no preliminary processing. In fact, long before the appearance of man, clays were molded by contact with minerals, plants, and animals, and they supply many records of value in geology and other sciences.

In the Bible,[1] the manufacture of brick is mentioned several times. It is of interest that the first occasion was in relation to the building of the tower of Babel, after the account of Noah's flood. Thoroughness of burning, the substitution of brick for stone, and the use of mortar are emphasized. Later, the supplying of straw to the brick makers was discontinued. It was ordered that the men must obtain their own supply, but produce as many bricks as before. This use of straw is the earliest mention of reinforcing a plastic, that has come to my attention.

Objects made of clay have proved very valuable to the archaeologist.[2]

> Archaeologically speaking, pottery and flint implements are the foundation of all knowledge. Pottery is the commonest of all objects and practically indestructible; fire or water will not destroy it; and if a pot is broken the shards can be put together, and the form identified. Pottery, since about 6000 B.C., has been so common that even the poorest person might have at least one vessel buried with him. It is bulky, heavy, and fragile, and therefore travels badly, so that it is rarely carried any distance from the place of making.

Even in modern times, clay, in addition to that made into bricks, is used as a building material. In January, 1943, it was announced that a contract had been let in Chungking, China, for a 75,000 dollar clay building,[3] to be used as the headquarters of the Office of War Information.

The term, *ceramics*, is applied to the technology of clays. This covers a wide range of products, arising from the use of clays of various compositions, and subjected to various heat treatments. As the intensity and duration of applied heat increase, the terms *drying, firing, baking* and *vitrifying* are employed. When surface moisture has been removed from clay by drying, usually plasticity will return with the addition of more water. Some clays develop considerable strength on drying. For example, three clays of a series of nine, studied in the Philippines,[4] developed tensile strengths in excess of 300 pounds per square inch, when dried at 220 to 230° for several days, and the strength of one reached 447 pounds.

Sun-dried clays are used extensively. The adobe brick made in the United States, Mexico, and some other countries where dry climates prevail, are sun-dried. There is evidence that the brick in the Great Wall of China,[5] in the vicinity of Shanhaikwan, is sun-dried. When clays are exposed to higher temperatures, water of composition is expelled, and plasticity does not return. Common clay products are fired to a red heat, and vitrified products result from still higher temperatures. Under water, clays do not harden. The removal of water is the first step in the hardening process.

Common Limes

It seems likely that, after clays, limes were among the first cementitious materials. Although clays can be used without any processing before the addition of water, in the case of limes, the application of heat must precede the addition of water. The starting point for the manufacture of lime is calcium carbonate, $CaCO_3$. The most widely distributed mineral having this composition is *calcite*. Just as kaolin is the base of clays, so calcite is the base of limestones, from which limes are made. It may be considered that the composition of limestones varies in one of two directions, or both, from pure calcium carbonate. First, there is a series starting with calcium carbonate and ending with magnesium carbonate. The halfway point, so to speak, is dolomite, $CaMg(CO_3)_2$, which contains calcium carbonate and magnesium carbonate in equal molecular proportions. A stone consisting almost completely of calcium carbonate is called a *high-calcium limestone*, and one that contains an appreciable quantity of magnesium carbonate is called a *dolomitic limestone*.

Limestones frequently contain varying quantities of silicon, aluminum, iron, and some other elements. For simplicity, we shall refer to lime manufacture on the basis of pure calcium carbonate. When this substance is heated to more than 1500° it is decomposed into calcium oxide or *quick lime* and carbon dioxide,*

$$CaCO_3 \rightarrow CaO + CO_2.*$$

When quick lime comes into contact with water, it is *slaked* or hydrated, forming calcium hydroxide, or hydrated lime. The chemical reaction is

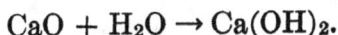

$$CaO + H_2O \rightarrow Ca(OH)_2.$$

In the presence of water, calcium hydroxide is a cementitious material. A mixture of one volume of hydrated lime, two or more volumes of sand, and sufficient water to produce the desired plasticity, is known as *mortar*. It is used for holding together brick, stone, and other masonry units. Hydrated lime hardens because of the evaporation of water and crystallization. Further hardening is caused by absorption of carbon dioxide from the air

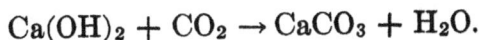

$$Ca(OH)_2 + CO_2 \rightarrow CaCO_3 + H_2O.$$

Formerly, it was taken for granted that carbonating extended entirely through a mass of mortar, but now it is known that this is primarily a surface phenomenon and usually not complete.

For the dolomitic limes, the chemical reactions and uses are much the same as for the high-calcium limes.† Corresponding to the compounds CaO, $Ca(OH)_2$, and $CaCO_3$, there are MgO, $Mg(OH)_2$ and $MgCO_3$. The series containing silicon, aluminum, and other elements will be discussed under hydraulic limes.

---

* Sometimes, the term *carbonic acid* (H_2CO_3), or *carbonic acid gas*, is substituted incorrectly for carbon dioxide. When a carbonate and a strong acid react, carbonic acid is formed, which decomposes into carbon dioxide and water. When a calcium or magnesium carbonate is calcined, carbon dioxide is evolved —not carbonic acid.

† It may be mentioned here that the term *lime* is frequently used incorrectly, which causes some confusion. As has been explained, lime denotes either the oxide or the hydroxide of calcium. Sometimes, however, the term is used as a substitute for calcium, as lime carbonate, lime sulfate, etc.

Hydraulic Limes

We come now to a very important characteristic of cementitious substances. Clays and common limes do not set—that is, they do not lose their plasticity—under water. After setting, clays are not very resistant to water, unless they have

FIG. 2:1. JOHN SMEATON—CIVIL ENGINEER
Early Developer of Hydraulic Limes

been exposed to relatively high temperatures. Common limes are not very resistant to water, even after surface carbonization.

Hydraulics may be defined as the science dealing with the action and use of water. A hydraulic cementitious material is one which can set under water. The term was not well chosen, but it has been so widely used that any attempt to substitute another term would encounter much opposition. John Smeaton is usually given credit for developing the extensive use of hydraulic limes. One cannot proceed very far with the study of cement before coming across the name Edystone (Smeaton's spelling). This is a group of gneiss rocks in the English Channel, off the coast of historical, picturesque Cornwall,[6] about fourteen miles south south west of the Plymouth breakwater. The latitude is 50°10′49″ north; and the longitude, 4°15′53″ west.[7]

For many years, this has been the site of lighthouses. Previous to 1756, these were frame structures. In that year, Smeaton was employed by Parliament to build a lighthouse which, it was hoped, would be permanent. It was well known that common limes would not harden under water, but, generally, it was believed that a high-calcium lime was superior to one that had a lower calcium content. Smeaton carried on a series of investigations. He found that lime made from limestones, relatively high in clay content, could harden under water. There was a deposit of such limestone at Aberthaw. It is known now that the hydraulic characteristics of such limes are due to the silicates they contain. A lighthouse, 72 feet high, was constructed of masonry, the mortar of which was made with Aberthaw hydraulic lime.

Figure 2:2 shows the title page of Smeaton's book, on the construction of the lighthouse.[8] Figure 2:3 is a map from the same book. Shown on this map, less than 100 miles north east of Edystone, is Portland.

Not only did Smeaton develop a cementitious material for his purpose, but it is said that modern lighthouse construction began with his work. A new masonry lighthouse, 133 feet high, was completed in 1882.

Modern hydraulic hydrated lime is defined in the ASTM Specification, C 141–61 as follows:

A

NARRATIVE OF THE BUILDING

AND

A DESCRIPTION of the CONSTRUCTION

OF THE

EDYSTONE LIGHTHOUSE

W I T H   S T O N E:

TO WHICH IS SUBJOINED,

AN APPENDIX, giving fome Account of the LIGHTHOUSE on the SPURN POINT,

BUILT UPON A SAND.

BY JOHN SMEATON, *CIVIL ENGINEER*, F.R.S.

The MORNING after A STORM at S.W.

An Appearance and Technical Reference

LONDON:

PRINTED FOR THE AUTHOR, BY H. HUGHS;

SOLD BY G. NICOL,

BOOKSELLER TO HIS MAJESTY, PALL-MALL. 1791.

FIG. 2:2. TITLE PAGE OF SMEATON'S BOOK (1813)

An hydraulic hydrated lime is the hydrated, dry cementitious product obtained by calcining a limestone containing silica and alumina to a temperature short of incipient fusion, so as to form sufficient free lime (CaO) to permit hydration, and at the same time leaving unhydrated sufficient calcium silicates to give the dry powder, meeting the requirements herein prescribed, its hydraulic properties.

Natural Cements

Natural cements are hydraulic cementitious materials, and they have much in common with hydraulic limes. Compounds of silicon, aluminum, and calcium are present in each raw material, and only one raw material is required for each product. Previous to calcination, the only preparation required for a raw material is reduction to a suitable size. Although the principal use of hydraulic limes is for mortar, natural cements are used primarily for concrete.

Modern natural cement is defined in ASTM C 10–54 as:

The product obtained by finely pulverizing calcined argillaceous limestone. The temperature of calcination shall be no higher than is necessary to drive off carbonic acid gas.

Meade [9] outlines the manufacturing process, in general, as follows:

The rock is blasted down from the face of the quarry, broken by hand with sledges into sizes suitable for the kiln, loaded on dump cars, and elevated to the mouth of the kilns. Here the rock is dumped into the kiln alternately with coal, a layer of rock and then a layer of coal. The charging is kept up continuously during the daytime but hardly ever at night. As the charge works its way down through the kiln it becomes calcined and the larger portion of its carbonic acid driven off. When it reaches the base of the kiln it is drawn and conveyed to the grinding machinery. The kilns used for the manufacture of natural cement are usually made of iron plates riveted together and lined with firebrick. They are circular in shape, upright, and their average dimensions are about 16 feet in diameter by 45 feet in height.

The clinker is usually ground by buhr-stones, the fine material in many mills being separated from the coarse by passing over screens, so placed as to allow the fine particles to go to the storehouse and to return the coarse ones to the grinders. The buhr-stones are preceded by crushers or crackers to reduce the clinker to a suitable size for them to handle. In some instances, ball and tube mills and Griffin mills have been installed in natural cement plants, particularly where these plants also make portland, but the clinker from these kilns is usually so soft as to be easily ground by buhr-stones.

The manufacture of natural cements in the United States began in 1818. During the next twelve years, several hundred thousand barrels were made. Then the output began to increase very rapidly. The maximum annual output was nearly 10 million barrels in 1899. By 1911, the annual output was less than one million barrels.

Pozzolana Cements

Pozzolana cements were developed and used extensively by the Romans. The name *pozzolana*, which today, is spelled in a variety of ways, was derived from Pozzuoli, a town in Italy a few miles from Naples and Mount Vesuvius. This material is of volcanic origin, containing compounds of silicon, aluminum, and some other elements. The Romans found that a hydraulic cementitious material could be made by mixing hydrated lime and a finely ground pozzolana.

The term is applied to other siliceous materials which can be substituted for the original pozzolanas. Shales, trass, and slag from blast furnaces may be mentioned. Sometimes, a clay or shale acquires pozzolanic properties when it is subjected to heat treatment below the fusing point. Sometimes, the addition of a relatively small amount of another substance, such as sodium hydroxide, is necessary. There is a group of cements called *portland pozzolana* in which portland cement is substituted for hydrated lime. These will not be discussed here.

Pozzolana cements are complex, and their classification is difficult. It is probable that they were the first hydraulic cements developed. On this basis, they properly could have been mentioned before hydraulic limes or natural cements. They are mentioned at this point because the order is on the basis of manufacturing processes. Pozzolana cement is the only one of the five cementitious materials discussed for which two component materials are required.

It is not known whether pozzolana cement represents a necessary step in the development of portland cement. When Smeaton made his hydraulic lime, his principal contribution was a simple method for manufacturing a cementitious material, to some extent equivalent to a pozzolana cement, using only one raw material, of a type that is widely distributed.

In his study of pozzolana, Palmieri [10] prepared Table 2:1. He stated that compounds of sodium and potassium are always present in pozzolana.

In 1810, Edgar Dobbs was granted a patent for the manufacture of a product that has been called an artificial Roman cement. An important step in the process was the calcining

TABLE 2:1

COMPOSITION OF SOME POZZOLANAS

Sample Number	Silica (SiO_2)	Alumina (Al_2O_3)	Ferric Oxide (Fe_2O_3)	Calcium Oxide (CaO)	Magnesium Oxide (MgO)	Sodium Oxide (Na_2O)	Potassium Oxide (K_2O)
1	45.50	15.12	12.05	9.33	3.59	3.78	2.15
2	46.24	19.26	11.18	9.82	2.84	2.66	2.77
3	64.71	20.03	4.98	4.39	0.96	1.77	3.07
4	48.58	32.72	9.63	7.16	–	0.74	0.74
5	47.10	21.90	8.01	9.91	1.56	2.56	6.01
6	30.00	24.00	23.30	13.00	1.00	–	–
7	45.50	15.125	12.050	9.336	3.595	3.782	2.154
8	52.77	17.00	4.84	3.28	1.35	4.94	7.65
9	47.10	21.90	8.01	9.91	1.56	2.56	6.01
10	56.20	19.30	8.70	2.90	0.25	2.90	3.43
11	54.16	19.14	3.51	5.12	1.05	2.89	9.35
12	48.00	17.53	5.09	11.19	3.60	2.56	8.47
13	45.00	14.00	12.00	3.80	4.70	4.10	1.40
14	57.02	18.85	6.90	3.18	0.65	3.51	2.78
15	44.00	10.50	29.50	10.00	Trace	–	–
16	44.50	16.50	15.00	10.00	3.00	–	–
17	42.50	15.50	12.50	9.50	4.40	–	–
Max	64.71	32.72	29.50	13.00	4.70	4.94	9.35
Min	30.00	10.50	3.51	2.90	Trace	0.74	0.74

Analyses 1 to 5, inclusive, are from G. Gallo, *The Microscopic Study of Pozzolanic Mortar.*

Analyses 6, 7, 10 are from G. Giorgis and U. Alvisi, *Natural and Artificial Pozzolanas.*

Analyses 8 and 9 are from Professor O. Rebuffat, *Studies upon the Technical Analysis of Pozzolana.*

Analysis 11 is by Professor O. Rebuffat.

Analysis 12 is from G. Giorgis and U. Alvisi, *Contribution to the Study of Pozzolana and Its Technical Value.*

Analysis 13 is by L. J. Vicat.

Analysis 14 is from F. Ferrari, *Of Special Cements in Connection with the Action of Aggressive Waters.*

Analyses 15 to 17, inclusive, are by L. J. Vicat.

of a calcium carbonate to which clay had been added. The product may be considered a pozzolana cement, in which the siliceous material is added before calcination rather than after calcination. Also, it has some features in common with hydraulic limes, and natural cements.

Early Portland Cement

The events that led to the development of portland cement could be discussed without limit. The brief summary that has been presented, together with the references at the close of this chapter, and in Chapter 18 should be sufficient.

TABLE 2:2

COMPARISON OF SIX TYPES OF CEMENTITIOUS MATERIALS

Type	Min. Num. of Raw Materials	Required Steps in Manufacturing Process	Principal Compounds Responsible for Characteristics	Hydraulic or Non-hydraulic
Clays	1	None	Aluminum silicates	N
Common limes	1	Calcination	Calcium hydroxide	N
Hydraulic limes	1	Calcination	Calcium hydroxide, Silicates	H
Natural cements	1	Calcination Grinding	Silicates	H
Pozzolana cements	2	Calcination of lime-stone, mixing with ground pozzolana	Silicates	H
Portland cements	1	Grinding and blending raw materials Clinkering Grinding the clinker	Silicates and Aluminates	H

Table 2:2 shows a comparison of the five types of cementitious materials as they have been discussed, together with modern portland cement. In Table 2:3 there are some typical analyses.[9] Chapter 13 should be consulted from time to time. To many persons outside the cement industry, the word *portland* is confusing. Its significance is explained in that chapter.

In 1824 Joseph Aspdin, a stonemason, living at Leeds, England, was granted a patent for a new type of cementitious material. The color of the product after hydration reminded him of the limestone of the Isle of Portland. Because of this, he named the product *portland cement*. Portland, therefore, is the name of a product made by a given process, somewhat analogous to Bessemer steel. The product is manufactured by individual companies, just as Bessemer steel is manufactured by individual companies.

TABLE 2:3

COMPOSITION OF SOME LIMES AND CEMENTS

Material	Silica (SiO_2)	Ferric Oxide (Fe_2O_3)	Alumina (Al_2O_3)	Calcium Oxide (CaO)	Magnesia (MgO)	Sulfur Trioxide (SO_3)	Carbon Dioxide (CO_2)	Water (H_2O)
High-calcium lime	0.15		0.85	98.01	0.45	–	0.55	–
High-calcium lime	0.52		0.24	97.14	1.28	–	0.96	–
High-calcium lime	0.25		0.15	97.46	0.73	–	1.41	–
Dolomitic lime	1.61		0.17	57.44	40.36	–	0.41	–
Dolomitic lime (hydrated)	0.34		0.18	45.37	31.20	–	3.02	20.07
High-calcium lime (hydrated)	0.38	0.08	0.06	72.59	0.74	–	2.10	23.11
Hydraulic lime	31.10	2.15	4.43	58.38	1.09	0.60	1.28	–
Hydraulic lime	19.05	0.55	1.60	65.10	0.65	0.30	–	12.45
Natural cement	29.92	4.78	11.23	36.50	11.93	–	5.42	–
Natural cement	27.75	4.28	5.50	35.61	21.18	0.50	4.05	–
Portland cement	22.56	2.85	7.44	62.73	1.99	1.46	–	–
Portland cement	21.29	2.72	7.64	63.48	1.53	1.77	–	–

A fundamental difference between natural cements and portland cements lies in the selection and processing of the raw materials before they enter the kilns, as is shown in Table 2:2. On this basis, portland may be considered a synthetic cement. It is not surprising that the portland cement of Aspdin's day was very crude, by our present standards. With the aid of various improvements, however, it has had the merit of retaining the lead in its field, without serious challenge from competing products for many decades.

Although there are numerous cements recommended for special purposes, in general, those manufactured in relatively

FIG. 2:3. MAP FROM SMEATON'S BOOK ON EDYSTONE LIGHTHOUSE

Table 2:4

PRODUCTION OF PORTLAND CEMENT IN UNITED STATES

Year	Barrels	Bbl per Capita	Value
1875–1879	82,000	0.001	$246,000
1880	42,000		126,000
1881	60,000		150,000
1882	85,000		191,250
1883	90,000		193,500
1884	100,000		210,000
1885	150,000		292,500
1886	150,000		292,500
1887	250,000		487,500
1888	250,000		487,500
1889	300,000		500,000
1890	335,500	0.005	704,050
1891	454,813		967,429
1892	547,440		1,153,600
1893	590,652		1,158,138
1894	798,757		1,383,473
1895	990,324		1,585,830
1896	1,543,023		2,424,011
1897	2,677,775		4,315,891
1898	3,692,284		5,970,773
1899	5,652,266		8,074,371
1900	8,482,020	0.11	9,280,525
1901	12,711,225		12,532,360
1902	17,230,644		20,864,078
1903	22,342,973		27,713,319
1904	26,505,881		23,355,119
1905	35,246,812		33,245,867
1906	46,463,424		52,466,186
1907	48,785,390		53,992,551
1908	51,072,912		43,547,679
1909	64,991,431		52,858,354
1910	76,549,951	0.83	68,205,800
1911	78,528,637		66,248,817
1912	82,438,096		67,016,928
1913	92,097,131		92,557,617
1914	88,230,170		81,789,368
1915	85,914,907		73,886,820
1916	91,521,198		100,947,881
1917	92,814,202		125,670,430
1918	71,081,663		113,730,661
1919	80,777,935		138,130,269
1920	100,023,245	0.90	202,046.955
1921	98,842,049		180,778,415
1922	114,789,984		207,170,430
1923	137,460,238		257,684,424
1924	149,358,109		264,046,708
1925	161,658,901		278,524,108
1926	164,530,170		277,965,473
1927	173,206,513		278,854,647

TABLE 2:4 (*Continued*)

Year	Barrels	Bbl per Capita	Value
1928	176,298,846		275,972,945
1929	170,646,036		252,153,789
1930	161,197,228	1.31	228,779,756
1931	125,429,071		140,959,906
1932	76,740,945		82,021,723
1933	63,473,189		85,583,916
1934	77,747,765		116,921,084
1935	76,741,570		113,372,182
1936	112,649,782		170,415,302
1937	116,174,708		168,835,208
1938	105,357,000		153,977,226
1939	122,259,154		180,893,208
1940	130,216,511	0.98	190,078,068
1941	164,030,559		246,621,914
1942	182,781,184		283,237,028
1943	133,423,788		200,103,216

Here are some data [11] for 1960:

Production	328,715,000 bbl
Barrels per capita	1.8
Number of establishments	170
Wage Earners, average for year	41,000

In 1963, the production in the United States and Puerto Rico was 352.5 million barrels. According to a preliminary report, mill shipments in the United States (exclusive of Puerto Rico) reached a new high in 1964—356 million barrels of portland cement.

large quantities are portlands, or they contain a portland and one or more other materials.

In commemoration of the centennial of the invention of portland cement, a bronze tablet, prepared by the Portland Cement Association with the cooperation of the British Cement Makers' Federation, was unveiled at Leeds, on September 6, 1924. In commenting on the ceremony, the late F. W. Kelley, then president of the Association, stated that there was on display,

the original Aspdin patent inscribed with India ink upon parchment, to which was attached a wax seal six inches in diameter and bearing the impress of the Royal Seal of George IV. (See Frontispiece.)

Developments in United States

Many years passed after the invention of portland cement before the industry became important in the United States.

The first cement was manufactured in 1875. From 1875 to 1879, the total production was 82 thousand barrels. Table 2:4 shows that the annual production increased from 42 thousand barrels in 1880 to more than 176 million barrels in 1928. The 1928 production was the peak until 1942, when well over 182 million barrels were manufactured.

The names of cement companies have much in common. Most of them are geographical in origin. A few were suggested by history or mythology, and the remainder are general. None is named for the owners; see list in Chapter 24.

In Table 2:5 there are listed some of the important events in the cement industry. Each of these has had an important

TABLE 2:5

SOME OUTSTANDING EVENTS IN CEMENT INDUSTRY

Before history: The first use of clays.
Before history: The first use of common limes.
1756: Smeaton's development of hydraulic lime, and the Edystone lighthouse.
1818: The first natural cement manufactured in the United States.
1824: Aspdin's patent for portland cement.
1868: The first reinforced concrete.
1873: Invention of the rotary kiln.
1875: The first portland cement manufactured in the United States.
1880: The first use of the barrel pulverizer, the prototype of the tube mill.
1886: The first rotary kiln made in America.
1887: The discovery of tricalcium silicate by Le Chatelier.
1891: The first concrete highway, built at Bellefontaine, Ohio.
1898: The first waste heat boiler.
1902: Organization of the Association of Portland Cement Manufacturers. (In 1916, this became the Portland Cement Association.)
1903: The construction of an eighteen story reinforced concrete building in Cincinnati.
1905: Organization of the National Association of Cement Users. This became the American Concrete Institute.
1905: The first use of the Bates valve bag and packing machine in the cement industry.
1906: The first publication of the results of the silica-alumina-calcium oxide system by the Geophysical Laboratory.
1911: Application of the Cottrell precipitator to cement manufacture.
1914: The formulation of the water-cement ratio law.
1920: Invention of the Fuller-Kinyon pump.
1924: The first high-early-strength cement manufactured in Germany on a commercial scale.
1925: The first use of slurry filters.
1935: The first soil-cement secondary road.

influence on some phase of plant design, manufacturing processes, or the development, or the utilization of the product. It is probable that some of the dates are approximate, rather than exact. The literature does not contain much information of this type.

Since 1935, the advance of the industry has been due largely to the further development in existing techniques in the manufacture and use of cement, rather than to outstanding events.

For example, the size of rotary kilns has advanced beyond all expectations. A unit 760 feet long and 25 feet in diameter is planned. Likewise, one-kiln plants, rather than multiple-kiln plants, are increasing in popularity. Such transitions bring about many operating problems.

At a number of plants, computers are in use, although these do not provide any basic advance in technique. A manager of a plant having computers remarked that he still must have kiln operators. But the future of these units in the industry is really promising.

Prestressed concrete, known for a long time, has reached a rapid development in the last few years.

Interest in astronomy, particularly in Mars, once led me to some speculation as to what cement and concrete industries on that planet might be like. Some material taken from a partly-completed paper on the subject was used in a book, *Chemistry and the Cement Industry*, written for the Universal Atlas Cement Company in 1933. The matter was forgotten until a section heading, "Canals of Mars," appeared in a book that was published after the manuscript for *Portland Cement Technology* was completed. This book, *The Magic Powder*,[12] is a history of the Universal Atlas Cement Company and, to some extent, of the industry in general. The term *magic powder* has a history of its own, having been applied to portland cement for many years.

REFERENCES

1. *Genesis*, 11:3; *Exodus*, 1:14, 5:7–11; *Isaiah*, 9:10, 65:3.
2. *The Popular Educator*, 1 (3): 191 (1938).
3. *The Chicago Tribune*, January 4, 1943.
4. J. C. Witt, *Philippine J. Sci.*, 11, Sect. A (5): 210 (1916).
5. J. C. Witt, *Philippine J. Sci.*, 12, Sect. A (5): 257 (1917).

6. Herbert Corey, *Nat. Geogr. Mag.*, 46: 653 (1924).
7. *Brown's Nautical Almanac*; Brown, Son, and Ferguson, Ltd., Glasgow, 1935.
8. John Smeaton, *Narrative of the Building and a Description of the Construction of the Edystone Lighthouse;* Longman, Hurst, Rees, Orme, and Brown, London, 1813.
9. R. K. Meade, *Portland Cement;* Chemical Publishing Co., Inc., Easton, Pa., 1930.
10. Mario Palmieri, *Concrete, Cement Mill Section,* 43(5):42; (7):39 (1935).
11. Census of the United States, 1960.
12. E. J. Hadley, *The Magic Powder;* G. P. Putnam's Sons, New York, 1945.

Chapter 3

RAW MATERIALS

The selecting, proportioning, and blending of raw materials have not received so much attention as have some other phases of manufacturing. The importance of these phases, however, is receiving more and more recognition. Many improvements have been made, and more may be expected as the industry progresses.

It has been found in studying various kiln feeds, in laboratories and in experimental plants,[1] that two feeds having identical chemical analysis may have widely different clinkering characteristics.

Some important developments in raw materials have come about within the last few decades. For many years after the invention of portland cement the only raw materials were natural calcium carbonates, such as limestone and marl, and clays or shales. In the early nineties, Jasper Whiting and J. G. Bergquist, blast furnace superintendent and chief chemist, respectively, of the old North Works of the Illinois Steel Company, began a search for some means of utilizing blast furnace slag. Slag was considered a waste material of the worst type. Not only did it have no commercial value, but the disposal of it was difficult and costly.

In his studies, Bergquist happened to compare the analysis of slag with that of portland cement. He noticed that slag contained all the elements present in portland cement, the most important difference being that the cement contained a higher percentage of calcium.

The years of development that followed established slag as an important raw material, although the technique required differs from that of other raw materials. Many millions of barrels of cement have been made from mixtures of slag and limestone.*

The next important addition to raw materials came with the beginning of the manufacture of special portland cements. Previous to 1920, the manufacture of a cement seldom required more than two raw materials. Some times two materials could be proportioned to produce a raw mix containing silicon, aluminum, iron, and calcium within the desired limits. With the introduction of high, early-strength, and other special cements, however, three or more materials were frequently required, one of which had a high silicon content, another, a high iron content, and so on. Later developments have shown that it is not always economical to limit the number of raw materials to two, even when this can be done.

Formerly, a raw mix of the desired composition could be obtained only by blending the raw materials, as received. Today, by the application of flotation,[3] another procedure is available. In some cases, a raw material may be separated into its components, and a portion of one or more of these components can be rejected. This greatly increases the flexibility of raw mix preparation and permits the utilization of some raw materials that would otherwise be unsuitable for cement manufacture.

The raw materials discussed in this chapter include only those employed in the manufacture of clinker. The use of gypsum is explained in Chapter 9.

Selection of Raw Materials

Cement contains four essential chemical elements: silicon, aluminum, iron, and calcium. Usually, these are reported as oxides. The other elements present, of which the number is not limited, are unessential, and their quantities are relatively small. Although, ordinarily, both aluminum and iron are present, only one of these is actually required. The iron content

---

* Much of this information was obtained from oral statements by Bergquist, and by men who were acquainted with him and with Whiting. The occasion for this was a request for me to prepare a paper on "The development of portland cement manufacture from blast furnace slag,"[2] one of a series on manufacturing process developments in the Chicago district.

of white cements is only a few tenths of one per cent, and
cements have been made which contain practically no aluminum. The limits in composition of more than one hundred
commercial cements are shown in Chapter 15.

In selecting each raw material, consideration should be
given to:

Composition	Quantity
Uniformity	Location, Topography
Physical characteristics	Transportation
Overburden	Unit cost

Composition:

On the basis of composition, the requirements for one or
more raw materials to be used in producing a clinker are as
follows:

a) The essential elements must be present;

b) It must be possible to proportion these materials so that
the essential elements may be present within the desired limits;

c) The nonessential elements must not be present in undesirable quantities.

Uniformity:

If a raw material does not possess a fair degree of uniformity, its use requires so many additional analyses and
mixing operations that it should be avoided, if any better
material can be located.

Physical Characteristics:

Some physical characteristics are important. One of these
is resistance to grinding, or the so-called *grindability*. Another
is the degree of *homogeneity*. A third is fairly uniform *hardness*
throughout.

The following special points should also be considered:

If it is an aggregate of particles varying in hardness, does
it tend to segregate after the material is ground?

If it is a material that is to be blasted, in a quarry or mine,
does it shatter easily, or with difficulty?

If it is clay or shale, to be used in a wet process plant,
does it contain an undesirable quantity of colloidal material?

Overburden:

Other factors being equal, the cost of quarrying increases rapidly as the thickness of the overburden, or waste material, increases. The physical characteristics of the overburden are important considerations. In some deposits, the raw materials can be obtained economically only by mining.

Quantity:

The soundness of an investment in a cement plant demands that the supply of the primary raw materials be adequate. This is calculated on the basis of the expected annual output, and the expected years of operation, together with a liberal allowance for contingencies.

Location, Topography, Transportation:

These three factors are mentioned together because, to a considerable extent, they are interdependent. The location of the principal raw materials deposits, and the location of the market for the product must be taken into account in deciding where a plant is to be placed. Every advantage and disadvantage due to freight rates, transportation lines, time required for deliveries, and the like, must be evaluated carefully. In the location of some plants, the availability of raw materials has been given the preference. In other cases, the location of the market has been given first consideration. In still others, the plant is near the market and a supply of one material, whereas another material must be shipped hundreds of miles. In fact, there are examples of almost every conceivable combination of circumstances.

Unit Cost:

The resultant of all the items that must be taken into account in the selection of raw materials is the unit cost. This must be reasonable if the investment is to be sound.

Primary, Secondary, Tertiary Raw Materials

The listing of raw materials in three divisions is an arbitrary classification which has been adopted for convenience in this discussion. Quantities of materials are stated on the basis of theoretical clinker, not taking into account losses from dust leaving the kiln; nor gains due to coal ash. On this basis, 376 pounds of clinker require 376 pounds of ignited raw materials.

Primary materials:

Primary materials are those that form more than 85 per cent of the weight of a clinker. In most cases, they can be used without the addition of any other materials.

Secondary materials:

The secondary materials are those that form less than 15 per cent of a clinker. Their function is to modify the percentage of silicon, aluminum, or iron that a clinker would contain if only primary materials were used in manufacturing it.

In Table 3:1, there is a list of some primary and secondary raw materials. The principal element of the clinker in the raw material is indicated.

TABLE 3:1

PRIMARY AND SECONDARY RAW MATERIALS

Raw Material	Primary or Secondary	Principal Element
Limestone	P	Ca
Cement rock	P	Ca, Si, Al, Fe
Marl	P	Ca
Oyster shell	P	Ca
Calcium carbonate (by-product)	P	Ca
Clay	P	Si, Al, Fe
Shale	P	Si, Al, Fe
Slag	P	Si, Al, Fe
Sand	S	Si
Sandstone	S	Si
Bauxite	S	Al
Diaspore	S	Al
Iron ore	S	Fe
Mill scale	S	Fe
Pyrites cinder	S	Fe

Tertiary materials:

Tertiary materials are employed for some purpose other than making clinker; for example, to decrease the water content of slurry, to increase the output of grinding equipment, or decrease power requirements; also catalysts, or fluxes, for facilitating clinker production. Many tertiary materials are proprietary products. They may have been developed especially for the cement industry, or they may have been used first in some other industry. New items appear from time to time.

Table 3:2 shows the analysis of some raw materials that have been used in widely separated plants for the manufacture of cements.

TABLE 3:2

COMPOSITION OF SOME RAW MATERIALS
USED FOR MANUFACTURE OF CEMENTS

Per Cent

Material	SiO_2	Al_2O_3	Fe_2O_3	CaO	MgO
Limestone	0.78	0.28	0.26	53.69	0.66
Cement rock	13.77	5.00	1.61	41.49	2.06
Marl	0.15	0.27	0.19	54.70	0.88
Oyster shell	1.50	–	–	53.80	0.25
Clay	54.24	29.12	5.72	6.75	–
Tufa	53.76	22.88	6.84	5.41	1.88
Blast furnace slag	37.37	11.81	0.65	45.50	2.75
Iron ore	9.02	4.76	72.30	4.76	1.17
Bauxite	2.00	58.83	21.99	0.22	0.03
Diaspore	21.94	55.87	9.13	–	–
Sand	61.35	18.46	4.82	5.22	2.72

Sources of Raw Materials

There are no general statements concerning the sources of raw materials. A raw material may be quarried, mined, or dredged from property belonging to the company—at the site of the plant, or at a long distance from the plant; or it may be purchased. It may be produced especially for the manufacture of cement, or it may represent a surplus from some other industry, such as screenings from blast furnace stone, or from concrete aggregates. A raw material may be a natural product, such as limestone or shale; or it may be a by-product from the manufacture of some commodity, such as alkali, or sulfuric acid.

It is probable that the tonnage of limestone used in manufacturing cement is greater than that of any other raw material. For this reason, limestone will be taken as an example, and followed through the typical operations from the original deposits to the crushers. Some stone deposits are far from homogeneous. The stone mentioned as a single raw material may be in reality a series of stones, requiring careful selection and blending. Frequently, shale, or clay, is obtained at the stone quarry, from the overburden, the floor, or intermediate strata. In this chapter, the raw materials will be followed through the primary and secondary crusher stages, ready for proportioning, which will be explained in Chapter 4.

The quantities of raw materials [5] used in the manufacture of clinker in 1660, are listed in Table 3:3.

TABLE 3:3

RAW MATERIALS USED IN MANUFACTURE OF CEMENT
IN UNITED STATES AND PUERTO RICO

(In 1000 Short Tons)

Cement rock	19,917
Limestone (including oyster shells)	66,823
Marl	1,224
Clay and shale [a]	9,657
Blast furnace slag	1,269
Gypsum	1,146
Sand and sandstone (including silica and quartz)	2,690
Iron materials [b]	774
Miscellaneous [c]	66
Total	103,566

[a] Includes fuller's earth, diaspore, and kaolin, for making white cement.

[b] Includes iron ore, pyrite cinders and ore, and mill scale.

[c] Includes fluorspar, pumicite, pitch, red mud and rock, hydrated lime, tufa, calcium fluoride, sludge, air-entraining compounds, and grinding aids.

Prospecting for Limestone

The existence of a suitable deposit may be known in advance. A quarry may be in operation, supplying stone for aggregates or lime manufacture, for example, or it may be necessary to start with the search for a quarry. In any event, the only safe procedure is to make a thorough, independent investigation. After a decision has been made as to the general locality, the next step is to assemble all the available data on deposits in this locality, including bulletins by the United States Geological Survey and the State Survey. Sometimes information can be obtained from the records of wells, railway and highway cuts, excavations for foundations, and the like. There should be a reconnaissance. It may also be advisable to take preliminary samples at outcrops, and analyze these samples.

Surveying and Core Drilling

Surveying:

After the preliminary tests have been completed, and options on the land have been obtained, some surveying is in order. This includes surveying the entire area, and locating points at which there is to be core drilling. North-south lines are laid off at intervals of one hundred feet. This is followed by a series of east-west lines at the same intervals. The elevation is measured at each point of intersection of the two sets of lines, and stakes are driven. The elevations may be measured from an arbitrary datum, which later may be referred to the elevation above sea level, if this is found desirable.

Core drilling:

Unless the company investigating a limestone deposit has high-grade drilling equipment, and men with considerable experience to operate it, the best procedure is to let a contract for the work. There are many drilling contractors who are well equipped to handle such assignments. If, as is usual, rapid completion of the work is desirable, sometimes more than one crew can be placed in the field. Usually, the quotation is for a definite amount per foot of drilling, taking into account the diameter of the core, with the minimum footage specified.

When arrangements with a contractor have been completed, attention must be given to a number of items such as the following:

Size of core:

A number of factors are involved in selecting the size of the core. While the cost of drilling varies directly with the

diameter of the core, it is important that the diameter be
sufficiently large for the purpose. It is well to consult the
contractor on this point.

Distance between holes:

The most important factor is the uniformity of the deposit.
If the preliminary information is limited, a good procedure
is to drill a few holes at 100-foot intervals and, on the basis
of the analyses, decide whether longer or shorter intervals
would be satisfactory for the remainder of the drilling.

Depth of holes:

For a rather thin deposit, it is customary to drill entirely
through it. For other deposits, the usual procedure is to
specify the depth to which the drilling is to proceed.

Water supply:

If provision for an adequate supply of water is not made
in advance, there may be serious delays in the progress of
the work. Temporary pipe lines may be laid, or water may
be transported by trucks.

Core boxes:

To avoid delays, a sufficient supply of core boxes and
covers must be on hand. It is the duty of the drilling contractor to place the cores in these boxes together with a
record of the number of the hole and the depths represented
by each core. In addition, he submits a report of each day's
drilling, showing the hole numbers, and the feet of drilling.

Delivery of cores:

Arrangements are made to have the cores that are obtained each day delivered without delay to the laboratory
where the analyses are to be made.

Analysis of cores:

The most representative sample is obtained by crushing a
portion of a core, riffling or quartering it until the desired
quantity of material is obtained, and then pulverizing it for
analysis. This procedure is not recommended, however, because much information on the location of the material in
the original deposit is lost. For example, if it were divided to

make an analysis for every 5 feet of core, and the results
indicated that additional analyses of the stone were desirable,
the needed samples could not be obtained. A better procedure is to remove a longitudinal portion of a core for
analysis, replace the remainder in the box, and preserve the
boxes, together with the complete record, for future reference.

To facilitate study, the analytical results are arranged
according to the depths at which the samples were obtained.
For a given line of drill-holes, a profile drawing is made
showing the thickness of each stratum of the quarry from
end to end, or side to side. This may be repeated for as many
lines of north-south holes, or east-west holes as may be desired. From the large mass of data that is obtained from
such an investigation, an almost unlimited number of calculations and graphs may be made. It is customary to analyze
cores only for silicon, aluminum, iron, calcium, and magnesium. Some representative samples, however, should be
examined for minor elements and compounds, including sodium, potassium, sulfate, and phosphate. Deposits investigated today may supply raw materials to a plant for several

TABLE 3:4

TESTS FOR MINOR ELEMENTS AND COMPOUNDS
IN RAW MATERIALS

Element or compound	Limestone	Shale
Mercury	—*	—
Lead	—	—
Bismuth	—	—
Cadmium	—	—
Copper	—	+
Arsenic	—	—
Antimony	—	—
Tin	—	—
Manganese	+	+
Strontium	+	—
Titanium	+	+
Chloride	—	—
Sulfate	—	—
Phosphate	+	+
Total alkalies, calculated to Na_2O	0.45 per cent	2.82 per cent

* + Indicates presence; — indicates absence.

decades. During this period, some minor elements may be found to be more important than they are believed to be at present, or less important. Table 3:4 shows the results of qualitative tests on two raw materials.

Faults in the strata of the limestone may interfere seriously with the prospecting procedure, thus curtailing the scope of the information obtainable. In some localities, the presence of faults is indicated early in the drilling schedule. In others, there may be little or no indication of faults until cores have been analyzed and the preparation of graphs has been started.

Quarry Operation

The remainder of this chapter, and all of Chapters 4 through 10, are concerned primarily with operation. The operation of a quarry is difficult. Hard, heavy materials are handled at various levels by massive equipment. Much maintenance is required and there are many hazards, some of which are caused by weather conditions. While this chapter was in preparation, temperatures in Chicago as low as $-15°$, and reports from one of the plants that stone could not be unloaded because of the ice surrounding it, made very realistic the influence of weather conditions on the operation of a quarry. Briefly, some of the principal procedures are:

Stripping:

Stripping means the removal of the overburden, or waste material. Depending on the depth and nature of the overburden, various types of equipment are used, such as bulldozers, scrapers, drag-lines, and power shovels. The stripping may be sufficiently ahead of the drillers to be out of their way, or it may be completed a year or two in advance. The work may be done by the quarry crew, or by contract.

Primary drilling and blasting:

The drills used in quarrying operations can be classified in a number of ways. They may be *hand-held*, or *mounted* (wagon drills); they may cut by impact or by rotation; they may be powered by electricity, compressed air, gasoline, or steam.

Several types of explosives are employed, including: straight nitroglycerol dynamites, extra (ammonia) dynamites, gelatin dynamites, high-count ammonium-nitrate dynamites, cored ammonium nitrate dynamite, semi-gelatins, and nitro-starch explosives.

An objective in primary drilling and blasting is to avoid, as far as possible, secondary blasting which is likely to cause expensive delays. A number of variables must be taken into account, and because these may be combined in many ways, the work is complex. Some of these factors are: [4]

1. Height of face
2. Hardness and uniformity of rock
3. Dip of beds
4. Prevalence of open bedding seams and joints
5. Dip of quarry floor
6. Size and depth of drill holes
7. Arrangement and spacing of drill holes
8. Number of holes shot at one time
9. Size of charge
10. Position of charges in drill holes
11. Type of explosive used
12. Method of firing shots
13. Method of loading rock
14. Size of shovel
15. Size of crusher

Secondary blasting:

It frequently happens that after the primary blasting, there are some pieces of stone that are too large for transporting, or for entering the primary crusher. These must be reduced in size by drilling and secondary blasting.

Loading and transporting:

In most quarries, material is loaded by the use of one or more power shovels. The distance from quarry to plant may be less than a mile, or it may be many miles. For short distances, the equipment in common use is the quarry car, or tractor-and-trailer combination. The use of tractors and trailers seems to be increasing. For distances of intermediate length, such as several miles, or in regions in which the terrain is rough, aerial tramways may be employed. Conveyor belts, several miles in length, are in common use. For relatively long distances, the stone may be transported by water or by rail. When it is to be shipped by rail, it is customary to load the stone directly into gondola cars.

Crushing

The crushers are installed at the quarry, at the plant, or at some intermediate point. Probably the most common arrangement is to locate the crushers near the raw materials storage. The size reduction of the stone, after the primary and secondary blasting, starts with the crushers, and proceeds until nearly all of it will pass through a 200-mesh sieve. Usually, there are primary and secondary crushers, one or

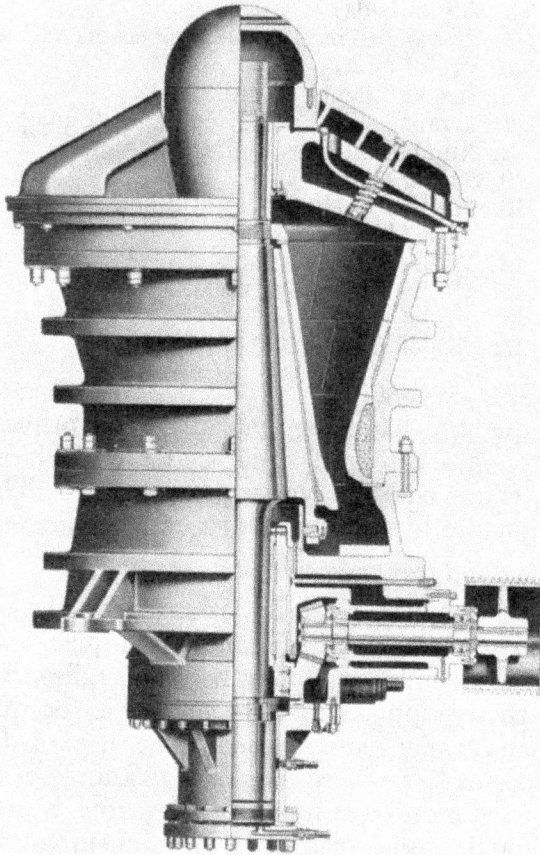

FIG. 3:1. GYRATORY CRUSHER WITH HYDROSET MECHANISM
(*From* Allis-Chalmers)

more of each. As primary crushers, *jaw* and *gyratory* types are commonly used (see Fig. 3:1). The *swing-hammer* type is often used for secondary crushing.

As an example of procedures for handling raw materials, an installation may be briefly outlined. Limestone arriving in gondola cars is dumped into a large gyratory crusher. The stone leaving the crusher, approximately 6-inch maximum size, is reduced to approximately ½-inch maximum in a hammer-mill. It then goes by conveyor belt to a storage building 75 by 497 feet in dimensions. By means of an automatic tripper the stone is distributed and blended during the discharge from the belt. In addition to stone, the building provides storage for shale (the other primary raw material), secondary materials, gypsum, and coal. There is a traveling crane by which all materials are reclaimed.

REFERENCES

1. J. C. Witt, *Concrete, Cement Mill Section*, 40(10):35 (1932).
2. J. C. Witt, *Chemical Bulletin*, 18(7):202 (1931).
3. C. H. Breerwood, U. S. Patent 1,931,921 (1933).
4. *Pit and Quarry Handbook;* Pit and Quarry Publishing Co., Chicago, 1961.
5. *Minerals Yearbook;* U. S. Bureau of Mines, Washington, D. C., 1961.

Chapter 4

PROPORTIONING
RAW MATERIALS

Sometimes it is possible to manufacture clinker from a single raw material, but usually two or more raw materials are required. In this way, the proportioning of raw materials becomes necessary, and it is one of the most important procedures at a cement plant. This chapter has to do primarily with calculations for proportioning.

The simplest proportioning problem is one in which there are only two raw materials, and the nature of the materials is such that when the calcium content of the raw mix is maintained within certain limits, the silicon, aluminum, iron, and magnesium are satisfactory. Even in this case there must be a fair degree of accuracy, because if the calcium content is too high, the cement will be unsound; and if the calcium is too low, the strength of the cement will not meet specifications.

There are some changes in weight in the raw materials in a kiln before they reach the clinkering zone, due to calcination of the carbonates, and the vaporization of water. These must be taken into account in proportioning calculations. It has been found most convenient to calculate the composition of each raw material to the ignited basis as the first step. When this is done, the total weight of the materials is approximately equivalent to the weight of the final product. As will be explained later, it is best to calculate each material, and the product, on the four-constituent basis.

Some changes take place in the weight of the kiln contents, which cannot be calculated with even a fair degree of accuracy. Examples are the addition of ash, when coal is the fuel, and dust loss. By making parallel determinations of calcium oxide in the kiln feed, and in the clinker, as frequently as experience indicates that this is necessary, correction factors may be established, by which the calcium oxide content of the kiln feed, necessary to produce a clinker having the desired CaO content, may be calculated. If it is considered necessary or desirable, similar correction factors may be calculated for some of the other oxides. When the composition of the clinker, and of the gypsum are known (see Chapter 7), the composition of the cement may be calculated, and verified by analysis.

Suppose it is required to make 100 pounds of clinker, containing a given percentage of calcium oxide, from limestone and shale containing known percentages of calcium oxide. Most of the quantities in the initial equations become zero. The remaining quantities are as follows:

q = per cent CaO in calcined limestone
p = per cent CaO in ignited shale
v = per cent CaO in clinker
z = pounds of calcined stone required

$$z = \frac{100(p - v)}{p - q}$$

All percentages are expressed as decimals.

Example: A calcined limestone contains 96 per cent CaO; an ignited shale, 4 per cent CaO; and the desired clinker, 65 per cent CaO. Substituting in the equation above,

$$z = \frac{100(0.04 - 0.65)}{0.04 - 0.96} = 66.3 \text{ pounds}$$

Thus, the preparation of 100 pounds of clinker requires 66.3 pounds of calcined limestone and 33.7 pounds of ignited shale.

Starting with the next paragraph, and continuing for several pages, a paper on *The Proportioning of Materials in Process Engineering* [1] has been reproduced. It will be noted that the procedures were developed in connection with the manufacture of cement, but they were stated in general terms, so that they could be used in other industries.

In all process industries in which two or more raw materials are used, proportioning is an essential step. It is usually one of the first steps, preceded only by the obtaining of the raw materials, and by such preliminary treatments of each material as crushing, drying and grinding. When the materials are proportioned, each should be as nearly homogeneous chemically and physically as conditions permit.

Proportioning should be followed by adequate blending, directly after the materials come together. This should not be left to chance, but should be given as much care, in plant design, as any other operation. Although, in general, additional blending takes place at successive points in a process, this should not be considered a substitute for positive blending at the beginning. Incorrect proportioning and blending tend to increase production costs and to result in products that are irregular in composition.

Calculations for the proportioning of materials seem to have received less study than other types of calculations, or other details of process industries. At least, this is the impression I have received from general observations and from the literature. Because some calculations are long and tedious, there is a temptation to substitute various types of approximations. I have developed a procedure by which proportioning calculations may be made easily and rapidly without sacrificing accuracy.

Most of the work was completed in 1935. At the time, there was no intention of preparing a paper for publication. Later the literature was consulted, including journals, and books on mathematics, engineering, and manufacturing. Nothing closely related to the work was found, but two published references [2,3] and one unpublished reference [4] are cited. It is probable, however, that other work along the same lines has been carried out, and some of it may have been published. The nature of the subject matter renders difficult its location in the literature.

Proportioning problems cover a wide range, from simple to complex, depending on a number of factors. The materials may be either solids or fluids. The proportioning equipment may be either gravimetric or volumetric, and either batch or continuous. The procedure to be described is applicable to all classes of materials and proportioning equipment, but, for simplicity, this description will be restricted to solid materials, proportioned by weight. Three cases are considered.

Case 1: Two or more materials are proportioned to prepare a mixture of desired percentage composition. The constituents of each material, chemical changes, and chemical or mechanical losses or gains in weight are not taken into account. An example is the preparation of a mixture of dry sand and dry gravel to contain 40 per cent of the former and 60 per cent of the latter.

Case 2: Two materials, having a constituent in common, are proportioned to prepare a product containing a desired percentage of the constituent. Chemical changes and chemical or mechanical losses or gains in weight may or may not be taken into account.

Case 3: Three or more materials are proportioned to prepare a product containing a number of constituents. Each material contains one or more of the constituents of the product. Chemical changes and chemical or mechanical losses or gains in weight may or may not be taken into account.

In these three cases, *chemical changes* refer to any chemical reactions that may take place between or among the materials in a mixture, during a manufacturing process. Chemical or mechanical losses or gains refer to any change in the weight of one or more of the materials. Examples are loss of moisture or combined water, absorption of water, oxidation, and reduction. Also, there may be losses due to leaking or dusting, and gains due to contaminating substances. There are variations, frequently due to irregularities in the composition of the materials, irregularities in the composition and quantity of fuels, and changes in temperature, pressure, humidity, and velocity.

In the solution of a proportioning problem, the quantities of the material required, represent the only variables. All other quantities must be considered constants for the time being. Losses and gains that can be calculated should be taken into account in advance. No attempt should be made to take into account losses and gains that may be irregular, or concerning which sufficient information is not available. When a process is in operation, corrections may be introduced by comparing repeatedly the calculated theoretical results with the results actually obtained, and making adjustments on this basis from time to time.

Although the procedures and formulas discussed in this paper were developed in connection with the manufacture of cement, it was evident that they are applicable to other manufacturing processes requiring the proportioning of materials. Consequently, they are stated in general terms so that they may be used in connection with other industries, and so that one need not be familiar with the details of cement manufacture to be able to follow the discussion readily. For convenience, the principal examples to which reference is made in discussing the calculations are taken from the cement industry.

Case 1 is so elementary that no comments are considered necessary. Case 2 also is simple. Anyone with a working knowledge of algebra and of the fundamentals of the industry in which he is interested, can prepare and solve equations that will give him the desired results. The calculations, however, require some little time.

In 1925, I became interested in what may be called the finer points of proportioning. At a cement plant, two raw materials were proportioned to obtain a product containing a desired percentage of calcium oxide. Frequent determinations of the calcium oxide in each material were made. However, the information could not be utilized to full advantage because there was not sufficient time available for making the needed calculations. This situation was remedied by preparing tables covering the full range of calcium oxide in each of the raw materials, and in the product, in steps of 0.25 per cent. The tables were voluminous (containing almost 35,000 terms). Their preparation required the equivalent of one man's time for several weeks, although every mathematical short-cut that could be found was employed, including some procedures mentioned in a former paper.[5] It is estimated that the tables have eliminated the solution of many thousands of individual equations—the answers to which would have been required for the accuracy in proportioning obtained by the use of the tables. This is an example of how it may pay to develop procedures by which simple calculations may be made in advance, so that their results are available indefinitely. The original investment pays large dividends in time, accuracy, and general satisfaction.

Formerly, many cement plants used only two raw materials. As a rule, proportioning involved only the increasing of one or the other of the two, to keep the calcium-oxide content within desired limits. Case 2 calculations were sufficient. Occasionally a third material was needed to modify the silica ratio, for example, but the permissible range of composition was wide and usually approximate results were satisfactory.

In 1929, for the first time, I found it necessary to use four raw materials to make a product of a definite composition. The occasion was the placing in operation of an experimental plant.[6] The calculations required were long and tedious. Later, shorter procedures were found and finally the formulas explained in this paper were developed. This experience proved to be valuable, because within a few years the use of three or four materials at commercial plants became prevalent, because of diversification of products in the cement industry.

It is evident from the statement of Case 3, that as the number of required constituents in the product increases, the calculations become more complex. Although, theoretically, the number of constituents of the product does not determine the number of materials required, practically this seems to be the case. The formulas have been prepared to cover a maximum of four constituents in the product and in the four materials. Of course there is no limit to the number of materials or to the number of required constituents in the product. It should be possible to develop formulas for any number of materials and constituents, but the formuals would become complex.

In some cases, formulas for more than four materials or constituents would be helpful, but the need for them has not arisen with sufficient frequency to justify the labor of preparing them. A four-material problem of the type of Case 3 ordinarily is solved by setting up four equations in four unknowns, and then solving the equation. The practice has been to use numerical coefficients for the unknown quantities, which means that each problem must be solved individually. Regardless of the number of problems, it has been necessary to repeat the algebraic procedures of substitution, transposition and elimination for each problem.

It seemed that it should be possible to set up the original four equations using literal coefficients, and then solve these equations to obtain a value for each unknown in terms having general rather than particular values. Substitution of numerical values in the formulas should give the results sought. In other words, the formulas in effect would represent the common solution of such problems. There are many engineering formulas of this general type.

The best way to explain the procedure is to follow through an example:

Suppose it is required to manufacture a cement clinker of the following four-constituent composition:

1) Silica (SiO_2)   = 22.51%
2) Alumina (Al_2O_3)   =   4.19
3) Ferric oxide (Fe_2O_3) =   3.32
4) Calcium oxide (CaO) = 69.98

Such an analysis, which represents the major constituents of clinker, is obtained from a complete analysis by dividing the percentage of each major constituent by the sum of the four major constituents.

Four materials are available. The complete analysis of each is given in Table 4:1, and the calculated four-constituent analysis is given in Table 4:2. The latter eliminates, for the purpose of the calculation, the minor constituents, and in the

TABLE 4:1

ANALYSIS OF MATERIALS

Constituent No.

Material No.	1 SiO_2 %	2 Al_2O_3 %	3 Fe_2O_3 %	4 CaO %	MgO %	MnO %	Volatile %	Undetermined %
1 Sand	76.58	4.28	2.00	4.82	2.77	–	5.55	4.00
2 Slag	38.33	11.29	2.05	44.22	2.44	0.83	–	0.84
3 Limestone	0.67	0.31	0.20	54.18	0.90	–	43.48	0.26
4 Mill scale	0.00	0.00	99.58	0.00	–	–	–	0.42

Table 4:2

FOUR-CONSTITUENT ANALYSIS OF MATERIALS

Material No.	%, Constituent No.			
	1 SiO_2	2 Al_2O_3	3 Fe_2O_3	4 CaO
1 Sand	87.34	4.88	2.28	5.50
2 Slag	39.97	11.77	2.14	46.12
3 Limestone	1.21	0.56	0.36	97.87
4 Mill scale	0.00	0.00	100.00	0.00

case of the limestone, the carbon dioxide which will be lost in the kiln. *Mill scale* is a by-product of steel mills. The sample mentioned here has a total iron content of 99.58 per cent, calculated to Fe_2O_3. Much of the iron is present as Fe_3O_4, which in the kiln becomes oxidized to Fe_2O_3. Consequently, it is assumed for the purpose of calculation that the material is 100 per cent Fe_2O_3. The constituents of each of the materials, and of the product, are listed in Table 4:3.

Table 4:3

CONSTITUENTS OF MATERIALS AND OF PRODUCT

Material No.	%, Constituent No.			
	1	2	3	4
1	87.34	4.88	2.28	5.50
	a	e	i	n
2	39.97	11.77	2.14	46.12
	b	f	j	p
3	1.21	0.56	0.36	97.87
	c	g	k	q
4	0.00	0.00	100.00	0.00
	d	h	m	r
Product	2251	419	332	6998
	s	t	u	v

NOTE: The letters a, b, c, etc., are the algebraic quantities that represent the constituents.

The initial equations are developed as follows:

Let

a, b, c, d = percentage of the first constituent of each material, respectively

e, f, g, h = percentage of the second constituent of each material, respectively

i, j, k, m = percentage of the third constituent of each material, respectively

n, p, q, r = percentage of the fourth constituent of each material, respectively

s, t, u, v = 100 times the percentage of each of the four constituents of the product, respectively

x, y, z, w = the pounds of materials numbers 1, 2, 3, and 4 required for 100 pounds of the product, respectively

Then

$$
\begin{aligned}
s &= ax + by + cz + dw \\
t &= ex + fy + gz + hw \\
u &= ix + jy + kz + mw \\
v &= nx + py + qz + rw
\end{aligned}
$$

Although these are first-degree equations, their solution (for x, y, z, and w) involves a large number of binomials, the handling of which becomes rather complex. For example, the value of z expressed in terms of a, b, c, etc., involves 96 terms.

To simplify the solutions, the following substitutions are made:

Let

$$
\begin{aligned}
A &= a - d; & B &= b - d; & C &= c - d; & D &= s - 100d; \\
E &= e - h; & F &= f - h; & G &= g - h; & H &= t - 100h; \\
I &= i - m; & J &= j - m; & K &= k - m; & L &= u - 100m.
\end{aligned}
$$

Taking advantage of the fact that

$$
\begin{aligned}
s + t + u + v &= 100 \\
x + y + z + w &= 100
\end{aligned}
$$

the values for x, y, z, and w are found to be

$$
z = \frac{(DFI + AHJ + BEL) - (DEJ + BHI + AFL)}{(CFI + AGJ + BEK) - (CEJ + BGI + AFK)} \tag{1}
$$

$$
y = \frac{(DE - AH) - (CE - AG)z}{(BE - AF)} \tag{2}
$$

$$
x = \frac{D - By - Cz}{A} \tag{3}
$$

$$
w = 100 - (x + y + z) \tag{4}
$$

These equations represent the *materials formulas*.

The 22 intermediate equations are not shown. In such equations, it is customary to substitute original quantities, such as $a - d$ for A. In this work, however, it has been found convenient to leave the formulas in the form as shown, substituting the numerical values for A, B, C, etc., in a given problem to obtain the desired solution.

Typical Proportioning Calculations Employing Materials Formulas

The procedure is outlined so that it may be followed in detail. The calculations may be made without logarithms or with logarithms. The former has the advantage that only simple arithmetic is required, although the calculator must understand positive and negative quantities. The latter is much faster for one who can use logarithms with facility. As some of the quantities appear a number of times, only a few logarithms are required. It is desirable to employ six-place logarithm tables, not using proportional parts.

1. Calculate each material and the product to the basis of four major constituents by dividing each major constituent by the sum of the four major constituents in each case. Set down the percentage of each constituent of each material. Multiply by 100 the percentage of each major constituent of the product and set down the results:

Material No.	1(SiO$_2$)	2(Al$_2$O$_3$)	3(Fe$_2$O$_3$)	4(CaO)
		%, Constituent No.		
1 Sand	$a = 87.34$	$e = 4.88$	$i = 2.28$	$n = 5.50$
2 Slag	$b = 39.97$	$f = 11.77$	$j = 2.14$	$p = 46.12$
3 Limestone	$c = 1.21$	$g = 0.56$	$k = 0.36$	$q = 97.87$
4 Mill scale	$d = 0$	$h = 0$	$m = 100.00$	$r = 0$
Product	$s - 2251$	$t = 419$	$u = 332$	$v = 6998$

2. Set down the values for a, b, c, etc.; also, the values for $100d$, $100h$, and $100m$. Make the subtractions $a - b$, $b - d$, etc., designating the differences as A, B, C, etc.:

$$a = 87.34 \qquad s = 2251 \qquad g = 0.56 \qquad j = 2.14$$
$$d = 0 \qquad 100d = 0 \qquad h = 0 \qquad m = 100.00$$
$$A = 87.34 \qquad D = 2251 \qquad G = 0.56 \qquad J = -97.86$$

$$b = 39.97 \qquad e = 4.88 \qquad t = 419 \qquad k = 0.36$$
$$d = 0 \qquad h = 0 \qquad 100h = 0 \qquad m = 100.00$$
$$B = 39.97 \qquad E = 4.88 \qquad H = 419 \qquad K = -99.64$$

$$c = 1.21 \qquad f = 11.77 \qquad i = 2.28 \qquad u = 332$$
$$d = 0 \qquad h = 0 \qquad m = 100.00 \qquad 100m = 10000$$
$$C = 1.21 \qquad F = 11.77 \qquad I = -97.72 \qquad L = -9668$$

Calculations Without Logarithms

3. Set down the values for A to I. Make the multiplication $A \times F$, $A \times G$, etc.

$A =$	87.34	$B =$	39.97	$F =$	11.77
$F =$	11.77	$E =$	4.88	$I =$	-97.72
$AF =$	1027.99	$BE =$	195.05	$FI =$	-1150.16
$A =$	87.34	$C =$	1.21	$G =$	0.56
$G =$	0.56	$E =$	4.88	$I =$	-97.72
$AG =$	48.91	$CE =$	5.90	$GI =$	-54.72
$A =$	87.34	$D =$	2251.00	$H =$	419.00
$H =$	419.00	$E =$	4.88	$I =$	-97.72
$AH =$	36595.46	$DE =$	10984.88	$HI =$	-40944.68

4. Set down the values for D, FI, AH, J, etc. Make the multiplications $D \times FI$, $AH \times J$, etc.:

$D =$	2251.00	$DE =$	10984.88
$FI =$	-1150.16	$J =$	-97.86
$DFI =$	-2589010.16	$DEJ =$	-1074980.36
$AH =$	36595.46	$B =$	39.97
$J =$	-97.86	$HI =$	-40944.68
$AHJ =$	-3581231.72	$BHI =$	-1636558.86
$BE =$	195.05	$AF =$	1027.99
$L =$	-9668.00	$L =$	-9668.00
$BEL =$	-1885743.40	$AFL =$	-9938607.32
$C =$	1.21	$GE =$	5.90
$FI =$	-1150.16	$J =$	-97.86
$CFI =$	-1391.69	$CEJ =$	-577.37
$AG =$	48.91	$B =$	39.97
$J =$	-97.86	$GI =$	-54.72
$AGJ =$	-4786.33	$BGI =$	-2187.16
$BE =$	195.05	$AF =$	1027.99
$K =$	-99.64	$K =$	-99.64
$BEK =$	-19434.78	$AFK =$	-102428.92

5. Set down the values for DFI, AHJ, etc., and add each column separately:

$DFI =$	-2589010	$DEJ =$	-1074980
$AHJ =$	-3581232	$BHI =$	-1636559
$BEL =$	-1885743	$AFL =$	-9938607
Total $=$	-8055985	Total $=$	-12650146

$$CFI = -1392$$
$$AGJ = -4786$$
$$BEK = -19435$$
$$\text{Total} = -25613$$

$$CEJ = -577$$
$$BGI = -2187$$
$$AFK = -102429$$
$$\text{Total} = -105193$$

6. Substitute the numerical values of DFI, AHJ, etc., in Equations 1, 2, 3, and 4, and solve for x, y, z, and w. The results are

$$x = \frac{2251 - 39.97(27.77) - 1.21(57.73)}{87.34}$$

$$= \frac{2251 - 1109.97 - 69.85}{87.34} = \frac{1071.18}{87.34}$$

$$= 12.26 \text{ lb of material No. 1 (sand)}$$

$$y = \frac{-25610.58 - (-43.01)57.73}{-832.94} = \frac{-23127.61}{-832.94}$$

$$= 27.77 \text{ lb of material No. 2 (slag)}$$

$$z = \frac{(-8055985) - (-12650146)}{(-25613) - (-105193)} = \frac{4594161}{79580}$$

$$= 57.73 \text{ lb of material No. 3 (limestone)}$$

$$w = 100 - (12.26 + 27.77 + 57.73)$$
$$= 2.24 \text{ lb of material No. 4 (mill scale)}$$

Calculations With Logarithms

For convenience in finding the logarithms, the equations are given here in the order 2, 3, 4, and 1. For the same reason, the logarithms for I, J, K, and L are first down in connection with Equation 2, although they are not used until later. After the value for z has been found in Equation 1, it is available for use in Equation 2.

7. Set down the logarithm for each quantity, as indicated. For Equation 2:

Log D	= 3.352375	
Log E	= 0.688420	
Log DE	= 4.040795	
DE =		10958
Log A	= 1.941213	
Log H	= 2.622214	
Log AH	= 4.563427	
AH =		36590
$DE - AH$ =		-25605
Log B	= 1.601734	
Log E	= 0.688420	
Log BE	= 2.290154	
BE =		195.1

Log C	= 0.082785	
Log E	= 0.688420	
Log CE	= 0.771205	
CE =		5.905
Log A	= 1.941213	
Log G	= $\bar{1}$.748188	
Log AG	= 1.689401	
AG =		48.910
$CE - AG$ =		-43.000
Log I	= 1.989983	
Log J	= 1.990605	
Log K	= 1.998434	
Log L	= 3.985337	

$$\text{Log } A \ = 1.941213$$
$$\text{Log } F \ = 1.070776$$
$$\text{Log } AF = 3.011989$$
$$AF = \qquad 1028$$
$$BE - AF = \qquad -833$$

$$(DE - AH) = \qquad -25605$$
$$\text{Log } (CE - AG) \ = 1.633468$$
$$\text{Log } z \qquad = 1.761498$$
$$\text{Log } (CE - AG)z = 3.394966$$
$$(CE - AG)z = \qquad -2483$$
$$(DE - AH) - (CE - AG)z = \qquad -23122$$
$$\text{Log } (DE - AH) - (CE - AG)z = \qquad 4.363988$$
$$(BE - AF) = \qquad -833$$
$$\text{Minus log } (BE - AF) = \qquad 2.920645$$
$$\text{Log } y = \qquad 1.443343$$
$$y = \qquad 27.75$$

For Equation 3:

$$\text{Log } B \ = 1.601734 \qquad\qquad \text{Log } C \ = 0.082785$$
$$\text{Log } y \ = 1.443343 \qquad\qquad \text{Log } z \ = 1.761498$$
$$\text{Log } By = 3.045077 \qquad\qquad \text{Log } Cz = 1.844283$$
$$By = \qquad 1110 \qquad\qquad Cz = \qquad 69.87$$

$$(D - By - Cz) = 1071$$
$$\text{Log } (D - By - Cz) = \qquad 3.029789$$
$$\text{Minus log } A = \qquad 1.941213$$
$$\text{Log } x = \qquad 1.088576$$
$$x = \qquad 12.26$$

For Equation 4:

$$x = 12.26 \qquad y = 27.75 \qquad z = 57.74 \qquad x + y + z = 97.75$$
$$w = 100 - 97.75 = 2.25$$

For Equation 1:

$$\text{Log } D \quad = 3.352375 \qquad\qquad \text{Log } DE \ = 4.040795$$
$$\text{Log } F \quad = 1.070776 \qquad\qquad \text{Log } J \ = 1.990605$$
$$\text{Log } I \quad = 1.989983 \qquad\qquad \text{Log } DEJ = 6.031400$$

$$\text{Log } DFI \ = 6.413134$$
$$DFI = \qquad -2589000 \qquad\qquad DEJ = \qquad -1075000$$

$$\text{Log } AH \ = 4.563427 \qquad\qquad \text{Log } B \quad = 1.601734$$
$$\text{Log } J \ = 1.990605 \qquad\qquad \text{Log } H \quad = 2.622214$$
$$\text{Log } AHJ = 6.554032 \qquad\qquad \text{Log } I \quad = 1.989983$$
$$\qquad\qquad\qquad\qquad\qquad\qquad \text{Log } BHI \ = 6.213931$$

$$AHJ = \qquad -3581000 \qquad\qquad BHI = \qquad -1637000$$

$$\text{Log } BE \ = 2.290154 \qquad\qquad \text{Log } AF \ = 3.011989$$
$$\text{Log } L \ = 3.985337 \qquad\qquad \text{Log } L \ = 3.985337$$
$$\text{Log } BEL = 6.275491 \qquad\qquad \text{Log } AFL = 6.997326$$
$$BEL = \qquad -1886000 \qquad\qquad AFL = \qquad -9939000$$

$$(1) = -8056000 \qquad\qquad\qquad (2) = -12651000$$

$$
\begin{aligned}
\text{Minus } (2) &= -12651000 \\
\text{Numerator} &= 4595000 \\
\text{Log of numerator} &= 6.662286
\end{aligned}
$$

Log C	= 0.082785	Log CE	= 0.771205
Log F	= 1.070776	Log J	= 1.990605
Log I	= 1.989983	Log CEJ	= 2.761810
Log CFI	= 3.143544	CEJ =	−577.8
CFI =	−1392		

Log AG	= 1.689401	Log B	= 1.601734
Log J	= 1.990605	Log G	= 1̄.748188
Log AGJ	= 3.680006	Log I	= 1.989983
		Log BGI	= 3.339905
AGJ =	−4786	BGI =	−2187

Log BE	= 2.290154	Log AF	= 3.011989
Log K	= 1.998434	Log K	= 1.998434
Log BEK	= 4.288588	Log AFK	= 5.010423
BEK =	−19440	AFK =	−102430

$$(3) = -25618 \qquad\qquad (4) = -105195$$

$$
\begin{aligned}
\text{Minus } (4) &= -105195 \\
\text{Denominator} &= 79577 \\
\text{Log of denominator} &= 4.900788
\end{aligned}
$$

$$
\begin{aligned}
\text{Log of numerator} &= 6.662286 \\
\text{Log of denominator} &= 4.900788 \\
\text{Log } z &= 1.761498 \\
z &= 57.74
\end{aligned}
$$

The number of decimal places retained determines the accuracy of the calculations. Multiplications may be made with absolute accuracy by retaining all decimals, but this is not always true with divisions. A few trials will indicate how far decimals must be carried to give the desired accuracy. It has been found that in step 4, only whole numbers need be used, as a rule.

It is necessary to pay careful attention to negative signs. A single error will lead to an incorrect result for the entire calculations—unless there happen to be compensating errors. It has been found helpful, in the interest of accuracy, to place either a positive or negative sign before each quantity—in place of designating only negative quantities, as is customary.

When the calculations are proved, it is found that the result agrees exactly with the four-constituent analysis of the product given previously. The complete analysis of the product now may be calculated. It is:

$$SiO_2 = 21.78\%$$
$$Al_2O_3 = 4.05$$
$$Fe_2O_3 = 3.22$$
$$CaO = 67.71$$
$$MgO = 1.97$$
$$MnO = 0.23$$

Undetermined
(alkalies, volatile, etc.) = 1.04

The product represents what may be termed ideal clinker, that is, what would be obtained if there were no changes in the manufacturing process except those which have been taken into account in developing the formulas.

There are only two conditions under which results cannot be obtained with the formulas, taking into account the limitations that have been pointed out already.

1. It is not always possible to obtain the desired product with the materials available. In this case, it is necessary to make some change in the materials, or to be content with some modification in the product. When this is not apparent from general observation of the constituents, the calculation should be started in the usual way. Generally, one of the material weights will appear as a negative quantity. In any event, the results obtained cannot be proved, and, therefore, are obviously incorrect.

2. It may happen, that due to some duplication in the values for a to r, $BE = AF$, in equation [2]. This results in a zero value for the denominator and the equation cannot be solved. In this case, the value for y may be found from the equation

$$y = \frac{HI - EL}{FI - EJ} - \frac{GI - EK}{FI - EJ} z$$

A number of preliminary calculations may be made from which some information of interest may be obtained without completing the whole procedure. For example, referring to Table 4:3:

$$s/a = 25.77 \qquad\qquad t/f = 35.60$$
$$v/q = 71.50 \qquad\qquad u/m = 3.32$$

This means that the quantity of material No. 1 required for 100 pounds of product will not be greater than 25.77

pounds; of No. 2, not greater than 35.60 pounds; and so on. Such information might be worth while if one material were relatively expensive, or the supply limited.

There is some choice in selecting the major constituents of the materials. For example, if the iron content of all the materials were low, and the magnesia content of the product relatively high, MgO could be used in place of Fe_2O_3 in the calculations.

The use of the formulas is not restricted to four materials nor to four constituents. They may be employed for a smaller number of either materials or constituents, assigning zero values to some of the quantities, as indicated previously. It is difficult, however, to find two or even three materials that will give a product for which the percentage of each of four constituents is specified. This means that, within reasonable limits, as the number of available materials increases, the probability of finding a solution for a proportioning problem increases.

Under some conditions it is possible to use the formulas for more than four materials. Suppose there are five materials, each containing the same constituents, and it is satisfactory that the ratio of material No. 4 to No. 5 remain constant. Then, the analysis of a given mixture of the two may be calculated, and the result handled in the formulas as the analysis of a single material. Also, an additional constituent can be introduced, in a fifth material, if the fifth material consists of this constituent only, and the constituent is present in none of the other four materials.

Considerable information may be obtained readily by the use of a graph such as shown in Figure 4:1. As the scale of the graph increases, the accuracy of such information increases. The abscissa indicates weights of materials, and the ordinate indicates weights of constituents of materials. Horizontal broken lines indicate the pounds of each constituent in 100 pounds of the product. The graph shown here has been plotted for the example mentioned in this paper.

If the product is to contain 69.98 per cent of calcium oxide, at least one material must contain more than that percentage of calcium oxide. This is indicated on the graph by a calcium-oxide line of a material crossing the calcium-

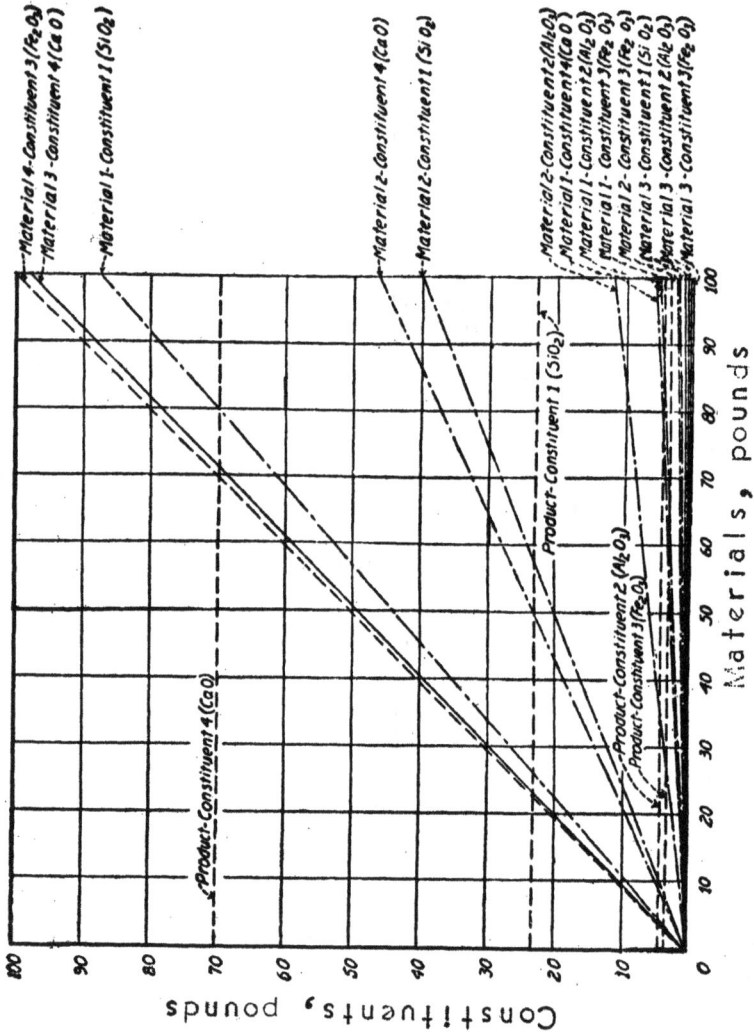

FIG. 4:1. MATERIALS AND CONSTITUENTS FOR 100 POUNDS OF PRODUCT

oxide line of the product. Only one such line crosses, that representing the calcium oxide in the limestone. In the same way it may be observed that only two materials, slag and sand, can supply sufficient silica; only one, sufficient alumina; and only one, sufficient iron. The points on the graph at which the diagonal lines cross the horizontal lines indicate how much of a given material would be required in order to supply all of a given constituent in the product.

On the basis of the maximum quantities of materials, a guess of the quantities required for the product may be made. The composition of the product resulting from such a proportion of the materials can be found from the curve, another guess may be made, and so on. Theoretically, the difference between the assumed product and the desired product will continue to diminish, so that if a sufficient number of trials is made, the correct result will be obtained. It is probable, however, that such a trial-and-error method would require more time than the actual calculation in most cases. It is necessary to prepare a graph for each product and set of materials. Time may be saved by providing forms in advance, so that to prepare a graph it is necessary only to draw a few straight lines.

In using the formulas for a series of proportioning calculations, the product or one or more of the materials remain the same. In such cases, some steps are common to the series and advantage of them may be taken to decrease the average work per calculation. The preparation of auxiliary tables and graphs will decrease, still further, the required work.

The first proportioning calculations for four raw materials required approximately 8 hours. By the use of the formulas, with logarithms, a set of calculations has been made in about 45 minutes. As stated previously, the work may be facilitated in various ways, such as by preparing calculating forms, and by the use of auxiliary tables and graphs. At a given plant, there may be some short cuts that are not applicable to the procedure in general. For one plant, some short cuts were found by showing the quantities representing the constituents of each raw material, and of the clinker in a separate color. When this was done, a number of relations that were not seen previously, at once appeared.

TABLE 4:4

A CALCULATING FORM FOR THE PROPORTIONING OF MATERIALS *

Raw Materials	1	2	3	4	5	6	7	8	9	10	11	12
	Constituents: Original Basis Pounds per 1000 Pounds				Sum of constituents	Reciprocal of sum	Constituents: Four-Component Basis, Pounds per 100 Pounds				Materials Required, 4-Component Basis	Materials Required, Original Basis
	SiO_2	Al_2O_3	Fe_2O_3	CaO			SiO_2	Al_2O_3	Fe_2O_3	CaO		
(1)	....						(a)	(e)	(i)	(n)	(x)	
(2) Shale							(b)	(f)	(j)	(p)	(y)	
(3) Stone							(c)	(g)	(k)	(q)	(z)	
(4)							(d)	(h)	(m)	(r)	(w)	
Clinker							$(s)^a$	$(t)^a$	$(u)^a$	$(v)^a$		

^{a} Multiply by 100

13

$A = a - \ldots$	$d = \ldots$
$B = b - \ldots$	$d = \ldots$
$C = c - \ldots$	$d = \ldots$
$D = s - 100d = \ldots$	$100d = \ldots$
$E = e - \ldots$	$h = \ldots$
$F = f - \ldots$	$h = \ldots$
$G = g - \ldots$	$h = \ldots$
$H = t - 100h = \ldots$	$100h = \ldots$
$I = i - \ldots$	$m = \ldots$
$J = j - \ldots$	$m = \ldots$
$K = k - \ldots$	$m = \ldots$
$L = u - 100m = \ldots$	$100m = \ldots$

14

$(+) DFI = \ldots$
$(+) AHJ = \ldots$
$(+) BEL = \ldots$
$(-) DEJ = \ldots$
$(-) BHI = \ldots$
$(-) AFL = \ldots$
$\qquad M = \ldots$

15

$(+) CFI = \ldots$
$(+) AGJ = \ldots$
$(+) BEK = \ldots$
$(-) CEJ = \ldots$
$(-) BGI = \ldots$
$(-) AFK = \ldots$
$\qquad N = \ldots$

16

$(+) DE = \ldots$
$(-) AH = \ldots$
$(-) CEz = \ldots$
$(+) AGz = \ldots$
$\qquad R = \ldots$

17

$(+) BE = \ldots$
$(-) AF = \ldots$
$\qquad S = \ldots$

18

$(+) D = \ldots$
$(-) By = \ldots$
$(-) Cz = \ldots$
$\qquad T = \ldots$

$$\frac{M}{N} = z \qquad \frac{R}{S} = y \qquad \frac{T}{A} = x$$

$$100 - x - y - z = w$$

Directions: Set down the analyses in columns 1–4. Add the quantities in each line and place the results in column 5. Find the reciprocal of each and place in column 6. Multiply each quantity in columns 1–4 by the corresponding reciprocal, and place the results in columns 7–10. Make the subtractions indicated in column 13, and the multiplications indicated in columns 14–18. By addition and subtraction, find the values M, N, R, S, and T. Find the values z, y, x, and w as indicated, placing them in the proper order in column 11.

* Prepared with the cooperation of George C. Wyatt, of the Marchant Calculating Machine Company (now Marchant Division, SCM Corporation), who pointed out the most direct sequences for making the required calculations.

Table 4:4 is a form for use with a Marchant calculator. In comparing Table 4:4 with the calculations by logarithms, it will be noted that some of the calculations have been omitted, and that the whole procedure is much more direct. The plus and minus signs in columns 14 to 18 indicate the usual sign of the products DFI, AHJ, and so on. It is well to become familiar with the calculating procedure as originally published, before following Table 4:4.

The uses of the proportioning formulas are not restricted to plant control. For example, they are helpful in studying a series of raw materials available at a given plant to select those that can be used most economically. The procedure is to calculate the cost of 4 pounds of each of the principal constituents of each material on the ignited basis—4 pounds being considered, for convenience, 1 per cent of the weight of a barrel of clinker. Graphs may be drawn to facilitate comparisons.

The accuracy of the proportioning of raw materials may affect each succeeding operation in the manufacturing process, and the uniformity of the product. If the cost per unit weight of the raw materials differs, as is usual, the minimum cost for raw materials will be obtained when the proportioning is most accurate. Other factors being equal, uniformity tends to reduce power for grinding raw materials and clinker, fuel for clinkering, and kiln refractories.

REFERENCES

1. J. C. Witt, *Trans. Am. Soc. Mech. Eng.*, 60:493 (1938); *Concrete, Cement Mill Section*, 46:265, 284 (1938).
2. R. K. Meade, *Portland Cement*, Chemical Publishing Co., Inc., Easton, Pa., 1930.
3. H. H. Steinour and H. Woods, *Rock Products*, 33(6):75 (1930).
4. L. A. Dahl, Unpublished manuscript, 1936.
5. J. C. Witt, *J. Ind. Eng. Chem.*, 12(6):591 (1920).
6. J. C. Witt, *Concrete, Cement Mill Section*, 40(10):35 (1932).

Chapter 5

PREPARING THE RAW MIX

In most chemical reactions the reacting substances are gases, vapors, liquids, or solutes. In each of these types the reacting particles are submicroscopic. They may be molecular, or less than molecular in size. The relation of surface to weight is very great. Also, because of the mobility of such small particles, the blending of the reacting substances is simple. The particles entering a cement kiln, on the other hand, are solids. The size is very large, in comparison with molecules, and the ratio of surface to weight is small.

In the preparation of raw mix the objectives are:

a) The raw materials shall be present in the desired relative quantities;

b) They shall be blended as completely as practicable;

c) The ratio of surface to weight shall be sufficiently great to permit the chemical reactions to reach equilibrium.

Theoretically, there is no limit to the decrease in particle size except the molecule. For a given set of conditions, however, there is an economic limit. This is the point at which the cost of further decrease in size is not justified. It is customary to grind raw mix to such fineness that at least 85 to 90 per cent by weight will pass through a 200-mesh sieve. The clear opening of such a sieve is 74 microns, or 0.074 millimeter, equivalent to approximately 0.0029 inch. For cement, the particle size is usually expressed in terms of *specific surface*. This unit has been applied to raw mix and to coal, but it has not been used very extensively, to date. The unit is defined in Chapter 9.

Number of Raw Materials

Generally, there are two or more raw materials, the number seldom exceeding four. For convenience, the preparation of raw mix will be discussed here on the basis of the two raw materials, limestone and shale.

Reclaiming the Raw Materials

Two procedures for assembling materials will be mentioned. 1) Under the storage receptacle for each raw material, there is a series of hoppers discharging through feeders to one or more conveyor belts. 2) The raw material is transferred by a crane from storage to feeders that supply mills.

Relative Quantities of Raw Materials

Materials may be proportioned by weight or by volume, or by some combination of these. Weight is the direct basis; volume, the indirect. If, for convenience or some other reason, materials are proportioned by volume, it must be kept in mind that volume is employed as a substitute for weight, and the relation of the two must be known. It is obvious that if any change in this relation takes place, an adjustment must be made, or an error results.

Choice of Process

One of the first decisions that must be made, in designing a cement plant, is whether the dry process or the wet process will be employed. This comes about from the fact that materials of certain moisture contents cannot be finely ground. The water must be either reduced to a low percentage, or it must be present in relatively large quantities. The choice of process affects primarily the preparation of the raw mix, but the influence is apparent in most of the departments of the plant. Although either process may be used in most cases, the problem is so important that very careful study is justified. An incorrect decision may jeopardize the success of the project. It is a matter of listing as completely as possible the advantages and disadvantages of both processes, and finding the resultant. Here are six of the items that are to be taken into consideration:

1. Water content of the raw materials
2. Water supply at the plant
3. Cost of fuel
4. Power supply
5. Dust collection
6. Uniformity of the kiln feed

Water Content of the Raw Materials:

Water content is one of the most important items. If the primary materials are wet when received at the plant, the wet process is suggested because the water must be removed at some point in the manufacturing process. Examples of such materials are marl, granulated blast furnace slag, and some clays. If the materials contain, in general, only surface water, the addition of large quantities of water, to be removed later in the process, may be difficult to justify.

Water Supply:

An adequate, dependable water supply is a prerequisite for the wet process. Also, there should be facilities for water disposal.

Cost of Fuel:

Relatively expensive fuel suggests the dry process, because when this process is selected the kilns consume less fuel.

Power Supply:

If the power is to be purchased, or is to be generated in direct-fired boilers, power is not an important consideration in the selecting of the process. If the power is to be generated in waste-heat boilers, the dry process is favored. In the wet process, the heat required to evaporate water in the kilns is not available for the boilers.

Dust Collection:

Less dust is to be expected in the wet process than in the dry process.

Uniformity:

Frequently, it is said that wet-process cement is more uniform than the dry-process product. Such a statement may be misleading. The wet process favors the preparation of a uniform kiln feed, because of the facility by which particles of materials suspended in water may be blended. Of two plants, designed without special attention to uniformity, the wet process probably will be the better. However, a dry-process plant can be designed to produce a cement sufficiently uniform for any commercial use. The weighing of dry materials, after preliminary grinding, is an advantage which the wet process does not have.

There are so-called semi-wet procedures, in which features of the dry and wet processes are combined. These have not become extensively established in the industry. An example [1] is the mixing of slurry with previously dried slurry.

The importance of each of these factors in the characteristics of the process is well understood by cement manufacturers, but they are not always evaluated in the same way. For many years, nearly all new plants have employed the wet process. An exception is the Lehigh (Alsen) plant,[2] in New York. The first plant on that site used the dry process; the second, the wet; and the third, the dry process. Possibly dry process plants may again predominate.

Generally, the total weight of dry raw materials required for a barrel of clinker is from 550 to 650 pounds. In this book, the weight of a barrel of clinker is considered identical with the weight of a barrel of cement, 376 pounds. Actually, the clinker required to produce a barrel of cement is a few per cent less than this amount, but the weight can be calculated for a given set of conditions only. If substantially all the calcium in a raw mix is present as the carbonate, and nothing is volatilized in appreciable amounts except carbon dioxide, the minimum total weight of raw materials for a barrel of clinker is approximately $376 + 297c$. Here, $c =$ the per cent of calcium oxide, expressed as a decimal, in the clinker. For example, if the clinker contains 65 per cent of calcium oxide, the required total weight of raw materials, in pounds, is $376 + 0.65(297) = 569$. This formula is not applicable when blast furnace slag is one of the raw materials. For each per cent of magnesia in the clinker, there is required an addition of approximately 4 pounds to the total raw materials.

Dry Process

The principal operations in the preparation of raw mix by the dry process are drying, blending, preliminary grinding and finishing grinding. There is no rule of order for these operations, but here are some sequences that have been employed. For the sampling, and the analysis of raw materials, see Chapter 14.

1. The materials are proportioned, dried, ground, and blended. (This has the undesirable feature that an allowance must be made for the moisture content of each material.)

2. The materials are dried separately. Then they are proportioned, ground, and blended.

3. Each material is separately dried, given a preliminary grinding, and stored. The materials are then proportioned, passed through the finishing mills, and blended. When the details of this procedure are controlled carefully, a raw mix, having a very high degree of uniformity, is to be expected.

Drying:

One of two procedures may be employed for drying raw materials. They are passed through dryers before entering the preliminary mills, or they are dried during the preliminary grinding.

There are various types of dryers. A common type is a rotary steel cylinder, lined with refractory brick, and inclined from the horizontal about 0.5 to 0.75 inch per foot. The diameter is from 5 to 8 feet, and the length from 50 to 80 feet. For a direct-fired dryer, the heat is supplied by the combustion gases from a stationary furnace at the lower end of the dryer. The furnace may be hand-fired, stoker-fired, or pulverized-coal-fired. Fuel oil or natural gas may be used. In some dryers, waste heat is substituted for some or all of the fuel. This type of dryer is simple in construction and operation, but, in general, the heat transfer is poor, and the thermal efficiency, low.

The use of rotary dryers has decreased because most of the recently built plants employ the wet process, and because

the application of the drying-during-grinding principle is increasing. When conditions are suitable, raw materials can be
dried in the preliminary mills. There is considerable heat
generated by friction. This may be augmented by waste heat
from the kilns, or by the installation of auxiliary heating
units.

In the drying process, the raw materials for cement have
certain advantages in comparison with some other materials.
For example, they may come into contact with combustion
gases without injury; the vapor pressure of the water present
is not lowered appreciably by the presence of any substances
in solution; it is not necessary to evaporate all the water.

The temperature of the materials leaving the dryers should
be no higher than has been found necessary to obtain satisfactory results. One reason for this is economy of fuel; another is that excessive heat in materials entering the mills is
not desirable. The following is quoted from a description of
the Lehigh (Alsen) plant,[3] at Cementon, New York:

> An 18-inch belt conveyor, 67-ft centers, then delivers the
> clay into a 7-ft, 6-in by 57-ft rotary dryer. This dryer is not
> intended to remove all the moisture from the clay, since the
> clay together with correct proportions of stone are later
> ground together in a mill that dries the material during the
> process of grinding. Just enough moisture is removed to
> facilitate the storing and handling of the clay.

Preliminary and Finishing Grinding:

In general, the same types of mills are employed in the
grinding of raw materials and clinker. Some examples will be
discussed in Chapter 9. Here, it is sufficient to say that in
preliminary mills, the raw materials, either separately or together, are ground to 20-mesh fineness. The mixed materials
then pass through the finishing mills, in which the fineness
is brought to 200-mesh.

Open and Closed Circuits:

In open-circuit grinding, a material passes through a mill
only once. In closed-circuit grinding, the product of a mill
goes to some classifying device such as a screen, or an air
separator and the coarse particles return to the mill. The

equipment and procedure, which are much the same for dry raw materials and for clinker, will be discussed in Chapter 9.

The product discharged from the finishing mills is known as *raw mix*. This may be sent directly to the kilns. Generally, however, there is some additional processing by which the raw mix is converted into kiln feed.

Collector Dusts:

The most troublesome class of materials handled at a cement plant comprises the collector dusts. Even the terminology is difficult. It is customary to install dust collectors in almost every department of a plant. The design of these collectors is based on various physical phenomena, of which inertia, filtration, sedimentation, and electrical precipitation are examples. The operation of collectors requires the removal of the dusts that have accumulated in them, and procedures for disposal must be developed. There will be mentioned here only the electrical precipitation type of collector, developed by Cottrell, and employed for the removal of dust from kiln gases (see Chapter 7).

Usually, this dust is returned to the process. It is equivalent to an additional raw material, and its effect on the composition of the raw mix must be taken into account. Dust may be added to one of the raw materials, or to the raw mix. In some installations, devices for introducing the dust directly into the kiln are provided.

Blending System:

At some plants silos are employed for blending raw mix. The number, diameter, and height of the silos vary greatly. They are designed to contain a supply of raw mix sufficient for the operation of the kilns for a given number of days.

That blending silos increase uniformity is not questioned. It may be said, however, that no matter how carefully a system has been designed and the operating schedule has been formulated, the results may not be according to expectations. One reason for this is that many variables are involved and it is difficult, if not impossible, to take all of them into account. Too much dependence may be placed on a blending system. When a decision has been made to construct the silos, there is a tendency to omit details in equipment for proportioning and mixing materials before they reach the silos, on the basis that the system will correct all irregularities. The result of this policy may be a blending system much larger than otherwise would be necessary, and a kiln feed that does not possess the desired uniformity.

Wet Process

In general, similar types of mills and similar operations are employed in both the dry and wet processes. In the wet process, however, there is an even greater variety in the schedules of operations. The most important single factor in determining the schedule is the physical characteristics of the primary raw materials as received at the plant. If these are limestone and shale with little plasticity, they may be proportioned by weight, or by volume, and ground, with the addition of water, in the preliminary and the finishing mills. If one material is a wet, finely divided clay, it is customary to mix it with additional water in a wet mill, forming a suspension, called a *clay slip*. This slip is added at a controlled rate to the limestone as it enters the preliminary mills. Various other sequences are in use. Regardless of the schedule, the raw mix produced by the wet process is a suspension of solids in water, called *slurry*. The solids are ground so that at least 85 per cent will pass through a 200-mesh sieve. Usually, the water content is 35 to 40 per cent by weight.

Preliminary and Finishing-Grinding:

As an example, the procedure at one plant will be outlined. The primary raw materials are limestone and shale crushed to ¾-inch maximum size. There are provisions for handling three secondary materials. After proportioning, the materials go to two *preliminators*, 9 feet, 6 inches in diameter, and 5 feet, 6 inches long. Each preliminator has a capacity of 30 to 40 tons an hour, and 35 to 45 per cent of the discharged material passes through a 200-mesh sieve.

Water is added to produce a mixture containing approximately 38 per cent of water. The product of these mills is pumped to two electromagnetic vibrating screens, measuring

4 by 6 feet. The tailings return to the preliminators. The fines are pumped to four tube mills, each 6 feet in diameter and 20 feet long. The discharge of these mills is the slurry. In studying the finishing-grinding of raw materials, in either the dry or the wet process, the proportion of each raw material coarser than 200-mesh should be known.

Collector dusts:

In the wet process, these materials cannot be handled in the same manner as in the dry process. The disposal of them will be mentioned in Chapter 7.

Blending System:

In the wet process materials can be blended with more facility than in the dry process. Sometimes, also, this is more greatly needed. This is true when it is not practicable to proportion the raw materials so accurately as in the dry process. The customary layout of a blending system includes a series of reinforced concrete silos, and such auxiliary items as valves, pipe lines, pumps, and stirring equipment. In modern installations, provision should be made for secondary, as well as for primary raw materials.

At a certain plant, there are eight principal silos, 25.5 feet in diameter, and 39 feet in depth; and four silos, 15 feet in diameter, and 39 feet in depth. All silos are equipped for stirring mechanically, and by compressed air; and for the circulation of slurry, from tank to tank in any sequence. The principal tanks are operated on a four-day schedule. The first two silos receive the slurry discharged by the tube mills. In the second two, the slurry is blended, the composition is adjusted, if this is desirable, and samples are taken. In the third two, the slurry is held for analysis, and any further adjustment in composition. The last two silos contain the kiln feed.

The auxiliary silos provide storage for slurries held for adjustments in composition. Also, they are available in emergencies, and during periods in which slurry for a different type of clinker is being prepared. The blending system was designed to provide the maximum flexibility that might be required by manufacturing procedures for some time to come.

The operation of the blending system is facilitated by the use of tables and graphs showing the volume of slurry, the weight of solids, the equivalent barrels of clinker, and other data, for each inch of depth. The weight-volume relation is affected by several items which can be evaluated and taken into account by employing correction factors.

The Solids Contents of Slurry

Table 5:1 was prepared for the plant just described. It is for convenience in handling slurry, particularly in combining slurry from two or more tanks, to obtain a desired composition. The values shown are on the basis that (1) one cubic

Table 5:1

IGNITED SOLIDS PER INCH OF DEPTH IN SLURRY TANKS

Total Depth (inches)	Ignited Solids (cwt) Principal Tanks	Auxiliary Tanks	Total Depth (inches)	Ignited Solids (cwt) Principal Tanks	Auxiliary Tanks
1	16.3	5.7	200	3268.6	1131.0
2	32.7	11.3	210	3432.0	1187.6
3	49.0	17.0	220	3595.5	1244.1
4	65.4	22.6	230	3758.9	1300.7
5	81.7	28.3	240	3922.3	1357.2
6	98.0	33.9	250	4085.8	1413.8
7	114.4	39.6	260	4249.2	1470.3
8	130.7	45.2	270	4412.6	1520.9
9	147.0	50.9	280	4576.0	1583.4
10	163.4	56.6	290	4739.5	1640.0
20	326.9	113.1	300	4902.9	1696.5
30	490.3	169.7	310	5066.3	1753.0
40	653.7	226.2	320	5229.8	1809.6
50	817.2	282.8	330	5393.2	1866.2
60	980.6	339.3	340	5556.6	1922.7
70	1144.0	395.9	350	5720.0	1979.3
80	1307.4	452.4	360	5883.5	2035.8
90	1470.9	509.0	370	6046.9	2092.4
100	1634.3	565.5	380	6210.3	2148.9
110	1797.7	622.0	390	6373.8	2205.5
120	1961.2	678.6	400	6537.2	2262.0
130	2124.6	735.2	410	6700.6	2318.6
140	2288.0	791.7	420	6864.0	2375.1
150	2451.5	848.3	430	7027.5	2431.7
160	2614.9	904.8	440	7190.9	2488.2
170	2778.3	961.4	450	7354.4	2544.8
180	2941.7	1017.9	460	7517.8	2601.3
190	3105.2	1074.5			

foot of slurry contains 60 pounds of dry solids, and (2) the
loss on ignition of the dry solids is 36 per cent.*

The contents of the tanks have been calculated from the
diameters (25.5 feet for the principal tanks, and 15 feet for
the auxiliary tanks), and the depth (39 feet, or 468 inches).
No corrections for any variations in diameter that may exist
nor for the space occupied by mixing equipment have been
made.

It is customary to blend slurries on the basis of volume.
Since, however, clinker composition can be calculated only
on the basis of the weight of ignited raw materials, calculations on the basis of volume are indirect. The blending of
two or more slurries by volume is permissible only when the
ignited solids content of each slurry differs within narrow
limits, or when correction factors are applied.

The weight of ignited solids is shown for each inch of
depth from 1 to 9, and then for each 10 inches. To obtain
the weight for a depth that is not a multiple of 10, two values
are added. For example, the weight for 18 inches in one of
the principal tanks equals the weight for 10 plus the weight
for 8 inches, or $163.4 + 130.7 = 294.1$. In the same way,
the depth (in whole inches) corresponding to a given weight
may be found.

Fractionating Raw Materials

A raw mix is usually prepared by a cumulative procedure.
In general, each raw material received at a plant contains
all the elements required. The minimum number of materials
needed to produce a mix containing the elements in the desired proportions are used. Stated in another way, the purpose of obtaining additional raw materials is not to introduce
an additional element, but to modify the proportions of the
elements already present.

Early in 1932, a study of these principles suggested an entirely different procedure. It should be possible to make a
raw mix from a single material by fractionating the material,

* If the slurry contains more or less than 60 pounds of dry solids per cubic
foot, or if the loss on ignition is more or less than 36 per cent, a correction-
factor chart is employed.

and then combining the fractions in the desired proportions. The first trial demonstrated, without difficulty, the correctness of this idea. The fractionation was made chemically, and was not suitable for manufacture on a commercial scale. It was evident, however, that in most cases complete separations would not be required, but that the preparation of concentrates would be sufficient. Various other procedures were reviewed, including fractional crystallization, flotation, and some other other procedures employed in ore dressing.[4]

In 1933, a patent on a cement manufacturing process involving flotation was issued to Breerwood.[5] Developments have been rapid, and today flotation equipment is in operation at a number of plants. An example is the plant of the Argentine subsidiary of the Lone Star Cement Corporation.[6] The raw materials in the locality are not suitable without preliminary processing.

In another patent,[7] a process is described which "has especially to do with the beneficiation of available natural raw materials, by a process in which grinding and particle size classification, among other means, are employed in combination to derive from the natural raw material at least two mixtures requiring little or no subsequent correction."

REFERENCES

1. L. S. Petersen, U.S. Patent 2 290 068 (1942).
2. Bror Nordberg, *Rock Products*, 45(7):32 (1942).
3. *Rock Products*, 45(10):54 (1942).
4. S. J. Truscott, *Text Book of Ore Dressing;* Macmillan Co., New York, 1923.
5. Charles H. Breerwood, U.S. Patent 1 931 921 (1933).
6. *Pit and Quarry*, 31(5):52 (1938).
7. Charles H. Breerwood, U.S. Patent 2 288 179 (1942).

Chapter 6

FUELS

For the preparation of clinker, to be discussed in Chapter 7, heat, in large quantities, is required. In this chapter, the source of heat for kilns and other equipment units, will be discussed.

Nearly all the heat at the plant is generated by the combination of carbon and hydrogen with atmospheric oxygen. When one pound of carbon is burned to produce carbon dioxide, 14,600 B.t.u. of heat are generated.[1] When one pound of hydrogen combines with oxygen to form water, 62,000 B.t.u. of heat result. One pound of sulfur burned to sulfur dioxide generates 4,000 B.t.u. The values for heats of combustion stated in various publications differ to some extent, but the values given here are in common use. Some reactions of carbon, hydrogen, and oxygen may be listed as follows:[2]

	Reaction	Calories (Kilogram)
1.	$C + O_2 \rightarrow CO_2$	$+ 94300$
2.	$CO_2 + C \rightarrow 2CO$	$- 41000$
3.	$2C + O_2 \rightarrow 2CO$	$+ 53300$
4.	$2CO + O_2 \rightarrow 2CO_2$	$+135300$
5.	$2H_2 + O_2 \rightarrow 2H_2O$	$+115500$
6.	$H_2O + C \rightarrow CO + H_2$	$- 31100$
7.	$2H_2O + C \rightarrow CO_2 + 2H_2$	$- 21200$
8.	$CO + H_2O \rightarrow CO_2 + H_2$	$+ 9900$

In the original table, the compounds are expressed in gram-moles; and the heat in mole-calories. For convenience here, the compounds are expressed in kilogram-moles and the heat in large (kilogram) calories. For conversion factors, see Chapter 24.

Although it is possible to burn elementary carbon or hydrogen for the generation of heat, such a procedure is not practicable for several reasons. Therefore, substances that contain carbon and hydrogen are selected. In this connection, these substances are known as *fuels*. The principal fuels are bituminous coal, natural gas, and petroleum distillates.

Illinois coals contain approximately 60 to 70 per cent of carbon, and 5 to 6 per cent of hydrogen.[1] Some coals contain less carbon and hydrogen, and others more. Most liquid and gas fuels belong to the aliphatic hydrocarbon series, the first member of which is methane, CH_4. The general formula for the series is C_nH_{2n+2}, in which n is the number of carbon atoms. As n increases, the percentage of carbon increases, and the percentage of hydrogen decreases. The heating value of a fuel is determined usually, and preferably, by the use of a calorimeter, as will be explained later. If, however, a fuel does not contain methane or carbon monoxide, its heating value may be calculated, approximately, by Dulong's formula:[1]

$$h = 14{,}600\,C + 62{,}000(H - 0.125\,O) + 4{,}000\,S.$$

where $\quad h =$ total heat of combustion,

$\quad C, H, O, S =$ the percentage, expressed as a decimal, of each of the four elements.

Fieldner[3,4] has presented many important and interesting data on fuels. One of the papers cited here was his inaugural address as president of the ASTM.

Sometimes the statement is made that although fuels are expensive, the air, which contains the oxygen required for combustion, is free. Air is free, on the *as is* basis. If we wish to remove impurities from it, however, humidify or dehumidify it, heat or cool it, or convey it from one location to another, equipment and power—both of which may be costly—are required. This is apparent from some statements in other chapters.

It may be pointed out also that although the production of heat is one of the most expensive operations at the plant, the cost of removing heat, when required, is not negligible. In some operations, portions of the heat are recovered, but in others the heat is lost completely. If all the heat generated could be utilized, a large reduction in the total fuel purchased would result.

Fuels in the Cement Industry

Solid, liquid, and gaseous fuels are used in the cement industry. The solid fuels consist almost exclusively of bituminous coals. In some localities, relatively small quantities of anthracite, coke, and petroleum coke are ground with the bituminous coal. The liquid fuels are petroleum products. Natural gas constitutes practically all the gaseous fuel, but some producer gas has been used.

Sources of Fuels

Fuels are generally purchased, rather than obtained on company-owned properties, as are raw materials. Probably at most plants, fuels cost more than any other purchased materials received. Occasionally, sites have been selected on or near coal deposits, but this does not happen often. As a rule, fuels are purchased in the open market, and solids or liquids are shipped by rail or water to the plant.

Storage of Fuels

It is desirable to stock fuels for emergencies that may arise in the production or transportation, but there are some disadvantages. With many coals, for example, certain precautions must be taken. The stock pile should have good drainage, all vegetation having been removed in advance. Coals that are not known to be satisfactory for the purpose should not be stored, and coals from two sources should not be mixed. Even when coal is stored for limited periods and under the most favorable conditions, consideration must be given to such items as (1) the possibility of fires in stock piles, (2) the gradual loss of heat by oxidation, (3) contamination of the coal during reclaiming, and (4) the cost of rehandling.

These disadvantages do not apply to oil, but this fuel has some others. Storage tanks must be provided, and as the consumption is high, the storage of a supply for more than a few days means a rather large investment. When an oil of relatively high viscosity is handled during cool weather, a higher temperature of the oil must be maintained. Usually, this is done by steam coils.

It is not customary to store gaseous fuel at a cement plant. It is recognized, however, that for any one of several reasons, a gas supply may be diminished or stopped completely for a time; consequently, at some plants there are arrangements to substitute coal or oil during an emergency. If conditions permit, the arrangements for substituting oil may include some storage tanks. The gas company may agree to pay the cost of substituting another fuel during a gas shortage.

Choice of Fuels

The selection of the fuel for the manufacture of cement may be easy or difficult, depending on the types of fuels available in the vicinity, the plant equipment, and the relative delivered cost of each type. In some localities, only one type is available economically. In others, two or three types may be obtained. If there is not much difference in the cost of the fuels, taking all items into account, an intelligent decision may require extensive data and calculations. There are advantages and disadvantages peculiar to each type. Some of these will be mentioned with reference to the operation of kilns, which require the major portion of all fuels in cement plants.

Coal:

Pulverized coal is the fuel that was used to operate the first rotary kiln, and today the coal burned at cement plants exceeds other fuels. It is widely distributed, and its delivery to a plant requires no special equipment. On the basis of net B.t.u., somewhat higher thermal efficiency may be obtained with coal than with gas. Although oil and gas may be burned in the form in which they are received, coal must be ground fine. In some cases, it must be crushed before it is ground, and, in general, the coal must be dried either before or during grinding.

Oil:

When oil is available at a price not greatly in excess of that of coal, on the net B.t.u. basis, it is often preferable.

There is no cost for preparation, and the control of the fuel is easier than when coal is used. As has been pointed out, however, the investment in storage facilities and, sometimes, the maintaining of temperature must be taken into account. In the manufacture of some special cements, in which color is particularly important, oil may be used in preference to coal, regardless of the additional expense.

Gas:

Natural gas is a very desirable fuel for cement plants. When the quality, quantity, and price are satisfactory, other fuels do not offer much competition.

Selection of Coal

After the type of fuel has been chosen, it is necessary to decide which source of supply, of those available, is most suitable. In general, this selection is most difficult in the case of coal, because there are many items that should be taken into account. In many respects, the cement manufacturer wants the same quality of coal as do other manufacturers—uniformity, high B.t.u., low ash, and low sulfur. From this point on, however, certain differences appear. Although high-grade coal is desirable, it is not absolutely necessary, and sometimes it is unobtainable. Cement is a low-priced commodity, and the quantity of coal required per barrel is relatively large. From these facts, it may be readily understood that in selecting coal, the cost at the mine and the freight are important considerations.

Obtaining a representative sample of coal is a complex procedure (see ASTM D 492–48).

There are two groups of analytical procedures for coal,[1] the ultimate and the proximate. The first includes the determinations of the chemical elements carbon, hydrogen, nitrogen, and sulfur, and the calculation of oxygen. The second group includes the determinations of moisture, volatile matter, and ash, and the calculation of fixed carbon. In the cement industry, ultimate analyses are made rather infrequently, except in connection with research (see ASTM D 271–58).

The most important determination in coal is that of the heat of combustion. This is the quantity of heat that is generated when a unit weight of coal is burned. It is expressed in calories per gram in the metric system, and in British thermal units per pound in the English system. The latter, B.t.u., is used in this book.

A B.t.u. may be defined as 1/180 of the quantity of heat required to raise the temperature of one pound of pure water at standard pressure from 32 to 212°. Or, approximately, a B.t.u. is the quantity of heat required to raise one pound of water one degree in temperature. The heat of combustion is determined by igniting a small quantity of coal in oxygen under pressure in a bomb calorimeter.

Additional determinations sometimes required include the analysis of the ash, and the softening temperature of the ash. The moisture content of coal may change considerably while it is at a plant, and this causes corresponding changes in the other constituents. Therefore, when an analysis is reported, it is customary to state the basis on which the results were calculated, for example, the *as-received basis, the dry basis*, or *the as-fired basis*.

As a rule, the result of a calorimeter determination is calculated to "the gross calorific value at constant volume, with the water in the products of combustion condensed to liquid at the temperature of the calorimeter. . . ." Net calorific value, net heat of combustion, at 68° refers to the result corrected for latent heat of vaporization, as follows:

$$N = B - 1030(H \times 9)$$

where N = net calorific value, net heat of combustion, in B.t.u. per pound,

B = gross calorific value, gross heat of combustion, at 68° in B.t.u.,

H = total hydrogen.

It is preferable to make all comparisons of fuel consumption on the basis of net B.t.u. This is particularly important when gas is to be compared with other fuels, because there is a considerable difference between the total and the net heat of combustion of gas.

Previous to 1934, there was no satisfactory basis for classifying coals, but now there is a standard specification for the classification of coals by rank; see ASTM D 388–64T. The basic scheme of classification is according to fixed carbon, and heat of combustion calculated to the mineral-matter-free basis.

In addition to composition and calorific value, the *grindability* of coal should be taken into account. The Hargrove machine is employed (ASTM D 409–51) to determine the relative grindability or ease of pulverizing of coals in comparison with a coal chosen as 100 grindability. The method is based on Rittinger's Law, which states:

> The *work done in pulverizing is proportional to the new surface produced.*

A prepared sample receives a definite amount of grinding energy in a miniature pulverizer, and the new surface is determined by sieving.

Studies of *ignitibility* of coal have produced many interesting results, which may have important effects on coal technology. The term has been defined by the American Society for Testing and Materials Committee D-5 on Coal and Coke as follows:

> The ignitibility of a fuel is that characteristic which determines the ease with which the fuel may be brought to a condition of self-supporting, active oxidation. The ignitibility is governed not only by the inherent chemical characteristics of the fuel that determine its rate of oxidation at various temperatures but also by the physical characteristics of the fuel and its surroundings that determine the rate at which it can be heated. Among these physical characteristics are (1) the size of the pieces of fuel, (2) the specific heat of the fuel, (3) the thermal conductivity of the pieces and the aggregate, (4) the rate at which the oxygen or air is brought into contact with the fuel, and (5) the rate of heat loss to the surroundings. Chemical and physical changes, such as melting and decomposition of bituminous coal, may occur as the fuel is heated.

Selection of Oil

When oil is the type of fuel chosen, the selection of the source of supply should not be difficult. There may not be

many sources available for a given plant. The most important
determination is the heat of combustion which is made by
the use of a bomb calorimeter. The relation between total
and net heat of combustion is the same as for coal. Consumption of oil is reported in terms of gallons per barrel of
clinker. The heat of combustion is calculated to B.t.u. per
pound of oil, and then to B.t.u. per barrel of clinker.

Selection of Gas

Many items may be included in a natural gas contract.
The most basic one is the heat of combustion. Several types
of calorimeters are available. Probably the type employed
more than any other at cement plants is the Junker. A
metered volume of gas is burned in the calorimeter, and the
temperatures of water entering and leaving the instrument
are observed. The water is weighed, and corrections are
made, including those for the temperature and pressure of
the gas. The heat of combustion is calculated, taking into
account a constant for the instrument.

Two ASTM designations should be mentioned: 1) Heat of
Combustion of Liquid Hydrocarbon Fuels by Bomb Ca-
lorimeter, D 240–64; and 2) Net Heat of Combustion of
Liquid Petroleum Products, Estimation of, D 1405–64.

A committee of the American Society for Testing and
Materials is studying gaseous fuels problems with the intention of writing a series of specifications. There are subcommittees on the collection and measurement of samples; the
determination of heat of combustion, density, water vapor,
and special constituents; and the complete chemical analysis.

Preparation of Fuel

The fuel-burning equipment at a cement plant may be
divided into four groups: (1) dryers, (2) kilns, (3) boilers,
(4) miscellaneous. The function of dryers has been mentioned
in Chapter 5. Kilns will be discussed in Chapter 7, and boilers
in Chapter 8. The miscellaneous equipment—which includes
such items as heating plants for office buildings, forges, and
occasional plant locomotives and steam shovels—consumes
only a small fraction of the fuel, and need not be discussed
here.

Oil and gas require no preparation. Only the coal that is burned in the pulverized condition receives any preliminary treatment. As the same pulverized coal is suitable for use in dryers, kilns, and boilers, this discussion will be restricted to the preparation of kiln coal.

Formerly, there was a separate department in the cement plant for preparation of coal. The coal as received was

FIG. 6:1. RAYMOND BOWL COAL MILL

crushed, if necessary, and fed to one or more dryers. The coal discharged from the dryers was reduced to the desired particle size in one or two stages. For one-stage grinding, a mill of the Fuller type was commonly used. For two-stage grinding, a ball mill followed by a tube mill was the usual arrangement. A Fuller mill may be classified as 1a, 2b, 3a, 4b, 5b; see Chapter 9. The ground product was distributed to the kilns and other coal-burning equipment.

Central coal systems are being displaced by unit mills, that is, a separate coal mill for each kiln or other piece of equipment for the operation of which heat is required. There are several types of unit mills.

Figure 6:1 shows a Raymond Bowl Mill. According to the manufacturer nearly 400 of these mills are in operation—firing kilns that produce cement, lime, magnesite, and other materials. Many of these units are employed for firing boilers in steam-generating power plants.

In comparison with central systems, unit mills for coal have certain advantages. The coal is ground as needed and discharged directly into the kiln, or other equipment for which the coal is required. There are no storage bins nor conveyors which are hazardous and from which the pulverized coal should be removed during shut-down periods to avoid fires. In general, the installations are cleaner and less noisy than a central system.

There are some disadvantages. A unit mill and its auxiliary equipment constitute a complex mechanism. Most adjustments and repairs require the stopping of the mill. Unless there is a spare unit mill, this means that the kiln must be stopped during this period, with the accompanying loss in production and in fuel.

The coal is ground customarily to a fineness of 80 to 90 per cent through a 74-micron (200-mesh) sieve. At some plants, the specific surface of coal is estimated by the use of an air-permeability apparatus. The quantity of coal ground by a unit mill should be sufficient to permit the kiln to which it is connected to operate at maximum capacity. As many factors influence the production of clinker and requirements for fuel, a unit mill having a capacity well in excess of the calculated requirements should be specified.

The heat required for the production of a barrel of clinker varies over a wide range. One million B.t.u. may be taken as an example for a dry-process plant. This is not stated as a mean nor an average value. This is equivalent approximately to 100 pounds of coal, having a net heat of combustion of 10,000 B.t.u. per pound; to 7 gallons of oil, having a net heat of combustion of 144,000 B.t.u. per gallon; or to 1,100

cubic feet of natural gas, having a net heat of combustion
of 900 B.t.u. per cubic foot.

At any cement plant, the major quantity of fuel is used
in operating the kilns. At a wet-process plant, for which the
power is purchased, nearly all the fuel goes to the kilns. The
cement industry is a large user of fuels.

In 1960, the types of fuel consumed, by number of plants
and by percentages, were as shown in Table 6:1.[5]

TABLE 6:1

CONSUMPTION OF FUEL IN CEMENT PLANTS

Fuel	Number of Plants	Percentages
Coal	67	33.7
Oil	9	4.2
Gas	37	18.1
Coal and Oil	23	15.0
Coal and Gas	21	
Oil and Gas	18	
Coal, Oil, Gas	7	

In 1960, the use of natural gas decreased 10% from 1959.
Coal and oil supplied 57% of the heat used, in comparison
with 55% in 1959. The 176 active plants used an average of
1.31 million B.t.u. for each barrel of cement produced.

REFERENCES

1. G. F. Gebhardt, *Steam Power Plant Engineering;* John Wiley
   and Sons, Inc., New York, 1928.
2. W. H. Walker, W. K. Lewis, W. H. McAdams, E. R. Gilliland,
   Principles of Chemical Engineering; McGraw-Hill Book Co.,
   1937.
3. A. C. Fieldner, *Proc. Am. Soc. Testing Materials,* 37, Part I
   (1937).
4. A. C. Fieldner, *J. Western Soc. Engrs.,* 48(3):127 (1943).
5. *Minerals Yearbook;* U.S. Bureau of Mines, Washington, D.C.,
   1961.

Chapter 7

CLINKERING

The selection, proportioning, and preparation of raw materials have been discussed in chapters 3, 4, and 5; and fuels, in Chapter 6. The raw materials and fuels are brought together in one or more rotary kilns. Here, by a process known as clinkering, a series of chemical compounds is synthesized. Collectively, these compounds are called clinker. The term burning is sometimes used as a synonym for clinkering, but this has no justification.

The kiln may be described largely by superlatives. Here are some examples. The kiln

is the largest equipment unit in the cement plant,
is the most expensive unit,
is the most important unit,
has the most functions,
has had the greatest increase in size,
is the greatest consumer of fuel,
is one of the most wasteful consumers of fuel.

The growth in the size of the kiln has been unique. The diameter has increased from 5 to 18 feet; and the length, from 25 to over 500 feet. In general, ratio of length to diameter has increased with the length of the kiln, but there is no definite relation. That is, kilns of the same diameter may differ in length, and kilns of the same length may differ in diameter. Most kilns have the same diameter, throughout. There are some kilns, however, of which some portions have longer diameters than other portions.

Development of the Rotary Kiln

From the invention of portland cement in 1824, until 1866, clinker was made by a batch process in vertical kilns. Soon after the rotary kiln came into use, it was realized that although the quality of the clinker was improved and the labor was decreased, there was a large increase in consumption of fuel. Decreasing the amount of fuel required for production of clinker has been the most important goal in the development of the rotary kiln, since it was invented. Some other factors have been the introduction of the wet process, increase in production, and improvement in the quality of the clinker.

For convenience in discussion, it is considered in this book that a *rotary kiln* comprises the steel shell, the refractory lining, any insulation that may be present between the lining and the shell, the front and back housings, the supporting structure, and the driving mechanism. All other items are classed as auxiliary equipment. A kiln and its auxiliary equipment constitute a *kiln system*. A kiln system and a waste-heat boiler, together with its auxiliary equipment, constitute a *kiln-boiler system*. The following is a list of some of the developments in connection with kiln-boiler systems that have resulted from the factors that have been mentioned:

1. Increase in size of kiln
2. Waste-heat boilers
3. Insulation of kiln
4. Pre-heating the kiln feed
5. Preliminary removal of water from the kiln feed
6. Pre-heating the combustion air
7. Recovery of radiated heat

The significance of each of these may be pointed out briefly.

Increases in length and diameter of kilns:

These changes have come about partly from calculations based on construction and operating experience, and partly from trial-and-error procedures. One of the greatest losses of heat is from the gases discharged from a kiln. It was evident that if the length of a kiln of given diameter were increased, a larger percentage of the heat in the combustion gases would be transferred to the feed before the gases were discharged into the atmosphere. This tended to increase the output of the kiln. Some additional fuel was required, which in turn called for an increase in diameter, and so on.

Many formulas have been suggested for calculating the output of a given kiln; or conversely, for calculating the dimensions of a kiln for a desired output. These have been based on such factors as the ratio of length to diameter, the area of the inner surface of the brick lining, and the net volume. Occasionally, a complex formula,[1] such as the following, appears:

$$\text{Output, in metric tons/hour,}* = 0.04552kVe^{-0.00045V}$$

where V = Volume of kiln, in cubic meters
k = 1.00, for dry process kilns
k = 0.96, for wet process kilns
e = base of Napierian logarithms

One kiln manufacturer announces that more than 100 variables are taken into account in designing a unit.

Waste-heat boilers:

The gases discharged from many kilns, particularly dry-process kilns, contain heat sufficient for the operation of boilers, notwithstanding all efforts that have been made to increase thermal efficiency. Waste-heat boilers, which are very important pieces of equipment for the conservation of kiln fuel, are discussed in Chapter 8.

Insulation:

Insulating block may be placed between the refractory lining and the shell of the kiln. In the insulated zone, thinner refractory brick is used, so that the block may be placed without increasing the total thickness of the lining. It may be said, therefore, that a portion of the refractory material is displaced by insulating material. The purpose, of course, is to decrease the quantity of heat that is transmitted through

---

* For conversion to barrels per cubic foot, see Chapter 24. So far I have not found a formula by which the production of a series of kilns can be correlated satisfactorily.

the lining and shell, and radiated from the outer surface of the shell.

The heat that can be conserved in this way can be calculated, but sometimes it is difficult to demonstrate the saving. The over-all value of kiln insulation has received much discussion. There are some disadvantages. The most important of these is that it may not be practicable to install insulating material in the hottest zone of the kiln, where it is most needed. Decreasing the quantity of heat transferred through refractory brick in this zone may decrease the durability of the brick.

Preheating the kiln feed:

Many devices and procedures have been developed for preheating the kiln feed. Usually, the source of the heat is the kiln gases. Some or all of the gases may be brought into contact with the feed. The result is that the feed enters the kiln at a higher temperature than it would otherwise. An example is described in a patent issued in 1942.[2]

Removal of water from the kiln feed:

When the kiln feed has been prepared by the wet process, the preliminary removal of water is particularly important. This is true because every pound of water in the slurry that enters the kiln must be evaporated, and the steam thus formed, must be superheated to the exit temperature of the kiln gases.

Preheating the combustion air:

Some years ago it was customary to pass the clinker discharged by a kiln through a cooler. Several types of coolers have been developed. One of the earliest types was a rotating cylinder inclined slightly from the horizontal. The types most in use at present consist essentially of vibrating screens or perforated plates. Both primary and secondary air may be preheated by being drawn through the clinker. Some of the heat leaving the kiln system may be recovered in this way.

Some other advantages may be claimed for clinker coolers. Rapidly cooled clinker is said to have a lower grinding resistance, and a lower autoclave expansion than clinker of the same analysis that has not been prepared in this manner.

Figure 7:1 illustrates an air-quenching cooler manufactured by Allis-Chalmers. The discharge pan, splitter gate, and diaphragm connection are shown at right.

It may be said that water-quenching coolers have some advantages, but in contrast to air-quenching coolers, the disadvantages are generally believed to predominate. Further research seems justifiable.

Fig. 7:1. Air-Quenching Cooler with Clinker-Breaker

Recovery of radiated heat:

Some of the heat radiated from the front end of the kiln may be recovered by installing a stationary steel jacket, and drawing air between the jacket and the kiln. Sometimes this hot air is utilized for drying coal.

Design of the Modern Kiln

The Allis-Chalmers wet-process kiln, installed at the Des Moines plant of the Marquette Cement Manufacturing Company in 1942, was the longest kiln in the Western hemisphere at that time.[3] Some of the principal data are as follows:

FIG. 7:2. WET PROCESS KILNS AT ADA, OKLAHOMA
(*From* Allis-Chalmers)

Diameter	11 feet, 6 inches
Length	475 feet
Approximate weight	1,428,600 pounds
Thickness of steel plate	$\frac{3}{4}$-inch except at the front end, where it is 1$\frac{1}{2}$ inches
Pitch, or slope	$\frac{7}{16}$-inch to linear foot
Supports	7 rings and 7 pairs of rollers
Girth gear	193-tooth, 3$\frac{1}{2}$-inch pitch, 20-inch face
Speed reducer	Falk, herringbone gears
Motor	150 horse power
Speed of kiln	20 to 80 revolutions an hour.

The kiln for most of its length, was lined with several types of fire brick, including magnesia, high-alumina, and low-alumina brick.

Two Allis-Chalmers kilns, of 12 by 450 feet, in the wet-process plant of the Ideal Cement Company at Ada, Oklahoma are shown in Figure 7:2.

In 1956, a kiln more than 557 feet long (see Fig. 7:3) was installed at the Barker plant of Loma Negra S.A., Buenos Aires. This kiln, which has a production in excess of 5000 barrels of clinker in 24 hours was manufactured by F. L. Smidth and Company, Copenhagen.

Auxiliary Equipment

A kiln system may include numerous auxiliary items, and these may require auxiliary items of their own. Some of these are essential, but the value of others, under some conditions at least, may be questioned. As the complexity of a system increases, there are more and more items that require adjustment and maintenance, and all this may cause delays in kiln operation.

A partial list of auxiliaries includes draft fans, draft controls, seal rings, feeders, slurry filters, chain systems, dust collectors, and clinker coolers. Slurry filters and clinker coolers have been mentioned in connection with heat conservation. The others will be discussed briefly here.

Draft fans:

When gases are discharged from the kiln directly into the atmosphere, either natural draft or fans may be employed depending on the general arrangement of the installation. When, however, the gases go through waste-heat boilers,

FIG. 7:3. KILN IN BUENOS AIRES

dust collectors, or both, induced draft fans are required. The gases consist of combustion gases, excess air, carbon dioxide, and small or large quantities of water vapor, depending on several factors. In any case, there is a relatively large volume of gases to be handled, and large-capacity fans are necessary.

Draft controls:

Drafts may be controlled either manually or automatically. The use of automatic equipment is increasing in the industry. Conditions arise, however, during which manual control adjustments are required.

Seal rings:

For several reasons, it is desirable that the volume of excess air entering a kiln system be as small as practicable. Seal rings are devices designed to minimize infiltration of air. A set of rings consists of a stationary ring attached to a kiln housing, and a ring attached to one end of the kiln. The two rings are kept in contact by springs or by some other mechanical arrangement. For satisfactory results, frequent inspection and, occasionally, rather frequent adjustments are necessary.

Feeders:

The rate at which feed enters a kiln is almost, or quite, as important as are the chemical and physical characteristics of the feed. In the dry process, there is a variety of installations, but in general, the feed is sent to a short screw conveyor from which it is distributed through a pipe, set in an oblique position, into the back end of the kiln. If one of the raw materials is plastic when wet, the addition of a relatively small quantity of water to the feed, as it enters the kiln, may be desirable. This tends to decrease the quantity of dust carried in the kiln gases, by nodulizing the feed to some extent.

In wet-process plants, the most common type of feeder is the Ferris wheel. This consists essentially of a wheel that rotates on a horizontal axis, and to which is attached at the periphery a series of buckets. A bucket is filled with slurry by entering a tank, in which the slurry is maintained at a constant level, and is discharged into the feed pipe for the kiln.

It is customary to synchronize the feeders and kiln rotation in both the dry and the wet process. One arrangement by which this can be accomplished is to install a small generator to be driven by the kiln. The generator supplies current to a motor that drives the feeder by means of a variable-speed device. The ratio of feeder-speed to kiln-speed may be varied, and when desirable, the speed of the feeder can be controlled manually.

Chain systems:

When water is not removed mechanically from the slurry before it enters the kiln, devices to facilitate the evaporation of the water in the kiln may be provided. One procedure is to introduce the slurry in the form of a spray, so that its surface of contact with the kiln gases is increased greatly.

The most commonly used device is a system of chains suspended in the upper end of the kiln. The chains are so arranged that as the kiln rotates, they become coated with slurry. This greatly increases heat transfer, and results in the evaporation of water by the utilization of heat, some of which would otherwise be lost. The use of chain systems is restricted to rather long kilns. If a kiln is not long enough, the chains cannot resist the heat to which they are exposed, and frequent replacements are required.

The design of a chain system is much more complex than it may appear to be on the surface. To obtain the most satisfactory results, the physical characteristics of the raw materials, particularly plasticity, must be taken into account. A system suitable for one group of raw materials may be far from suitable for another group. In the interest of heat conservation, the total surface of the chains, and the frequency of the change in direction of the flow of the gases should be maxima. If this is carried too far for a given slurry, however, the material does not move satisfactorily through the kiln, and delays and losses in production result.

Dust Collectors:

There are many types of dust collectors, and many locations in a cement plant in which these may be found. Because of the large quantities of finely ground solids that enter a kiln, the kiln gases may be heavily laden with dust. In many plants, most of this dust is recovered. Usually, but not always, it is returned to the kiln system. Several types of dust collectors are employed for this purpose, of which the Cottrell electrical precipitator is the most common. It is of interest that the physics experiments that suggested the development of the Cottrell precipitator [4] many years later, were started in 1824, the year that the Aspdin patent for portland cement was granted.

The operation of this type of dust collector may be outlined as follows [5]:

When a dust-laden gas or mist is subjected to a high-potential, unidirectional, ionizing, electrical field, the individual suspended particles, or droplets, become electrified and move to one electrode, on which they collect and from which they are removed either continuously or intermittently. The other electrode is usually of small radius of curvature to facilitate the production of the gas ions. In practice, this electrode is ordinarily a wire or edge, and is called the discharge electrode. The collecting electrode on which the suspended material is precipitated is usually a plate or a pipe. If the former, several discharge electrodes are suspended from insulators, vertically, between two collecting electrodes. A complete precipitator may be made up of as many of these units as are desired or needed. If the collecting electrode is a vertical pipe, a single discharge electrode is suspended vertically through its center. The dust-laden air passes upward through the pipe, and any number of these units may be assembled into a large unit.

The collecting electrodes are grounded, and so are safe to touch or to connect to other apparatus, such as pipes and conveyors. The discharge electrodes must be thoroughly insulated against the operating potential, which may be from 40,000 to 75,000 volts. This is usually obtained by transformers, whose secondary current is rectified or made unidirectional by means of a rotating pole-changer, or commutator, run by a synchronous motor.

Cottrell electrical precipitators are built to treat as little as a few hundred or as much as a hundred thousand cubic feet per minute, and some multiple-unit installations on cement-kiln gases and power-plant gases handle over a million cubic feet per minute. The collection efficiency can be made as high as desired, either on relatively coarse dust or on submicroscopic particles, by a suitable choice of type and size of equipment. The efficiencies of removal of suspended

matter obtained range from 90 per cent for ordinary applications up to over 99 per cent removal where it is essential
to deliver a particularly clean gas."

Functions of the Kiln

It has been stated that the kiln has more functions than
any other piece of equipment in a cement plant. The kiln
serves as each of the following:

1. Chamber for the combustion of fuel
2. Flue for gases and vapors
3. Conveyor for solids
4. Heat exchanger
5. Dryer
6. Calciner
7. Mixer
8. Apparatus for synthesis

Because of its many functions, the kiln cannot be designed
for the best results for any one function, so compromises or
additions must be made in favor of one or more of the others.
For example, functions 1 and 2 do not require rotation. As
a conveyor, the kiln need not be so elaborate. In the drying
zone, relatively low temperatures are sufficient, and the heat-
input is determined largely by the water content of the kiln
feed; in the calcining zone, higher temperatures and high
heat-input are necessary; in the clinkering zone, still higher
temperatures are required, but rather low heat-input is sufficient.

Heat

In this chapter, the subject of heat is discussed under four
subdivisions: (a) generation, (b) requirements, (c) losses, and
(d) recovery.

Generation of heat:

Heat is generated by the combustion of a fuel, that is, by
the combination of the principal constituents of the fuel with
the oxygen of the atmosphere. In our calculations, this is
considered the total heat-input of the kiln, although any
substance, the temperature of which is not absolute zero
($-460°$ F), contains heat. We shall assume that all substances enter a kiln at 70°. The heat required for clinkering

may be said to be, in round numbers, 1,000,000 B.t.u. per barrel of clinker (376 pounds).

Heat requirements:

The heat requirements for clinkering may be listed as follows:

1. Heat to raise the temperature of the kiln feed from 70 to 212°.

2. Heat to evaporate water from and at 212°, and to superheat the steam to the exit temperature of the kiln gases.

3. Heat to raise the temperature of the dry solids to the calcination temperatures of calcium carbonate and magnesium carbonate under the conditions in the kiln.

4. Heat to calcine the calcium carbonate and magnesium carbonate.

5. Heat to raise the temperature of the materials to the temperature necessary for the formation of the clinker compounds.

6. Heat to form the clinker compounds.

Some of the reactions are endothermic and some are exothermic. The algebraic sum of these quantities of heat is the total quantity required.

There may be some other requirements for heat, such as for the vaporization of combined water, and oxidation or reduction of minor constituents of the kiln feed. These vary considerably, but are relatively small in quantity. Therefore, they are not taken into account here.

Heat losses:

There are three principal sources of loss of heat:

1. Heat carried by the clinker when discharged from the kiln.

2. Heat in the kiln gases which contain not only gases, but vapor and solids as well.

3. Heat radiated from the kiln shell.

Heat recovery:

A portion of the heat lost by a kiln system may be recovered. The percentage of recovery depends on a number of factors, the most important of which are the auxiliary equipment provided, and the accuracy of the control of this equipment.

The recovered heat may be returned to the kiln system, or it may be utilized in some other manner. This has been mentioned previously in this chapter. The heat recovered from the clinker and returned to the kiln, in primary and secondary combustion air, reduces the heat necessary to be generated by combustion. This heat may be said to be deducted from the heat carried by the clinker leaving the kiln, and added to the heat-input.

Heat recovered from the shell of the kiln may be utilized to dry coal, as has been stated. This heat is not returned to the kiln system. If, however, a portion of the recovered heat has increased the temperature of the coal entering the kiln, it is evident that this quantity of heat is returned to the kiln system. The greatest recovery of heat takes place in waste-heat boilers, which are discussed in Chapter 8.

The distribution of heat in a rotary kiln is one of the most interesting and important factors in cement technology. The available information has been collected from many sources such as decades of kiln operation, tests, research, calculations, and the chemical, physical, and petrographic examination of samples taken at regular space intervals from kilns that have been stopped and allowed to cool before the contents have been removed. There are many variables and probably no two sets of data exactly alike have been collected. One particularly important variable is heat transfer.

Kiln Operation

The kiln operator is supplied with plant equipment, fuel, and raw materials that have been ground and proportioned. It is his responsibility to produce clinker. The four principal points that he must keep in mind are:

1. Quality of clinker
2. Rate of production
3. Economy of manufacture
4. Care of the equipment

These items are subject to a number of subdivisions, and they are inter-related to some extent.

Under the most favorable conditions, the operation of a kiln is difficult. The degree of difficulty is affected by a number of factors, such as the chemical and physical uniformity of the kiln feed and the fuel, the accuracy with which the entrance of the feed and the fuel into the kiln may be controlled, and the kiln instruments that are available. The farther these items are from the best practice, the greater the difficulty of operation. To some extent, good operation may compensate for irregularity of kiln feed, but if the chemical and physical characteristics of the feed are not within rather narrow limits, satisfactory clinker cannot be made, regardless of the care with which the kiln is operated.

The desired rate of production may range all the way from the minimum quantity of clinker that can be made when the kiln is properly operated, to the maximum output that can be obtained. As a rule, the highest thermal efficiency and the greatest output do not occur at the same time. The flexibility of kiln operation at a given plant increases as the number of kilns increases. When the number is relatively large, fluctuation in the demand for clinker may be adjusted readily, by increasing or decreasing the number of kilns in operation, and then adjusting the output of one or more of them. Such an arrangement permits kilns to be operated more nearly at the point of highest economy, and provides for the maintenance of kilns that are not in use.

When a plant has only one kiln, it is evident that all adjustments in the production of clinker must be made by modifying the output of this kiln. This may result in decreases in the economy of the operation, and it may interfere with the proper maintenance of the equipment. For a given annual output of clinker, the choice of the number of kilns to be installed—which is related directly to the sizes of the kilns—is one of the most important decisions that must be made. Sometimes emergencies arise during which kilns are operated intermittently rather than continuously. Such a schedule has many disadvantages, and is not put into effect if it can be avoided.

It has been indicated that one of the points an operator must keep in mind is the care of the equipment. As a rule, he is not responsible for general maintenance, but he must be on the lookout for emergencies. One example is the appearance of a red hot area on the shell of the kiln. This means that the lining is thin in that area, and it must be repaired. Sometimes this may be done by the application of a special mortar under pressure, and sometimes the kiln must be shut down, and allowed to cool, after which some of the lining is replaced with new brick.

Some operating procedures:

It will be assumed here that the demands for clinker permit a kiln to be operated at its maximum thermal efficiency, and that the kiln feed is reasonably uniform, chemically and physically. Some trial-and-error procedures may be necessary, but these are to be kept at the minimum. Three operating procedures may be mentioned:

1. Fuel is supplied to the kiln at the maximum rate compatible with complete, or nearly complete, combustion. Sufficient air is supplied, and the relation between primary and secondary air is maintained at the most satisfactory point. The kiln feed is adjusted to the maximum rate at which satisfactory clinker can be obtained. In discussing unit coal mill operation, Hardgrove [6] recommends the following:

 a. Primary air, about 2.5 to 3 pounds per pound of coal
 b. Air temperature, leaving pulverizer, 160–220°
 c. Velocity at burner tip, 6,000–8,000 feet per minute
 d. Fineness of pulverized coal, more than 80 per cent through 200-mesh
 e. Extremely uniform delivery of coal to the burner, which is best obtained by automatic feeder control and automatically controlled air temperature
 f. Very uniform control of secondary air to the kiln, which is best obtained by automatic draft control

2. Feed is supplied to the kiln at a rate found by experience to be the highest that is practicable. Fuel and air sufficient for the formation of clinker are introduced. Good results may be obtained by either of these procedures or by a combination of the two.

3. Automatic operation of kilns is attracting more and more attention. Certain automatic features are coming into rather extensive use, but complete automatic operation is hardly in sight (see Chapter 12). It is obvious that as the uniformity of the feed and fuel becomes more nearly constant, and as the facilities for manual control improve, complete automatic control will become more nearly feasible.

REFERENCES

1. T. Yosii, *Chem. Abstr.*, 33:9573 (1939).
2. L. S. Petersen, U.S. Patent 2 290 068 (1942).
3. R. C. Newhouse, *Rock Products*, 45(8):65 (1942).
4. F. G. Cottrell, *Ind. Eng. Chem.*, 3:542 (1911).
5. *Pit and Quarry Handbook*; Pit and Quarry Publishing Co., Chicago, 1961.
6. R. M. Hardgrove, *Trans. Am. Inst. Mining Met. Engrs.*, 148: 398 (1942).

Chapter 8

POWER

The four principal items of cost in the manufacture of cement are *raw materials,* in pounds per barrel; *labor,* in man-hours per barrel; *heat,* in British thermal units per barrel; and *electrical energy,* in kilowatt hours per barrel. Heat and electrical energy may be combined as total energy, thus reducing the principal items to three. Electrical energy is discussed in this chapter, with some reference to the other three items. The history of power development is long and interesting, and its application to the cement industry parallels closely the history of the industry as a whole.

The following dates, some of which are approximate, indicate the course of development of prime movers, generators, and motors:

Prototype of steam turbine	B.C.	130
Reciprocating engine, invention	A.D.	1601
Reciprocating engine, commercial use		1698
Steam turbine		1884
Boilers fired with pulverized coal		1895
Faraday's electromagnetic induction law		1831
Direct current generators		1835
Direct current motors		1849
Alternating current generators		1850
Alternating current motors		1888
Central power plants		1881

In the development of prime movers, there are some points that do not agree with what we are likely to take for granted on the basis of modern equipment. Here are some examples. It is probable that the first steam engine was a turbine. At least, a device that may be so classified was described by Hero in 130 B.C.[1] Condensing engines preceded non-condens-

99

ing engines. In the earliest form of cylinder-and-piston engines, one vessel was employed as both boiler and cylinder. From 1601 to 1763, there are records of various inventions. In 1763, James Watt made his first contribution to steam engines, and his name is familiar to all. It is said that he was the first to state that a horsepower is equivalent to 33,000 foot-pounds per minute.

The steam engine was used commercially for pumping water in 1698, for driving a steamboat in 1802, and for operating railways in 1829. It is of interest that John Smeaton built many large engines, making some important improvements in them, and that steam engines were employed at an early date for pumping water from mines in Cornwall. Thus, the development of steam engines and of portland cement have some points in common. In 1884, there was developed the first of the steam turbines which were to replace, very largely, steam reciprocating engines.

The history of electrical equipment is no less interesting. Starting with the work of Faraday, generators and motors in great variety, and many other types of electrical equipment, have been developed.

The application of power in the cement industry may be divided roughly into five periods:

1. Water-driven mill stones for grinding raw materials and clinker; hand labor for other operations
2. Mechanical power from reciprocating engines
3. Mechanical power from reciprocating engines, plus some electrical power
4. Transition from reciprocating engines to turbines
5. Complete electrification; steam from pulverized coalfired boilers drives turbo-generators and the power produced by the generators operates motors which drive the plant equipment units

The transitions represented by the third, fourth, and fifth periods have been made in the last few decades. At one time, it was customary to drive a long line shaft with a steam reciprocating engine, and to drive a series of mills by clutches, each of which was attached to the line shaft. In some cases, the steam engine was displaced by a large electric motor, the line shaft and the clutches being retained. Later, the line shaft and clutches were replaced by individual motors for the mills. This is typical of the change from mechanical to electrical drive. Today, the principal equipment units have individual motor drives, and this is true also of many of the auxiliary units. Electronics will probably soon become important in the modification and control of electrical equipment.

Total Power Consumption

The United States Bureau of Mines [2] reported the power consumption in 1960 as 7,426,000,000 kilowatt hours for the manufacture of 319,009,000 barrels of cement; or 23.3 kilowatt hours per barrel. Table 8:1 is taken from the same report.

The total power, exclusive of quarry operation, may be divided roughly into thirds, one third for grinding the raw materials, one third for grinding the clinker, and one third for all other purposes. An approximate value for power consumption, that may be remembered easily, is 20 kilowatt hours per barrel of cement. This is not to be considered an average. It is probable, however, that, at the majority of plants, the values are only a few units above or below twenty.

The present power consumption is the resultant of two series of events in the industry. The first series caused increases in power; and the second, decreases in power. Under the first series, there may be listed, the application of power to more and more unit processes, such as the change from vertical to rotary kilns; the introduction of power-driven crushing and grinding equipment; the conversion from mechanical to electrical drive; and the changes in manufacturing processes, such as increases in the fineness of raw materials, coal, and finished products.

The second series includes improvements in the design, operation, and maintenance of plant equipment units, including motors. Improvements in the generation of electric power have reduced the cost per kilowatt hour, rather than the number of kilowatt hours per barrel.

It is evident that, at a given plant, the consumption of fuel is affected by a large number of factors. Some of these

TABLE 8:1

ELECTRIC ENERGY USED AT CEMENT PLANTS IN UNITED STATES AND PUERTO RICO IN 1960

Process	Electric Energy Used						Finished Cement Produced Thousand (Barrels)	Average Electric Energy per Barrel Cement Produced (Kilowatt Hours)
	Generated at Portland Cement Plants		Purchased		Total			
	Active Plants	Million Kilowatt Hours	Active Plants	Million Kilowatt Hours	Million Kilowatt Hours	Percent		
Wet	25	615	100	3,537	4,152	55.9	186,370	22.3
Dry	29	1,222	67	2,052	3,274	44.1	132,639	24.7
Total	54	1,837	167	5,589	7,426	100.0	319,009	23.3
Per cent of Total Electric Energy Used	–	24.7	–	75.3	100.0	–	–	–

factors are: the grinding resistance of the raw materials, coal, and clinker; open- or closed-circuit grinding; dry or wet process; the plant equipment units; maintenance; and control.

Generation of Power

Electric power, almost exclusively, is employed in cement plants. Generators may be driven by steam turbines, reciprocating steam engines, water turbines, or internal combustion engines. The most common arrangement is a generator directly connected to a steam turbine (Fig. 8:1), the steam for

FIG. 8:1. STEAM TURBINE GENERATOR

which is generated in a boiler fired by pulverized coal. For convenience, in this section, the discussion will be limited to this combination. Alternating current is used almost exclusively. Usually, motor generator sets are installed to provide any direct current that may be required. The three sources of power are:

1) A separate power plant containing direct-fired boilers;
2) Waste-heat boilers;
3) Purchased power.

There are various combinations of these three basic arrangements. Here are some examples. If there is a waste-heat boiler installation, there may be in addition one or more direct-fired boilers. If most of the current is generated at the plant, some current may be purchased on a stand-by basis, or there may be an arrangement by which at times, some of the plant current is sold.

Frequently the choice is between waste-heat boilers and purchased power. The decision need not be permanent, but a change is not desirable. If waste-heat boilers are installed and power is purchased afterwards, there is a loss in the return on the investment; and if waste-heat boilers are installed to supply power formerly purchased, the installation usually is more expensive and inconvenient than if made originally.

It may be stated that "waste heat" is not a well-chosen term. When such heat is utilized, it is no longer waste heat. *Reclaimed* or *recovered* would be a much better term, but *waste heat* has become so firmly established in the industry that any attempt to change the nomenclature would encounter much resistance.

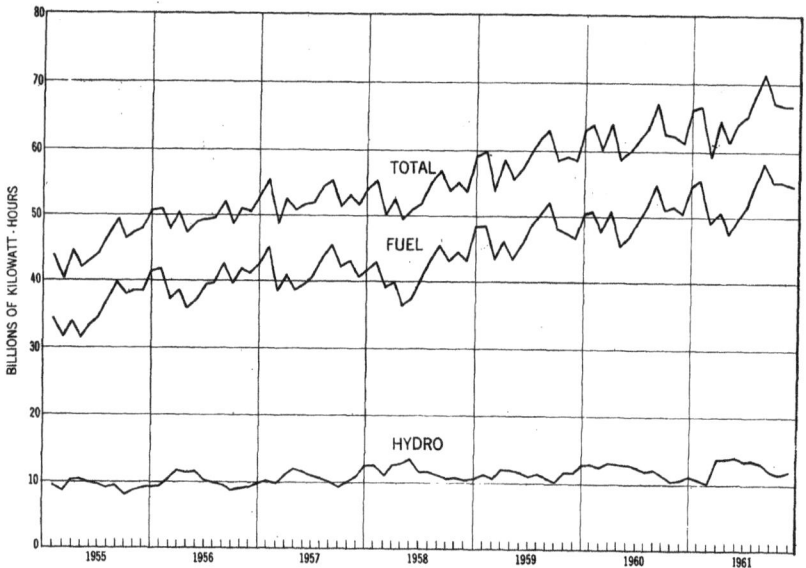

FIG. 8:2. MONTHLY PRODUCTION OF ELECTRICAL ENERGY
BY ELECTRIC UTILITIES

The production of electric energy by electric utilities is increasing greatly in the United States. The cement industry is largely dependent on this production. Here are some data, as of November, 1961; and see Figure 8:2.[3]

CONSUMPTION OF FUEL FOR PRODUCTION OF ELECTRIC ENERGY

Fuel Consumption

COAL CONSUMPTION by electric utility power plants was 15,937,039 tons in November 1961. This is an increase of 7.1 percent over the 14,878,553 tons consumed in November 1960. The November 1961 total was 2.4 percent above the 15,556,814 tons consumed in October 1961.

COAL AND ESTIMATED COAL EQUIVALENT of other fuels consumed in November amounted to approximately 23,492,000 tons, an increase of 7.7 percent over the preceding November, compared with an increase of 8.2 percent in total fuel generation. The indicated November rate, for combustion of coal only, was 0.857 pounds per kilowatt-hour compared with 0.862 pounds a year earlier.

FUEL OIL CONSUMPTION during November 1961 totaled 7,805,637 barrels. This is a decrease of 0.2 percent from the 7,822,705 barrels consumed in November 1960. November 1961 oil use was 21.1 percent above that for October 1961.

CONSUMPTION OF GAS during November amounted to 140,913,077 Mcf. This is an increase of 10.5 percent over the 127,514,618 Mcf burned in November 1960. November 1961 gas use was 15.6 percent below the 166,953,358 Mcf burned in October 1961.

TWELVE MONTHS' CONSUMPTION OF FUEL totaled 181,843,-593 tons of coal, 1,815,317,914 Mcf of gas, and 85,871,623 barrels of oil for the year ended November 30, 1961. These amounts represent increases of 3.9 percent for coal, 5.3 percent for gas, and 1.0 percent for oil as compared with corresponding totals for the 12 months ended November 30, 1960.

PRODUCTION OF ELECTRIC ENERGY

Electric Utility Production — Capacity

ELECTRIC ENERGY PRODUCTION by electric utilities in the United States totaled 66,669,369,000 kilowatt-hours in November 1961. This is the highest November production total of record and an increase of 8.6 percent over the 61,409,742,-000 kilowatt-hours produced during November 1960. The November 1961 production total was 0.3 percent below the 66,848,462,000 kilowatt-hours produced during October 1961.

WATER POWER plants of electric utilities produced 11,863,-
053,000 kilowatt-hours in November. This was 10.2 percent
above production from this source in November 1960. As a
proportion of the November total, water power increased
from 17.5 percent last year to 17.8 percent this year. Production by fuel-burning plants in November was 8.2 percent
above that for November 1960.

TWELVE MONTHS' PRODUCTION by electric utilities totaled
787,299,046,000 kilowatt-hours for the year ended November 30, 1961. This is an increase of 0.7 percent over the
782,039,419,000 kilowatt-hours for the 12-month period
ended a month earlier and 5.0 percent above the 749,819,489,-
000 kilowatt-hours for the year ended November 30, 1960.

INSTALLED CAPACITY of generating plants in utility service
totaled 178,816,125 kilowatts on November 30, 1961 according to reports received during December. This is a net increase of 1,162,061 kilowatts during the month. Occasionally
capacity changes are not reported promptly, therefore the
figure for any one month may reflect additions or retirements
for prior periods. The November capacity compares with
166,314,434 kilowatts reported in service on November 30,
1960.

Industrial Production — Capacity

ELECTRIC ENERGY PRODUCTION by power plants of industrial concerns, including the stationary plants of electric
railroads and railways, was 7,552,176,000 kilowatt-hours in
November. This is an increase of 10.1 percent over production in November 1960. Industrial generating capacity was
17,833,507 kilowatts on November 30, 1961.

COMBINED UTILITY AND INDUSTRIAL PRODUCTION was
74,221,545,000 kilowatt-hours in November, an increase of
8.7 percent as compared with November 1960. Combined
production for the year ended November 30, 1961 was
873,107,738,000 kilowatt-hours which was 4.2 percent above
the year ended November 30, 1960. Utility and industrial
generating capacity totaled 196,649,632 kilowatts on November 30, 1961.

Waste-Heat Boilers

An intelligent decision for or against waste-heat boilers
can be made only after a careful study of all the factors involved. There are likely to be more factors than are apparent
at first. Reduced to simplest terms, the basis for the decision

is the net cost of power during the period of operation of the plant. The wet process is much less favorable to waste-heat boilers than is the dry process, particularly if the slurry contains more than the minimum practical percentage of water. Long kilns are less favorable than short kilns, because, in general, there is less heat to be recovered from long kilns. Small plants, producing less than 3,000 barrels of clinker a day, are less favorable than large plants.

After such preliminary considerations have been disposed of, it is in order to study such items as the capital investment, depreciation, insurance, operation, and the maintenance of boilers, prime movers, generators, and auxiliary equipment. The power required to operate the auxiliary equipment must be deducted from the gross power to obtain the net power available for operating the plant.

Balancing the generation of power and the requirements of power introduces many problems. Let it be assumed that the kilns are producing the desired quantity of clinker. If the power generated is more than sufficient for the operation of the plant, and if the excess of power cannot be utilized in some way, nor sold, an uneconomical condition exists. If normal operation of the kilns does not generate sufficient power, it is customary to supply additional fuel. This increases the coal consumption per barrel of clinker, which causes some accounting difficulties, because, the boilers are no longer operating on waste heat exclusively. Such changes in fuel consumption may also affect the quality of the clinker, particularly if the fuel is coal.

Changes in manufacturing schedules may throw power production out of balance. For example, if the quantity of clinker ground is above normal, a power deficiency may result; and conversely, when the quantity of clinker ground is below normal—there may be an excess of power. Plant operators are familiar with such problems, and, frequently, satisfactory adjustments can be made.

Meade [4] gives the following account of the development of waste-heat boilers:

The attempt to utilize the heat of the kiln gases under boilers was first made, I believe, at the plant of the Nazareth Cement Company, Nazareth, Pa., in 1897 by Dr. Irving A.

Bachman. Dr. Bachman placed the boiler immediately over
the rear kiln housing so that the dust-laden gases entered
the first pass of the boiler directly after leaving the kiln. No
intervening flue was employed to collect even a portion of
the dust. Natural draft was used to carry the gases through
the boiler. It was found impossible under these conditions
to keep the boiler clean, and the dust accumulated so rapidly
that continuous operation of the kiln was impossible. After
encountering these difficulties, the plan was abandoned and
the boilers were taken away.

The late Professor R. C. Carpenter, of Cornell University,
installed waste-heat boilers to receive the exit gases from the
rotary kilns in the plant of the Cayuga Lake Cement Company. The installation here was somewhat better than that
at Nazareth, but much was still to be desired in the way of
facilities for keeping the boiler clean. In this plant, one boiler
of the Wickes vertical water tube type, of 3,000 square feet
of heating surface, was installed for each two kilns (6 feet by
60 feet) of the plant.

When these boilers were clean they gave very satisfactory
results. When operated on kiln gases, alone these boilers produced about 250 boiler-horsepower. The same general difficulties, however, were encountered at Cayuga Lake as at
Nazareth. The company, in spite of these, struggled for
quite a long period to handle the dust before finally abandon-
ing the waste-heat boilers. A similar installation by Professor
Carpenter at the plant of the Kosmos Portland Cement
Company near Louisville, Ky., had pretty much the same
history.

The efforts to use waste-heat boilers were abandoned after
these attempts, for some years. The late Mr. Spencer B.
Newberry, of the Sandusky Cement Company, however,
profiting by the experiences of the early attempts, placed
his boilers at a greater distance from the kiln with a sub-
stantial flue constructed between the kilns and the boilers.
As the importance of this began to be realized, other engineers
experimenting with the use of the waste gases for steam generation provided additional facilities for taking care of the
dust and for keeping the boilers clean. It was also found
advisable to employ induced draft rather than stacks.

Among the plants which installed successful waste-heat
boilers prior to 1915, when the subject received the almost
universal attention of cement manufacturers, may be mentioned the Louisville Cement Company, the Sandusky Portland Cement Company, and the Burt Portland Cement
Company. With these installations, about half of the steam
required to operate the plant was obtained. The boilers were
also generally so installed that they could be fired by hand

when the kilns were not operating, or a greater quantity of steam was desired than the kiln gases could produce.

While these installations were faulty they turned the attention of the cement mill engineers to the possibilities of the waste-heat boiler. It was soon realized that in the early installations due consideration had not been given to air leakage. Little or no importance was attached to the lowering of the temperature of the kiln gases by the infiltration of cold air at the feed end of the kiln. This condition occurred both in the boiler-setting itself and at the opening between the kiln and the flue. In the modern installations, air seals are placed on the upper end of the kiln shell to exclude the cold air at this point.

Figure 8:3 shows a waste-heat boiler installed by Combustion Engineering, Inc., in a cement plant in California. There are three of these units in operation at the plant.

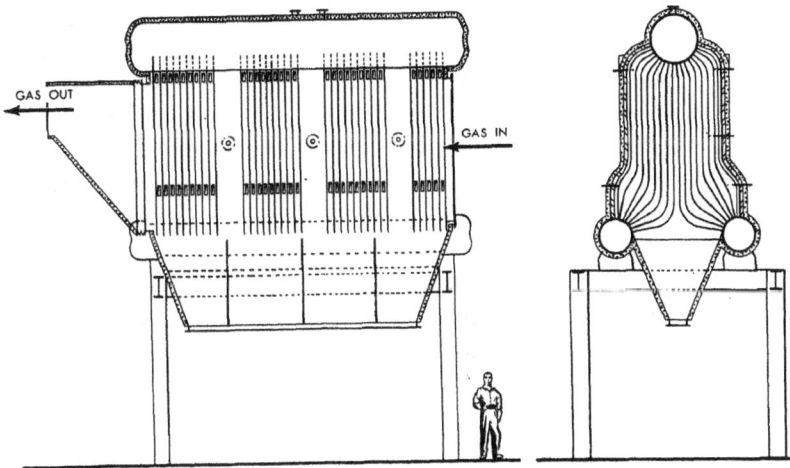

FIG. 8:3. WASTE-HEAT BOILER
(*From* Combustion Engineering, Inc.)

Purchased Power

Power problems are by no means confined to waste-heat boilers. Many arise in connection with purchased power. Some of the items that must be studied are:

Over-all cost of power
Demand charge

Energy charge
Any variation in supply from day to day, or from season to
   season
The nature of the contract
The ownership, location, and operation of transformers
Line losses as affected by the location of meters
Scheduling of plant operation to take advantage of off-peak
   loads

The percentage of industrial power supplied by public utilities has been increasing (Fig. 8:2). This is not surprising, because companies specializing in generation of power can be expected to operate their equipment at higher efficiencies.

Miscellaneous Applications of Electricity

Nearly all the electric current at a cement plant is utilized to drive motors. Notwithstanding this, the application of electricity is very extensive and varied. Transformers have been mentioned. Remote-control switches, controllers, magnetic pulleys, signal devices, indicating and recording instruments, and laboratory devices are used in great numbers. In fact, it is probable that examples of most of the applications of electricity can be found in the cement plant.

Today the principal sources of power in the cement industry are the same as for many years past. It is too early to predict the application of nuclear power. The present capacity (1964) of the Dresden Nuclear Power Station, near Chicago, is 180,000 kilowatts, which go into the energy pool of the Commonwealth Edison Company. The territory supplied by this company includes Oglesby, Illinois, where there are two cement plants. So, even now, some nuclear power may have gone into the manufacture of cement. At Dresden, it is considered possible that in two or three years, nuclear power can compete economically with coal in that locality. So many factors are involved, however, that each locality has its own problems.

The editor of *Power Engineering*, R. F. McCaw, generously supplied some of the information on power and fuels that appear in this chapter.

REFERENCES

1. *Mech. Eng.*, 66(5):352 (1944).
2. *Minerals Yearbook*, U.S. Bureau of Mines, Washington, D.C., 1961.
3. *Electric Power Statistics*, Federal Power Commission, Washington, November, 1961.
4. R. K. Meade, *Portland Cement;* Chemical Publishing Co., Easton, Pa., 1930.

Chapter 9

GRINDING THE CLINKER

There are a number of items that belong equally to Chapters 5 and 9. Duplication has been avoided by limiting the discussion of these items to this chapter. Here, only two primary materials and only the dry process are to be considered. The processing that clinker receives between the kilns and the preliminary mills varies. Some cooling and some blending always take place. In some plants, there are clinker-crushers, but some of the clinker produced does not require crushing before it enters the preliminary mills.

Of the unit processes in the manufacture of cement, fine grinding is second in importance only to clinkering. Also, it is second only to clinkering in the study that has been applied to it, and the developments that have resulted from this study. In fact, when we realize that clinkering, as we know it today, takes place only when the kiln feed is finely ground; and that modern cement is produced only when clinker is finely ground, it may be considered that grinding and clinkering are equally important.

Although no chemical reactions of major importance take place during the grinding of clinker, our information concerning some phases of it is surprisingly limited. One of the first steps in investigating a process or the operation of an equipment unit is to calculate the efficiency—thermal, mechanical, or some other type. The well-known formula is simple:

$$\text{Efficiency} = \frac{\text{Output}}{\text{Input}}$$

Whoever first applied this formula to a grinding machinery unit—a tube mill, for example—must have encountered considerable difficulty. Various values for efficiency have been reported, but until maximum and minimum values differ less widely than at present, or until some formula is developed on the basis of which differences can be reconciled to some extent, it is best to assume that the power efficiency of a tube mill is not definitely known. The power input can be measured, but the difficulty is with the numerator of the fraction.

Martin [1,2,3] and his coworkers have carried on extensive researches on the theory of fine grinding. Leighton Buzzard quartz sand was ground in a small ball mill. Some of the outstanding assumptions, calculations, results, and conclusions were:

> Imagine the quartz sand to be continually ground to particles the size of molecules. Then the work required to produce this result is the same as the work required to convert the silica into a gas, where each molecule is separated from the other. So that the work of grinding will be the same as the work required to volatilize the substance.
>
> A weighed sample of sand was shaken with five-normal hydrofluoric acid. The acid was then diluted, and the sand filtered out, dried and weighed. The loss in weight was assumed to be a function of the total surface.

This formula was developed:

$$W = B(S_2 - S_1)$$

Where W = the work input, in foot pounds
S_1 and S_2 = original and final surface of the sand, respectively, in square feet.
B = a constant for the mill.

> A relationship exists between the surface of the balls in the mill and the surface of the sand being ground, efficiency only being maintained when the surface of the sand exceeds the surface of the balls from 100 to 500 times.

Martin deduced that "in the case of quartz sand, the efficiency of an ordinary tube mill was of the order of only $\frac{1}{27}$th per cent."*

---

* It is recommended that anyone particularly interested in fine grinding read the original papers. It is difficult, if not impossible to abstract them adequately. There are more papers in the series than are indicated by the references given at the end of this chapter.

Clinker and Gypsum

The primary materials of cement are clinker and gypsum. The manufacture of clinker has been outlined in previous chapters. Most finely-ground clinker sets too rapidly for commercial use, after it has been mixed with water. A retarder, or, as it sometimes is called, a *negative catalyst*, is required for controlling the set. Various substances may be employed for the purpose, but for many years, only calcium sulfate and water have been recognized by specifications. The use of water for the purpose is not commercially important, at least in the United States.

Dry-process and wet-process clinkers are much the same, but usually the particle size of the wet-process clinker is the smaller of the two. This small size is favorable to heat transfer and to chemical reactions, and it tends to decrease the resistance to grinding.

Gypsum

The source of calcium sulfate is the mineral, *gypsum*. Its chemical formula, $CaSO_4 \cdot 2H_2O$, shows that it is a neutral calcium salt of sulfuric acid, containing two molecules of water of crystallization. The percentage composition of oxides in the pure substance is:

Calcium oxide	32.5
Sulfur trioxide	46.6
Water	20.9

Gypsum often contains such impurities as clay, silica, and calcium carbonate. In the United States, commercial deposits occur in New York, Michigan, Iowa, Kansas, New Mexico, and Oklahoma. In Canada, there are large deposits in Nova Scotia, and New Brunswick. The mineral is obtained by quarrying or by mining.

When gypsum is heated to approximately 500°, and not to exceed that temperature, plaster of paris $(CaSO_4)_2 \cdot H_2O$, results. When this is hydrated, the formula is again $CaSO_4 \cdot 2H_2O$.

Deposits of the mineral *anhydrite*, $CaSO_4$, are often associated with gypsum. There is a debate, of many years' standing, as to whether or not anhydrite can be used satisfactorily as a retarder. No decision will be stated here, but anhydrite should not be used unknowingly. Anhydrite may be distinguished from gypsum microscopically. If the content of sulfur trioxide in a shipment of gypsum is higher than 46.6 per cent, the presence of some anhydrite is indicated.

For several reasons, it is desirable that gypsum be free from surface moisture when it is received at the plant, and that it be stored under shelter. It is desirable also that the particle size of the gypsum be somewhat similar to that of the clinker. Large lumps are difficult to handle, and fine material favors segregation of the clinker-gypsum mixture. Ordinarily, clinker enters the preliminary mills practically dry. If it has been stored in the open, it may be dried to some extent by mixing it with clinker that still retains some heat from the kilns. If conditions arise by which clinker from an open storage must be ground without the addition of hot clinker, some arrangement for drying it must be made.

The temperature of clinker on entering the preliminary mills should not be any higher than can be avoided. In some Hercules mill tests, it was found that when relatively cool clinker entered the mill, the heat content was increased by friction; but that when hot clinker entered, the heat lost by radiation overbalanced the heat generated by friction.

For some reason, the sulfur trioxide present in the clinker has little or no value as a retarder. Usually, this amounts to only a few tenths of a per cent. When the kiln feed, or the fuel, is relatively high in sulfur, the sulfur trioxide in the clinker may exceed one per cent. This decreases the quantity of gypsum that may be ground with the clinker.

Proportioning Clinker and Gypsum

The proportioning of clinker and gypsum by weight is recommended strongly. The statements in Chapter 4 concerning proportioning and the formula apply here. We repeat the formula:

$$z = \frac{100(p - v)}{p - q}$$

in which q = per cent SO_3 in clinker

$\quad$ p = per cent SO_3 in gypsum

$\quad$ v = per cent SO_3 in cement

$\quad$ z = pounds of clinker required for 100 pounds of cement

All percentages expressed as decimals.

$\quad$ Example. A clinker contains 0.5 per cent SO_3, and a gypsum contains 43 per cent SO_3. It is desired to make a cement containing 1.50 per cent SO_3.

$$z = \frac{100(0.43 - 0.015)}{0.43 - 0.005} = 97.65$$

Therefore, the quantities required are 97.65 pounds of clinker and 2.35 pounds of gypsum. For one barrel of cement, 376 pounds, the requirements are: 367 pounds of clinker, and 9 pounds of gypsum.

When the complete analysis of the clinker and gypsum, and the relative quantities of the two are known, the analysis of the cement may be calculated. Probably the result will not be quite exact, because, during grinding, some moisture is likely to be volatilized from both clinker and gypsum, and there may be some loss of dust. For a given set of conditions, it is not difficult to develop correction factors by comparing calculated analyses with actual analyses.

It is desirable so to operate the mills that the temperature of the cement is as low as practicable. Less than 250° is recommended. When the temperature is higher than 250 to 300°, the clinker may have abnormal setting characteristics. Often, this is said to be due to the presence of alkali carbonates, or to partly dehydrated gypsum, but, in my opinion, these statements have not been demonstrated. The installation of a recording thermometer or pyrometer at the discharge area of a finishing mill is helpful.

Development of Grinding Equipment

One of the most important factors in the development of the cement industry has been the recognition of the importance of fine grinding. This recognition has led to the more intelligent utilization of the grinding equipment available. This utilization, in turn, has been an incentive for the design of better and better equipment, and so on.

The first power-driven mill of the industry was similar to the early grain mills. There were two massive buhr stones mounted horizontally, one above the other. The next device was called a *barrel pulverizer*, which appeared first in Cornwall, England, about 1880. It consisted of a cylinder, a few feet long, supported in a horizontal position and rotated. In it were placed some iron balls (the grinding media) and a batch of the substance to be ground. There is a type of laboratory mill of somewhat similar design in use today.

Following the barrel pulverizer, many types of mills have been designed that have found their way into cement plants. Some of the types have remained; others are obsolescent or have disappeared almost completely. It is of interest that probably most grinding media are spheres, a large range in sizes being employed. There is a group of cylinders, and there are some other shapes. Some of the characteristics of a mill, on the basis of which its over-all value is rated, are:

Original cost
Maintenance cost
Power requirements
Lubrication requirements
Output
Particle size distribution of the product
Vibration
Noise
Dust production

Types of Mills

Of the types of mills in use at present, the following are examples:

Preliminary Mills: Ball mill; Roll and ring mill
Finishing Mills: Tube mill
Combination Mills: Compartment mill

For convenience in discussing these mills, some definitions are in order.

1. The *body* of the mill includes the entire mechanism with the exception of the grinding media.
2. *Grinding media* are the cast iron or steel bodies inside a mill that are largely responsible for the grinding.
3. The *charge* is the total media.
4. The *feed* is the material to be ground, which is introduced into a mill.
5. The *product* is the ground material discharged from the mill.

A few of the characteristics of the mills may be listed as follows:

1. *Shape of grinding media*
   a) Spherical
   b) Cylindrical
   c) Other shapes
2. *Movement of media*
   a) Free
   b) Restrained
3. *Ratio of diameter to length*
   a) Greater than one
   b) Less than one
4. *Direction of the axis of the mill*
   a) Horizontal
   b) Vertical
5. *Class of grinding*
   a) Preliminary
   b) Finishing
   c) Combination

Ball mills:

A ball mill is a steel cylinder rotated in a horizontal position at low speed. The lining consists of iron or steel plates. The grinding media are iron or steel balls, usually not less than two inches nor more than four inches in diameter, when new. Referring to the list of mill characteristics, the following figures and letters apply to this type of mill: 1a, 2a, 3a, 4a, 5a. Since ball mills first appeared in cement plants, they have been modified in various ways, and types with special names have appeared, such as, the Kominuter and the Preliminator. The name *Preliminator* is applied to mills for both dry and wet grinding. Most of the product will pass through a 20-mesh sieve.

Roll and ring mills:

There have been several mills of the general roll-and-ring type; an example is the Hercules mill.[4]

Some of the characteristics of the Hercules mill are: 1b, 2b, 3a, 4b, 5a. Most of the product will pass through a 20-mesh sieve.

Tube mills:

The tube mill is designed for finishing, when the feed has received preliminary grinding. There are many modifications, but, in general, it consists of a horizontal rotary steel cylinder 5 to 8 feet in diameter and 20 to 27 feet long. The lining consists of iron or steel plates. The media are iron or steel balls not exceeding 1.25 inch in diameter, when new. Some of the characteristics are: 1a, 2a, 3b, 4a, 5b.

Compartment mills:

The compartment mill may be considered a combination of ball and tube mills. There may be 2, 3, or 4 compartments, containing grinding media in different sizes. In comparison with 2-stage grinding combinations, the compartment mill has the advantage that no intermediate conveying and feeding equipment is required. There are some disadvantages, however. The material proceeding directly from one compartment to another retains most of its heat. Also, it is difficult to determine the degree of grinding in each compartment. Some of the characteristics are: 1a, 2a, 3b, 4a,

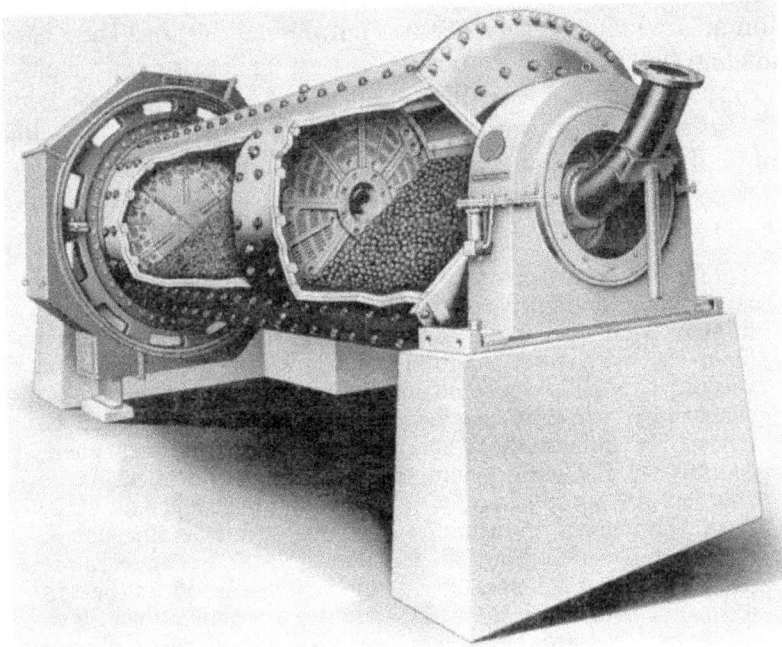

FIG. 9:1. TWO-COMPARTMENT MILL
(*From* Allis-Chalmers)

5c. A compartment mill, made by Allis-Chalmers Manufacturing Company, is shown in Figure 9:1.

Open and Closed Circuits

The fundamental difference between open circuit and closed circuit grinding is mentioned in Chapter 5. So far, in this chapter, only the open type has been mentioned. Closed circuits, however, are used extensively with preliminary, finishing, and compartment clinker mills. Furthermore, there are several types of mills that are operated in closed circuit only. Closed circuits will be discussed here only in connection with tube mills and air separators.

When a portion of the load of a tube mill has been reduced to 200-mesh fineness, this fine portion retards the grinding of the remainder. This is undesirable because it not only decreases the output, but also increases the power required per barrel of cement. It was recognized that the removal of the fines, as soon as formed, would facilitate the grinding operation as a whole. Screens are not employed, because they cannot handle 200-mesh materials so well as coarser materials.

Some years ago, so-called *wind-swept tube mills* were tried by Edison. They were not satisfactory, probably because some of the particles removed from the mill were too coarse. A type of air separator that has been used extensively may be described as follows: [4]

> The standard separator consists essentially of an upright steel plate, cylindrical outer shell terminating in a conical hopper which is of approximately 60-deg. slope, measured from the horizontal, in order to insure free discharge. The top of the shell is closed off except for the feed opening near the center, while at the lower extremity of the hopper an outlet for finished material is provided. An inner shell, roughly of the same general contour as the outer casing, is stayed to the outer wall in a concentric position, the inner cone terminating in an inclosed tailings spout leading out of the machine. The inner shell, which does not extend to the full height of the separator, is partially closed off at the top by means of a diaphragm plate having a circular opening at its center. This diaphragm may be a solid plate or it may be made up of segments adjustable from the exterior in order to vary the size of the opening which governs the velocity of the air passing through it.

A horizontal, steel-plate fan of the paddle-wheel type is placed immediately above the diaphragm and is mounted on a vertical shaft driven through bevel gearing. The lower end of this shaft carries a horizontal, circular distributor plate driven as a unit with the fan. A second plate, similar to the distributor plate and rotating with it, is usually employed above the distributor and this plate frequently carries additional blading.

In operation the feed opening allows the material to drop on the lower or distributing plate where it is spread and thrown off by centrifugal force, the larger and heavier particles being projected against the inner casing, while the smaller and lighter particles are picked up by the ascending air current created by the fan. These fines are carried over into the outer cone and deposited. Concurrently, the rejected coarse material drops into the inner cone and passes out through the tailings spout.

The air, after dropping the major portion of its burden, is again drawn up through the diaphragm opening, means being provided for the air return through the inner shell which is cut away for the purpose. In order to distribute the return air in a proper manner a series of guide vanes, or *louvres*, encircles the inner shell at this point. In certain machines these vanes are of the fixed type mounted in a radial position, while in others the vanes are pivoted and may be adjusted in unison by means of an external control. This adjustment allows the free area through the vanes to be varied and at the same time the vanes may be moved from a radial position to one approximately tangential to the circle of the inner shell.

The return air entering between the guide vanes into the inner shell encounters the falling oversize material through which it passes and, as a result, a further separation is made and an additional amount of finished material is recovered. Because the air is recirculated continuously, no external air is required for classification purposes. It is found in practice, however, that a relatively small amount of extraneous air is entrained in the feed and, consequently, a pressure is frequently built up in the separator, in which case the excess air may be vented off.

The major adjustment of the separator consists in establishing a proper fan speed and air velocity for the required classification, although various other adjustments are provided. When an adjustable diaphragm opening is employed, obviously any increase or decrease in the area of this opening has an immediate effect upon the velocity of the air at this point. The effect, however, upon particle sizing is not so pronounced as might be expected, for when this opening is reduced the diaphragm acts as a baffle and cuts out of the air stream some of the coarser particles which otherwise would tend to pass over into the finished product, because of the increased velocity.

In machines employing adjustable guide vanes these are usually pointed in the direction of the rotation of the fan, assisting in maintaining the swirling movement of the air. A moderate change in the position of these vanes from their normal setting produces no marked results, but it is worthy of note that a complete reversal of the vanes has the immediate effect of coarsing the finished product, in some instances as much as 20 points on the 200-mesh scale. Advantage may be taken of the fact particularly in cases where a separator is required to make intermittently two separate and distinct products of different degrees of fineness.

Where additional blading is carried on or near the distributor plate, the effect is naturally to augment the blast action of the fan and it is claimed also that these blades assist in projecting coarse particles in the feed to the outer reaches of the chamber, clear of the ascending air current. Other details and adjustments not universally adopted consist in raising or lowering the distributor plate, altering the blade area of the main blast fan and baffling fan blades and guide vanes.

Some advantages of air separators are:

1. Increased output for the same fineness
2. Decreased consumption of power
3. Decreased variation in particle size

Disadvantages are to be expected; some of them are:

1. Power is required for operating the separator.
2. There must be taken into account the cost of the separator, maintenance, and occasional loss in tube mill operating time because of repairs to separator.
3. For the same 200-mesh fineness, the strength of air-separated cement may be lower.
4. There is some segregation of the clinker and gypsum. This usually reaches equilibrium in time, however, the sulfur trioxide content becoming the same as it would be in closed circuit operation.

Opinions differ greatly concerning the over-all merit of air separators in general. As many factors must be taken into account, evaluation of a single installation may be difficult.

Importance of Surface Area

For many years it has been recognized that little information on the particle size of finely-divided solids is obtained by sieving. Measurements had been made by the use of standard microscopes, and it was known that, with the ultramicroscope, particles could be seen that were too small to be visible with ordinary equipment.

It was known that, other factors being equal, the chemical activity of a substance is directly proportional to the surface per unit weight. This led to the calculation of the surface of several pure substances, by various procedures. Martin determined the surface of sand by observing the rate of its reaction with hydrofluoric acid. Martin's methods are not applicable to cement, because cement is a mixture of compounds.

In 1927, I developed a procedure for estimating the surface of cements, in terms of square feet per barrel. Relative weights and surfaces of several particle-size groups were studied in some detail. The importance of surface estimations was evident, and it was apparent that the procedure could be applied to other finely divided solids, including raw mix and coal.

Since 1933, several types of apparatus for estimating the surface of cement in terms of square centimeters per gram have been developed. One of these is the Wagner turbidimeter, described in a paper entitled, *A Rapid Method for the Determination of the Specific Surface of Portland Cement.*[5] The following is an abstract of the paper:

> An apparatus and method are described for making rapid determinations of the specific surface and also the particle size distribution of portland cement. The apparatus, which is essentially a turbidimeter, consists of a source of light of constant intensity which passes through a suspension of the cement in kerosene and then into a photo-electric cell. The current generated in the cell is measured with a microammeter, and the readings afford a measure of the turbidity of the suspension. The relation between the turbidity of the suspension and the surface area of the suspended particles is calculated. Particle-size distribution is obtained by observing changes in turbidity as the particles settle from the suspension."

The following is an abstract of a paper by Klein,[6] describing a suspension turbidimeter:

This paper describes a portable apparatus for the rapid determination of the surface area of portland cement and other granular materials, in which determination the turbidity of a suspension of the material in castor oil is measured. The method does not involve Stokes' law, which is the basis of sedimentation methods, but consists in determining the mean effective cross-sectional area of the suspended material under static conditions, through the use of photoelectric equipment. Hence, no consideration of time or velocity is involved in the determinations, no standard samples are required, and the method is independent of the physical characteristics of the suspending medium. A theoretical constant, required for the analysis of a given material, is determined from an observation on a sieved sample of the material itself. A high degree of reproducibility is attained, and the method appears to furnish an accurate measure of the relative fineness of various materials.

An apparatus has been developed by Lea and Nurse [7] for estimating surface by air permeability. With any one of these devices, surface values are obtained by indirect calculations and may be considered relative only. Values obtained by any two procedures are not comparable. The development of a procedure for the direct determination of cement surface, even approximately, would be a worth-while contribution to the industry.

Tube Mill Output

As explained in Chapter 8, the power consumption of tube mills is large. Another item of expense in the operation of tube mills is the wear of grinding media and linings. Additional balls must be placed in a tube mill from time to time. Formerly, it was common practice to remove the media completely and screen them, rejecting all units of less than a given size. Now, it is customary to allow the balls to wear until they disappear. Calculations of the consumption of grinding media per barrel of cement show little agreement.

Much study has been devoted to increasing the output of tube mills, thereby decreasing the consumption of power per barrel. For a given material to be ground, some of the factors affecting output are:

1. The maximum size and the size distribution of the media,
2. The percentage of the internal volume of the mill that is occupied by the gross volume of the media,
3. The gross volume of the material to be ground in the mill.
4. The temperature and humidity of the contents of the mill.
5. The peripheral speed of the mill.

The objective is to find the optimum value for each of these factor for a given set of conditions.

For controlling the rate of feed to obtain the maximum efficiency, an ingenious electrical device called the *electric ear* has been developed.[8] In principle, this device reduces the rate of feed when the noise from the mill drops below a predetermined level, and increases the rate of feed when the sound from the mill is slightly above the desired level.

The most frequently mentioned law of size reduction is that of Rittinger. The basis of this law is the assumption that the work required is proportional to the increase in total surface. Another law is that of Kick. This is based on the ratio of average size of the particles before and after the reduction operation. According to Walker, neither law represents the facts accurately.[9]

There is a procedure for calculating the grinding resistance, the so-called *grindability*, of materials. This is the Hardgrove method, which has found its greatest application in the grinding of coal.

Particle Size of Cement

It has been pointed out that, theoretically, there is no limitation to the fineness of raw mix, but there is an economic limit. The situation is different with cement.

Although the object of grinding clinker is to facilitate chemical reactions between the clinker and water, there are some practical considerations that must be taken into account. Particles of cement can be so small that some hydration may occur in the package. This is commonly called *warehouse set*. Furthermore, the particles may be so small that the control of set is difficult. The present limitations of surface have resulted more or less from trial-and-error procedures.

Anderegg [10] found that when typical clinker was stored in water, in the presence of calcium sulfate, the depth of hydration, was approximately:

0.5 micron at 1 day
1.7 microns at 7 days
3.5 microns at 28 days
5   microns at 90 days

When the power input of a tube mill is plotted against the specific surface of the cement produced, the graph is not a straight line. As the surface increases, the rate of increase in power becomes greater. For example, the increase in power-input for surfaces of 2000 to 2500 square centimeters per gram, is greater than the increase for 1500 to 2000.

REFERENCES

1. Geoffrey Martin, E. A. Bowes, and J. W. Christelow, *Trans. Ceram. Soc. (British)*, 25:51 (1925–26).
2. Geoffrey Martin, E. A. Bowes, and F. B. Turner, *Trans. Ceram. Soc. (British)*, 25:63 (1925–26).
3. Geoffrey Martin, F. B. Turner, and Francis Linstead, *Trans. Ceram. Soc. (British)*, 26:34 (1927).
4. *Pit and Quarry Handbook;* Pit and Quarry Publishing Co., Chicago, 1956.
5. L. A. Wagner, *Proc. Am. Soc. Testing Materials*, 33(II):553 (1933).
6. A. Klein, *Proc. Am. Soc. Testing Materials*, 34(II):303 (1934).
7. F. M. Lea and R. W. Nurse, *J. Soc. Chem. Ind.*, 58:277 (1939).
8. Harlowe Hardinge, *Tech. Pub. 1076;* Am. Inst. Mining Met. Engrs., 1939.
9. W. H. Walker, W. K. Lewis, W. H. McAdams, E. R. Gilliland, *Principles of Chemical Engineering;* McGraw-Hill Book Co., New York, 1937.
10. F. O. Anderegg and D. S. Hubbell, *Proc. Am. Soc. Testing Materials*, 29(II):554 (1929).

Chapter 10

STORING, PACKING, LOADING, SHIPPING

The question a cement man most often hears is, "Why is cement called 'portland,' and is all 'portland cement' manufactured by the same company?" There are other frequent questions such as, "Why is cement sold by the barrel but delivered in sacks or bags?" "Why does a sack of cement contain 94, rather than 100 pounds?" The first question is answered in Chapter 2. The background (1870–1880) of a number of the customs in the industry is given by Lesley:[1]

..., It is a matter of considerable interest in connection with the history of the American portland cement industry to describe the methods by which this large importation was handled. In those early years the usual cargo ship was a wooden sailing vessel. Barques or full rigged ships were generally employed. These had no auxiliary power, and when sailing without regular cargo required ballast of sand, stone, or other material to stabilize the ship. These vessels differed entirely from the liners which carried expensive cargoes from Europe, and which, being constructed of steel or iron, had compartments that were filled with water ballast when necessary, and discharged when taking on cargo. The vessels in the cement-carrying trade were of the type known as tramp cargo ships. In most cases, they came to the United States to get export cargoes of grain or cotton. Such cargoes paid high revenue. Instead of non-paying ballast (as there was little inbound paying cargo to this country) they took on cement or other heavy material. The result was that freight from European ports to this country on cement was very low. In some cases, the American consignee was not only able to get his cement brought across the Atlantic free of

charge, but was actually paid by the ship as high as ten cents
a barrel for the unloading of it. This occurred at times when
the outgoing grain paid such good rates that the tramp ship,
in order to take advantage of the market, was obliged to get
the cargo she had carried over discharged rapidly, even if
necessary to pay for unloading.

The points of heaviest importations of foreign cements were
New York, Philadelphia, Charleston, Savannah, New Orleans, Galveston, and some Pacific Coast ports. In the last
case, not only did Belgian, German, and English cements
come to the Pacific Coast as ballast for grain ships, but also
cements from Japan and China.

As these ships had no regular sailing dates, and because
they were sailing vessels, had no definitely known time of
arrival, the successful marketing of foreign portland cement
required far-seeing calculations. A ship would be loaded in
London or Hamburg. Its actual arriving time was governed by
the elements, and was a most uncertain factor. The cements
would be consigned to various importers representing the
manufacturers, and one importer might have a consignment
of one brand another importer an equally large lot of another brand, both on the same vessel. Thus, it became a
contest of wits to dispose of the material promptly. This had
to be done to secure discharge from the ship at the earliest
possible moment so as to avoid storage, demurrage, and other
unnecessary expense. The result was that each importer was
constantly endeavoring to secure orders for shipments to
arrive, and when an excess of material would reach the harbor
on several vessels at the same time, they had to go on the
building material exchange, and find customers who would
take it off their hands at the earliest possible moment. The
business was uncertain as to possible profit and sometimes
entailed considerable loss. . . .

In the early years, all importations came in sailing vessels,
the cement being packed in barrels having a gross weight of
400 pounds. Almost every cargo had a portion damaged by
water, owing to leaky ships. Occasionally, the damage would
amount to a large percentage of the entire cargo. Freight
from London to New York, or Philadelphia, ran from 25 to
40 cents per barrel, fluctuating according to amount of ship
room available, occasionally dropping as low as fifteen cents
per barrel. Upon one occasion, I brought a cargo of 3,000
barrels from Hamburg to New York for the nominal sum of
one dollar. . . .

During the first decade, importation by steamer became
necessary as the demand could not be supplied by sailing
vessels alone, nor was their time of arrival sufficiently de-

pendable. The Alsen Company was the only one so far as I know that attempted shipments in bags. A few shipments were made by steamer to New York, in jute gunny bags containing 188 pounds of cement, two to the barrel, but they were too heavy to be conveniently handled and the experiment for general trade purposes was soon abandoned. However, the large quantity of Alsen cement used by the Florida East Coast Railway in building the viaduct toward Havana, was shipped in bags by steamer direct to Key West. In that case, the cement went directly from the steamer to the work with the minimum of handling, and the bags did not meet with serious objections, especially as they effected a considerable saving over cost of barrels.

The Alsen Company was the only one, to my knowledge, to make the experiment of shipping cement in sheet-iron drums. They were cheaper to make than a wooden barrel, but being perfectly straight, without any bilge, they were awkward to handle and were discontinued. . . ."

An interesting recollection of many years' dealing with all of these importers is how much they were governed in their methods by the manners and customs of European business men. In many cases, the letters that went out of their offices were hand-written by the principals, being copied in old-fashioned letter books. The typewriter had great difficulty in finding its way into the offices of these great importing conconcerns. . . .

Through the firm of Lesley and Trinkle, which, in 1874, began the shipment of natural cement in bulk to tidewater from the Cumberland cement works, the Coplay Cement Company was induced to do the same thing with its natural and portland cements; and for many years both Johnson and Wilson, and Lesley and Trinkle had large packing houses in Jersey City, and Philadelphia, respectively, to which the cement was brought in bulk from the mills, and there repacked in second-hand barrels which had come from the Rosendale and other mills shipping by water, and which, therefore, had no means of getting the barrels back. These barrels, then a waste product, were bought up at low prices, and in times of shortage were supplemented by second-hand barrels from other sources available in the two cities.

This account explains: Why the unit of quantity for cement is the barrel; why sacks have displaced barrels as containers for cement, and how bulk shipments originated.

It is probable that the 400-pound gross-weight barrel contained approximately 376 pounds of cement, and that this suggested the 188-pound sack, to be followed by the 94-pound sack. The change from barrels to sacks, the world over, was gradual, however. As late as 1919, all the cement manufactured in the Philippines was packed in barrels, and all the cement imported from China and Japan was received in barrels or in steel drums. At the plant of the Rizal Cement Company near Manila, the barrels were manufactured, starting with logs and strip steel for the hoops, at a cost of approximately one and one-half to two dollars each. Trials were made with various types of sacks, but, at the time, none could be found that would protect the cement from the high temperatures and high humidity.

One important factor in the continued use of barrels was that all Rizal cement was shipped by barge to Manila for distribution. Some of it was transferred one or more times to other vessels before it reached the consumer.*

We have not outlined the manufacture of portland cement from the raw materials to the finished product. The storing, packing, and loading, before the product is ready for the user, remain to be discussed.

It is customary to store cement in bulk, placing it in containers just before loading into freight cars, or other transportation units. In designing a stock house, the first question to be decided about storage is the total capacity. The second is the size of the storage units. There is no fixed ratio of the total storage capacity to annual plant capacity. In a few examples, taken at random, the total storage capacity ranges from 9 to 26 per cent, with 10 per cent predominating. Some of the factors that must be taken into account are:

1. Length of the shipping season
2. Length of the manufacturing season
3. General market conditions
4. Clinker storage capacity
5. Shipping facilities
6. Reservation of tested cement
7. Number of types of clinker and cement manufactured

---

* The use of barrels for Rizal cement was discontinued years ago. When I was in Manila in 1954, I was interested to see for the first time a bag of this cement in the grounds of the U.S. Embassy.

Types of Storage Buildings

Few of the characteristics of storage buildings are fixed. Many variations have appeared in the cement industry during the last few decades. Materials of construction have included wood, steel, and concrete. Rectangular and circular cross sections, and large and small units have been tried. In the early days of the industry, according to Meade,[2] a stock house was usually

> . . . a long low frame building, divided into bins by means of wooden partitions, so that each day's grinding could be kept separate. These bins usually held from 1,000 to 5,000 barrels and were arranged either on each side of a central aisle or else with an aisle on each side. The parts of the bins facing the aisles were stopped up by means of boards which might be easily removed, and below the floor of the aisles, ran screw conveyors to the packing room, which was usually at one end or a large room in the middle of the stock house. The screw conveyors were covered with boards, except in front of the bins where gratings 3 or 4 feet in length were placed. The cement was usually brought in from the grinding mills by an overhead screw conveyor, from the trough of which spouts ran to the middle of the bins. The openings in the trough leading into the spouts were closed by iron slides or gates so that the cement might be run into any bin desired at any time. When it was desired to open a bin the bottom plank was removed from the front of the bin and the cement was allowed to run into the screw conveyor, through the grating. When it ceased to run of itself, a scraper, which consisted of a flat iron plate, about 6 inches by 18 inches, from the middle of which a long handle projected was introduced and all of the cement which could be pulled through the opening conveniently was drawn into the conveyor, after which the remainder of the boards were taken down, and the rest of the cement was drawn into the conveyor, either with the scraper or wheeled by barrows to the grating.

All of this has led to a considerable degree of stabilization. No changes in the general arrangement are in sight. The form of stock house accepted today consists of a series of vertical hollow reinforced concrete cylinders, cast integrally, and having a common foundation. Usually, there is more than one row of units, arranged in a rectangle. The ratio of the width to the length of this rectangle is frequently between 0.5 and 1.0. The height, diameter, and capacity of the individual silo are decided for each project. Theoretically, the most economical ratio of height to diameter may be calculated by maxima and minima, but other considerations may suggest some modification of this ratio. Each space surrounded by the walls of the four individual silos may be utilized, and is called an interstice, or a star bin. Sometimes, additional walls are built, so that the walls of two outside adjacent silos may be utilized.

When concrete silos first came into use, it was customary to construct them with flat bottoms. As all the cement could not flow from such a silo, it was necessary to remove the last of the cement by hand shoveling. The capacity was listed under two heads, total and residual. It was considered that this practice was justified on the basis that the residual cement was a substitute for concrete that otherwise would

TABLE 10:1

CAPACITY OF A SILO PER LINEAR FOOT OF DEPTH

Internal Diameter of Silo in Feet	Floor Area of Silo in Square Feet	*Capacity of Silo Bin per Foot of Depth in Barrels of Cement (376 lb)	Floor Area of Star-Shaped Bin Between Each Four Silos in Square Feet	* Capacity of Star-Shaped Bin for One Foot of Depth in Barrels of Cement
25	490.9	121.5	134	33.2
26	530.9	131.4	145	35.9
27	572.6	141.7	156	38.6
28	615.8	152.4	168	41.6
29	660.5	163.4	180	44.6
30	706.9	175.0	193	47.7
31	754.8	186.8	206	51.0
32	804.3	199.1	220	54.5
33	855.3	211.7	234	57.9
34	907.9	224.7	249	61.6
35	962.1	238.1	263	65.1
36	1017.9	252.0	278	68.8
37	1075.2	266.1	294	72.8
38	1134.1	280.7	310	76.8
39	1194.6	295.7	326	80.8
40	1256.6	311.0	343	85.0

* Weight of cement in bins 93 pounds per cubic foot or 4.04 cubic feet per barrel of cement.

have been required. Although some silos with flat bottoms may still be in use, many consider that the additional cost of conical- or hopper-bottoms is justified, taking everything into account. One important factor is that with the increase in the number of types of cement, it may not be practicable to reserve a silo for only one type. The complete removal of one type and the filling with another type may take place frequently.

Table 10:1, prepared by Meade,[2] shows the capacity of a silo for each foot in depth. Experiments have shown, however, as might be expected, that the weight of a cubic foot of cement increases as the depth of cement in a silo increases. That is, the bottom one-foot layer has the greatest weight per cubic foot. The weight per cubic foot is affected by some other factors, one of which is fineness.

Conveying Equipment for Storing and Reclaiming Cement

Formerly the cement was conveyed to and from the stock house largely by some combination of bucket elevators, screw conveyors, and belt conveyors. Today, however, the Fuller-Kinyon system is used extensively. Meade [2] discusses this as follows:

The Fuller-Kinyon system is now being used at a number of plants for handling cement. This system requires somewhat more power to operate than do elevators, belt and screw conveyors, but its evident advantages such as absence from dust, maintenance and attention required for long lines of screw conveyors, flexibility as regards arrangement, ease of extension to any length desired and in any direction and saving in space required give it the preference over other forms of conveyor yet devised for moving ground cement. Nearly all of the most recently built mills are employing this system for conveying cement from the mill into the stock house bins. One new mill is employing for placing cement from the stock house bins into the packing bins a portable Fuller-Kinyon pump. The latter is mounted on a truck operating on a track under the bins, connection is made with the bin and with the pipe line leading to the packing bin by means of removable connections.

Packing and Shipping

In 1960, 82.5% of the cement was shipped in bulk, almost doubling the 42% in 1942. There was a large increase in paper over cloth bags. Table 10:2 shows shipments [3] from producing plants to purchasers; shipments to distributing plants are not included.

TABLE 10:2

SHIPMENTS OF CEMENT IN 1960, FROM PLANTS
IN UNITED STATES AND PUERTO RICO
IN BULK AND IN BAGS

(Thousand Barrels)

Type of Transportation	In Bulk		In Bags *		Total Shipments	
	Quantity	Percent	Quantity	Percent	Quantity	Percent
Truck	119,680	46.5	27,515	50.3	147,204	47.1
Railroad	130,416	50.6	27,112	49.5	157,528	50.4
Boat	7,030	2.7	74	0.1	7,104	2.2
Used at Plant	417	0.2	39	0.1	456	0.2
Total	257,552	100.0	54,740	100.0	312,292	100.0
Percent of total	82.5	—	17.5	—	100.0	—

* Including paper and cloth.

From one silo, or from two or more silos simultaneously, cement is conveyed to a packing machine by one or more equipment units. Several types of packing machines [4] have been developed. Reference here is to model 150-FC St. Regis Bag Packer.

The operator puts the empty bags to the spouts, pushing with his free hand the button that releases the scale to the filling position. The filling of the bag is done in two steps: the first, in which the material flowing through the wide aperture streams into the bag under high pressure, and the second, in which the material is dribble-fed through a reduced aperture to complete the filling to the predetermined weight. Just before that weight has been reached the dribble-flow position of the mechanism comes into operation by the automatic lifting of the rear end of the scale beams and by means of the oil and spring pressure operated piston, the stroke being adjustable. The retarding of the dribble feed is set according to the outputs required. An automatic electrical timer delays release of the bag clamp for 2 or 3 seconds

after the weighing is completed, thus allowing the material
to settle in the bag and the dust to be vented from the filling
tube. When the scale has traveled to its lowest position,
three functions are automatically performed: (1) the feed
mechanism is entirely closed; (2) the bag is released from
its holder; and (3) the bag chair, the center of gravity of
which is slightly in front of its pivot point, provides a smooth
discharge for the bag direct to the conveyor belt. The hourly
output of the machine when handling cement is 1,200 bags
(300 bbl).

In order to provide uniform head of material over bag-
packing machines, a small auxiliary hopper is frequently
installed directly over the packer.

The sacks or bags are taken from the packing machines by
belt conveyors. At most plants they are hand-trucked to box
cars or to motor trucks, but at some plants they are loaded
by conveyor systems. Bulk shipments are discussed later.

Containers for Cement

The displacement of barrels and steel drums by sacks and
bags as containers for cement has been mentioned. At one
time, jute sacks were used extensively, but for many years
Osnaburg cotton cloth has been employed exclusively.

In most respects, cloth sacks are very satisfactory containers. They may be used repeatedly, probably making from
six to ten round trips from the plant to the cement user.
However, the use of cloth sacks requires a considerable investment and expense for the cement manufacturer. Large
numbers are in stock and in transit. Returned sacks must be
counted, sorted, cleaned, and repaired before they are ready
for the text trip, and, of course, there is a percentage of loss.

Multi-Wall Paper Bags

When the multi-wall paper bag was first offered to cement
manufacturers, it was not received with much favor. In recent years, however, it has advanced rapidly, and today
paper bags of this type outnumber cloth sacks. Paper bags
make only one trip. Therefore, the user of cement is saved
the annoyance of returning them to the plant, and the manufacturer is saved the expense of reconditioning them. The

use of paper bags, however, slightly increases the cost of the container.

As cement valued at millions of dollars has been shipped in these containers, the quality of paper, like so many other items, is important to the cement manufacturer. ASTM lists seventy-two methods in the D division for the testing of paper. Three of these methods are listed here.

CONDITIONING PAPER FOR TESTING: D 685–44

This method covers the procedure for conditioning specimens of paper and paper products prior to testing.

Standard condition shall be that obtained in a circulating atmosphere maintained at a relative humidity of 50 ± 2% and a temperature of 73.4 ± 3.6F ...

TENSILE STRENGTH: D 828–60

Tensile strength is a fundamental property and belongs to the group of properties associated with paper manufacture ...

BURSTING STRENGTH: D 918–49

Bursting strength is defined as . . . the hydrostatic pressure required to produce rupture of a circular area of the material under test 1.2 in. in diameter. . . .

Monographs on paper were published by the Society in 1951, 1959, and 1962.

Changes of Cement in Storage and in Containers

When cement is exposed to the atmosphere, there may be absorption of water vapor and carbon dioxide. The extent and importance of this absorption are influenced by a number of factors. Some of these are:

1. Ratio of the exposed surface to the quantity of cement
2. Circulation of the air
3. Fineness of the cement
4. Pressure

Possibly the chemical composition is another factor. Moisture may be absorbed independently of carbon dioxide, but when there is no moisture present, either in the cement or in the air, no carbon dioxide is absorbed. In any case, the effect of carbon dioxide may be considered unimportant.

Under the usual conditions in storage, the surface of the cement exposed to the atmosphere is relatively small. Only a thin layer at the top of the cement in a silo is subject to change during storage, and this is negligible as far as the cement as a whole is concerned.

Conditions change materially when cement is placed in cloth sacks or paper bags. During packing, there is some aeration, and the ratio of exposed surface to weight is increased greatly. Although the cement is not exposed to the atmosphere directly, as in a silo, there may be considerable penetration of moisture through the walls of the container.

Bulk Shipments

In many respects, the most satisfactory way to ship cement is in bulk. The cost of sacks and bags, and the labor of trucking packages of cement are eliminated.

Usually, bulk cement is taken from the silos in the same manner as is cement to be packaged. From a convenient point on the conveying system, the cement is transferred to the freight car or other carrier. The cement may or may not be weighed on the way to the car, but in either event, the weight of the cement may be obtained by weighing the car before and after loading. There are some limitations to the shipment of cement in bulk. The user must be able to handle relatively large quantities of cement without delay, or he must have facilities for storage in bulk.

As already explained, the handling of cement in bulk is not a recent development. For a number of years the percentage of cement shipped in bulk was not large, but there has been a rapid increase in the last few years. Several reasons for this exist. One of these is the industry of ready-mixed concrete. Another reason is that at times the scarcity of cotton and paper favors shipping in bulk.

From 1942 to 1960, rail shipments decreased from 85.6 to 50.6 per cent, while truck shipments increased from 12.6 to 46.5 per cent.

Distributing Plants

One of the important developments in the cement industry has been the construction of distributing plants. There was a time when all shipments to dealers and to consumers originated at producing plants only. Then the idea of a distributing plant presented itself. Anyone that follows the journals in the cement and concrete field knows how rapidly the number of such plants is increasing.

As the construction, operation, and maintenance of plants, and shipments to distributing plants are expensive, the many factors involved must receive careful consideration. As with many other problems in the industry, there is much difference of opinion concerning the over-all economy.

At first bulk cement was shipped in box cars only. At present a large variety of equipment is available. There are hopper cars; piggybacks; special trucks and trailers; ordinary, self-loading and self-unloading barges; and special self-unloading ships. The ships are improving in speed, capacity, and unloading facilities. The extensive development of equipment for handling bulk and the increase in bulk shipments (see Table 10:2) have much in common.

REFERENCES

1. Robert W. Lesley, *History of the Portland Cement Industry in the United States;* International Trade Press, Chicago, 1924.
2. R. K. Meade, *Portland Cement;* Chemical Publishing Co., Inc., Easton, Pa., 1930.
3. *Minerals Yearbook;* U.S. Bureau of Mines, Washington, D. C., 1961.
4. *Pit and Quarry Handbook;* Pit and Quarry Publishing Co., Chicago, 1961.

Chapter 11

PLANT DESIGN

In previous chapters, the principal equipment units for each department have been mentioned. There remain to be discussed in this chapter the plant as a whole, the basis for the selection of the major equipment, and some of the minor equipment, the use of which is not restricted to any one department.

It has become customary to express plant capacity in terms of millions of barrels of cement per year, as a million-barrel plant, a 2-million barrel plant, and so on. The cost of a plant is expressed in terms of dollars per barrel of annual capacity.

The plant may be designed by the engineering department of the cement company with the assistance of other departments, by the company that manufactures the principal equipment, or by a consulting engineering organization. In some cases, the first plant of a cement company is designed by consultants, and the successive plants by the company's engineering staff. There are various combinations of these general procedures, and for almost any combination conceivable, some example could be found.

It has been stated that almost every profession and every trade are represented in the cement and concrete industries. The activities of the groups are so interrelated that they cannot be separated on the basis of the various steps of a given project. The work that is directed by engineers and geologists, however, may be outlined in a very general way by the following cycle:

1. The geologist locates the raw materials;
2. The civil engineer surveys the land, designs the buildings, and superintends their construction;
3. The mechanical and electrical engineers design the plant, exclusive of the buildings, and install the equipment;
4. The mechanical and chemical engineers operate the plant;
5. The civil engineer uses the cement in his next construction project, obtaining from the geologist any information in his field that is required.

Locating the Plant

There are four groups of investigations and calculations that are required in locating the plant. These need not be undertaken in the order in which they are listed, and in fact they need not be carried on entirely separately.

1. There is a market survey of the district in which the construction of a plant is being considered. What are the requirements for cement at present, and what is the probable increase? What plants supply the cement now? What are their capacities, locations, transportation facilities, and freight rates? What advantages would the proposed plant offer in comparison with existing plants? Formerly, it was customary to give first consideration to the sources of raw materials and fuels, shipping the cement relatively long distances, if necessary, to the market. The tendency now is to locate a plant as close as practicable to the principal market, even if this requires bringing raw materials and fuels relatively long distances to the plant. Frequently, the principal limiting factor is the cost of transportation. In addition, there is a tendency to build more plants, with less capacity per plant.

2. Tentative plant sites are selected, and searches for primary raw materials are started. Deposits near tentative sites are preferable, but conditions may justify the selection of deposits hundred of miles distant. The sources of secondary raw materials are also investigated.

3. Options are obtained on plant property and on raw materials property. The extent, composition, and uniformity of the raw materials deposits are determined by drilling, pre-

ferably by core drilling, if the raw materials are suitable for this.

4. The plant sites available are listed and all the information that can be obtained concerning each is checked. The items that must be taken into account include transportation facilities for raw materials and finished products, water supply, drainage, fuels, power, climate, labor, and living conditions. If both urban and rural sites are available, the long list of advantages and disadvantages of a site are topography, the load-bearing value of the ground, and the total area available.

It is customary to prepare a flow sheet and to calculate the unit cost of each product, as soon as information sufficient to permit this has been collected. As the project develops, and more information becomes available, an improved flow sheet and revised calculations are made, and so on, until a definite decision, for or against the plant, is possible.

Pre-construction Work

After the site has been selected, much work and planning are required before the start of construction. For example, there are surveys for plat and contour maps, and explorations for foundations.[1,2] The ground on which a foundation to carry a heavy load is to be placed, should be investigated by drilling even if it appears to be solid bed rock or hardpan. Serious failures have occurred when this was not done.[1]

Another pre-construction activity is the calculation of the approximate quantities of raw materials, fuels, intermediate products, and finished products, taking into account transportation facilities to and from the plant, and the handling of materials inside the plant. Closely related to the required quantities of each material is the provision for storage between two successive steps in the manufacturing process. It is evident that the output of equipment unit A will not be exactly sufficient to supply unit B at all times, that unit B will not supply unit C uniformly, and so on.

It may be found desirable to prepare several sets of preliminary drawings. The final drawings may be started whenever conditions justify, but they cannot be completed until all major decisions concerning buildings and equipment have been made. The completion of all drawings before the start of construction is good practice, and the most satisfactory results, and over-all economy can be expected when this procedure is followed. Sometimes the demand for a plant is so urgent that construction is started long before the plans are completed, and the draftsmen are just in advance of the construction. In fact, there have been projects in which construction was, at times, well in advance of the drawings.

The Selection of Equipment

In designing a plant for manufacturing any product, it is convenient and customary to select standard equipment as far as this can be done, designing special equipment only when there is no standard unit available for a given operation. In the cement industry, little or no special equipment is necessary. Equipment manufacturers as a class not only provide for every requirement of a plant, but also sometimes develop equipment in advance of extensive demands for it. Starting with a quantitative flow sheet as nearly complete as conditions permit, there are a few items to be taken into account in selecting equipment. Basically, these items are simple, but actually each is complex and has numerous corollaries. Frequently, judgment must be exercised, sometimes based on data, and sometimes based on assumptions. These items are

1. Equipment units that have been installed in the most recent, most successful plants
2. Avoidance of any difficulties that have been experienced at these plants
3. Developments in equipment that have been announced since these plants were placed in operation
4. Study of the chemical and physical characteristics of the proposed raw materials

In some projects, the last item has received less attention than any of the others, resulting in disappointments, delays, substitutions, and financial losses that could have been avoided. The following is an example of the procedure that has been outlined in this chapter from the first consideration of a plant project to the selection of the principal equipment. The example is taken from experience and observation, and does not represent any one project.

A market survey has indicated that commercial conditions are favorable to the contsruction of a plant, having an annual capacity of 1.5 million barrels, in or near a certain city. A plant site with excellent transportation facilities has been located within the city limits. The site possesses all the requirements listed previously. There are limestone and shale deposits within a few miles of the plant site. Core drilling and preliminary calculations have been completed. Supplies of secondary raw materials, and fuels are available. Quotations on power have been received. They indicate that probably it would be more economical to generate power than to purchase it. Assumptions and preliminary calculations indicate the approximate average quantities of materials to be handled and processed daily for the manufacture of 4800 barrels of cement. The plant is to be in operation 350 days per year.

Material	Pounds per day	Tons per day	Tons per hour
Limestone	2,160,000	1080	45
Shale	720,000	360	15
Raw mix	2,880,000	1440	60
Coal	480,000	240	10
Air	4,800,000	2400	100
Clinker	1,804,800	902	38
Gypsum	62,400	31	1.3
Cement	1,867,200	934	39

In the limestone and shale quantities, allowances are made for the volatilization of moisture and carbon dioxide, and for mechanical losses. The weight of a barrel of clinker is considered to be 376 pounds. It will be noted that the weight of air required greatly exceeds the total of the raw materials and the coal.

If these approximate data were given to a group of engineers and each were asked individually to design a plant, it is unlikely that any two plants would be alike, but it is probable that some of the plants would have certain features in common. My tentative list of equipment for such a plant would include:

2 10 x 250-foot kilns (dry process)
2 Unit coal mills
2 Clinker coolers
2 Waste heat boilers, one connected to each kiln
Preliminary mills for raw materials, with auxiliary heating equipment, if necessary
Scales for proportioning raw materials
Tube mills for raw materials
Scales for proportioning clinker and gypsum
Preliminary mills for clinker
Tube mills for clinker
Stock house with a total capacity of 150,000 to 300,000 barrels

With further development of the project, some change in the size of the kilns might be found desirable. The final choice of kilns would determine to a considerable extent the size of the coal mills and the clinker coolers. The final decisions on the characteristics of the waste heat boilers, such as the pressure, capacity, and the like should not be made until the approximate power requirements of the plant are known.

Distribution of Principal Equipment

After the principal equipment units have been selected, the distribution, or placing, of these units is the next problem. A number of items must be taken into account, but not all of these may be applicable to the same plant. Some of these items are:

Land area available
Shape of area
Topography
Drainage
Transportation facilities
Length of conveyor lines
Provision for expansion of plant

Usually, the available area and the shape of the land are not important factors, particularly in suburban and rural districts. In cities, however, it may be necessary to design a plant for a very limited area. The plant of the Peerless Cement Corporation in Detroit is an example. In some localities, topography is an important consideration, and this may influence greatly the design of a plant.

Provisions for expansion may have a major influence, and increase materially the cost of the project. Sometimes market conditions make it desirable to increase the capacity of a plant for which no provision for expansion had been made. Such a situation is likely to increase the over-all cost of the project, and it may be undesirable in a number of ways. On one occasion the president of a cement company asked me to design a plant in which each department could be expanded without limit.

In addition to the major equipment units there must be such minor equipment as intermediate storage facilities, feeders, and materials-handling equipment. The details of these must be worked out as the design of the plant proceeds. The intermediate storage facilities have been previously mentioned.

Feeders are mechanical devices, the function of which is to regulate the supply of materials to equipment units, such as kilns, mills, and scales. There are many types of feeders, the designs being based on the application of various principles. Of course, each type has its advantages, disadvantages and limitations. As a group, they may be considered indispensable to cement manufacture. Sometimes feeders are substituted for scales for proportioning materials. In the experience of the writer, such substitutions have not given satisfactory results.

Materials-handling Equipment

In many modern plants, the raw materials, coal, clinker, and gypsum are taken from storage by cranes. From this point on, it is customary to provide equipment to convey materials to and from each major equipment unit. Before selecting the conveying equipment, the engineer must know the physical properties, and, in some instances, the chemical properties, of each material to be handled, such as:

> Weight per cubic foot
> Particle size
> Moisture content
> Temperature
> Flowing characteristics

These properties are to some extent interrelated. For example, the flowing characteristic is affected by particle size, moisture content, and temperature. Probably there are other factors such as the coefficient of friction, and the adsorption of air. In general, raw mix does not flow so readily as cement, but hot dust from a Cottrell precipitator flows so rapidly that its control is difficult.

Information on mean or average properties of the materials is not sufficient. The properties should be known at every transfer point, that is, in and out of a dryer, in and out of a tube mill, and so on.

There is a wide variety of materials-handling equipment. Examples are:

Screw conveyors
Belt conveyors
Vibrating conveyors
Pan conveyors
Rubber-tube conveyors
Shaker conveyors
Drag chain conveyors
Fuller-Kinyon systems
Airveyors
Bucket elevators
Pivoted bucket carriers

Each type of materials-handling equipment is made in various sizes. Information on speed, capacity, power requirements, wear resistance, and the like may be obtained from the manufacturer. The advantages and disadvantages of most types of equipment are generally known in the industry.

All the information outlined in this section, and any additional information available, are needed for selecting the best materials-handling equipment for a plant.

Types and Distribution of Buildings

There is nothing standard about the types of buildings nor the distribution of them. Nearly every kind of material has been used, from wooden frames, covered with galvanized steel, brick, and stone, to reinforced concrete. In general, utility, rather than architecture, has been the most important factor in designing and locating plant buildings, but this does not mean that architecture has been neglected entirely.

The buildings at some cement plants compare favorably with the buildings of other industries. Although a cement plant has much in common with other plants, there are certain characteristics that are to some extent distinctive:

1. Large tonnages of heavy materials are handled.
2. These require rugged, massive machinery. We may consider that a building is constructed around machinery units, rather than that machinery is placed inside a finished building.
3. There is only one by product of any consequence; this is carbon dioxide from the decomposition of limestone. Usually, carbon dioxide is allowed to escape, but it has been used in the manufacture of dry ice, and for the treatment of slurry to inhibit hardening.

Aside from the carbon dioxide, and some minor mechanical losses, all the raw materials entering a plant leave as cement. That is, there are no by-products nor waste products resulting from such processes as filtration, crystallization, or distillation.

Although, as it has been stated, there is nothing standard about the types or distribution of buildings, there are certain trends in the industry which influence them. At one time, practically the only equipment for conveying materials from one building to another was screw and belt conveyors. It was economical to place buildings in close proximity, and it was easy to trace the conveying lines from one building to another. The introduction of the Fuller-Kinyon system soon had its effect on the distribution of buildings. Many of the restrictions on location have been removed. Other changes have resulted from the trends towards larger equipment units, plants of smaller capacity, and changes in power transmission.

Figure 11:1 shows the general plan of the Permanente plant at Waianae, Oahu, Hawaii.

Rockwood has made a study of the design of cement plants. The following is taken from a portion of his paper.[3]

We find that many American plants follow one of three primary designs, or occasionally, combinations of two. The primary designs might be described as (1) straight-line; (2) parallel; (5) cross-T.

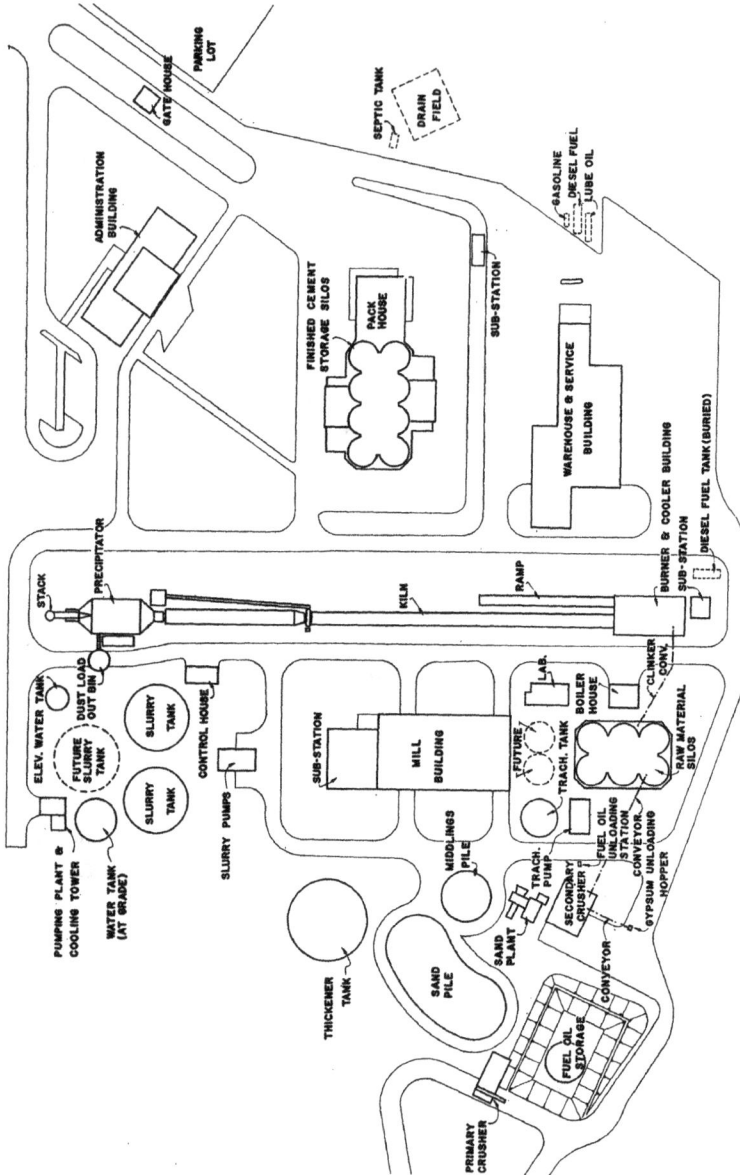

FIG. 11:1. GENERAL PLAN OF PLANT IN HAWAII

The principle of the straight-line design is to assemble the plant on a single long axis, with raw materials fed into one end and cement discharged at the other. Raw mills are at one end of the kilns and the finish mills at the other. Raw storage and clinker storage are widely separated.

The parallel design consists, usually, of assembling the various elements of the plant, (a) raw and clinker storage, (b) mills, (c) kilns, and (d) packing plant and silos on four parallel axial lines. Frequently, in this design, the axes of the mills are placed at right angles to the axis of, and backed up against, the raw storage bins or building, so that the mill hoppers are in the storage building. Sometimes the raw mills are in a separate building, but more frequently both raw and finish mills are in the same building, placed near the center of one side of the storage building, which has raw material for approximately half its length and clinker and gypsum the other half. The whole storage building is served by one or two bridge cranes, depending on the size of the plant, which feeds or feed the hoppers of both raw and finish mills.

In the cross-T (or H) design, not only the raw and finish mills back up at right angles to the axis of the raw and clinker storage, but the kilns as well, so that clinker is merely elevated from the coolers over the wall into storage.

REFERENCES

1. L. C. Urquhart, *Civil Engineering Handbook;* McGraw-Hill Book Co., Inc., New York, 1934.
2. ASTM D 1143–61T; D 1194–57.
3. Nathan C. Rockwood, *Rock Products,* 47(8):73 (1944).

Chapter 12

EXPERIMENTAL ENGINEERING

As employed in this book, the term *experimental engineering* * may be defined as the study of plant equipment in operation. Although most of the tests, measurements, readings, and observations take place at or near the equipment, much supplementary office and laboratory work is required. It is necessary, as a rule, to send samples to laboratories for chemical, physical, and microscopical examination. Also, considerable time must be devoted to calculations, the preparation of graphs, and the study and correlation of results.

Experimental engineering has been one of the most helpful subjects in the study of portland cement technology.

A paper on *Instrumentation and Control in the Textile Industry is of interest here.*[1] Although the raw materials, finished products, and equipment in the textile industry are very different from those in the cement industry, the two industries have much in common in experimental engineering procedures.

Objectives of Experimental Engineering

It would be difficult, if not impossible, to list all the objectives of experimental engineering, but some of them may be mentioned:

1. Control of operation
2. Efficiency

---

* Experimental engineering first came to my attention in connection with a course in the subject at Armour (now Illinois) Institute of Technology. The text was Carpenter's *Experimental Engineering*. A revised edition bore the names of Carpenter and Diedericks, as co-authors. Since then, there have been changes in the title and authors.[2]

3. Quality of products
4. Improvement in equipment
5. Choice of new equipment
6. Design
7. Research

The first three items have much in common, and can be discussed together. In many instances, improving the quality of a product involves improving the control of one or more of the plant equipment units. Sometimes, but not always, this results in increasing the efficiency of the unit.

By indicating the good and bad points of a unit, experimental engineering data provide sound bases for installing improvements for existing equipment, and for choosing the most suitable new equipment for handling the available raw materials, or for the processes peculiar to a given plant. This applies also to designing an extension to an operating plant, or designing a completely new plant. The field for research in experimental engineering itself, or for research in which experimental engineering is one of the essentials is unlimited. Research has been very important in the design of the instruments now in use, and no doubt research that is now under way will suggest many new instruments.

Instruments and Unit Measurements

Although a wide range of test instruments and equipment is employed in the industry, and some of the tests performed are elaborate, the whole system is based on relatively few types of instruments and measurements. There are three principal classes of instruments: indicating, recording, and integrating.

1) An *indicating* instrument provides information at the time of observation only; for example, an ordinary mercury thermometer or a pressure gauge.

2) *Recording* instruments supply continuous information for a given period of time, usually twenty-four hours, in the form of a record chart, which is usually operated by a clock mechanism.

3) *Integrating* instruments, such as steam flow meters, show totals in addition to continuous records of indi-

vidual values. In general, recording and integrating instruments are much more desirable than indicating instruments.

All units are derived from the centimeter-gram-second system. That is, all measurements are based on one or more of the independent variables length, weight, and time.

It may be well to outline a few of the fundamental measurements required in experimental engineering practice.

Weight:

In general, the weight of gases, liquids, and solids may be determined by weighing, or by calculating from the volume. The former is the more desirable on the basis of accuracy, but often it is inconvenient and involves considerable expense. In general, weighing is employed almost exclusively with solids, frequently with liquids, and seldom with gases. The type of scale is determined primarily by two factors— capacity and desired accuracy. For many plant tests, the ordinary type of platform scale is satisfactory.

Temperature:

Temperature determinations of gases, liquids, and solids are frequently required. The type of equipment is chosen largely according to the temperature range. For lower temperatures, thermometers may be employed, either the familiar type filled with mercury or other liquids, or the gas-filled type, in which temperature is determined indirectly from pressure. When temperatures exceed a few hundred degrees, some type of pyrometer is required. Base metal thermocouples may be used up to approximately 2000–2500°. For still higher temperatures, noble metal thermocouples, or radiation, or optical pyrometers, are employed.

Pressure:

The choice of instrument is based primarily on the range of pressure to be measured. Most of the instruments are of the indicating type, but there are many of the recording type in use. From low to high pressures, successively, the instruments commonly used are (a) draft gauge, (b) water manometer, (c) mercury manometer, (d) pressure gauge. The term *draft* is applied to pressures that differ only slightly from atmospheric pressure, occurring in such places as ducts, stacks, furnaces, dryers, and kilns. Draft is usually expressed in *inches of water*, whereas pressures of steam and compressed air are expressed in *pounds per square inch*.

Volumes:

Volumes of gases and liquids may be determined by taking the dimensions of their containers, but, in general, volumes are calculated from the size of a pipe or flue, through which a gas or liquid is passing, and the velocity of flow. There are several instruments for determining the velocity of gases, one of the most important of which is the *Pitot tube*. Frequently, it is important to know the analysis, and the dust and moisture content of gases, as well as the volume and temperature.

Electrical measurements:

Many of the instruments that have been mentioned are operated electrically; therefore, information on volts, amperes, ohms, watts, and efficiencies of motors is frequently required.

Large-Scale Tests

By making a number of unit tests, either simultaneously or over a given period, and combining and correlating the results, it is possible to perform such extensive tests as heat balances on dryers, kilns, and boilers; efficiency tests on dust collectors, and so on.

Some of the fundamental information needed in connection with the more important operating units is as follows:

Crushers and mills:

Particle size of materials, entering and discharged
Power consumption
Capacity
Temperature of materials entering and discharged, in the case of finishing mills

Dryers:

Capacity
Fuel consumption

Draft

Fineness of coal, when pulverized coal is used

Combustible matter in ash, in the case of hand-fired and
stoker-fired dryers

Temperature and moisture content of materials discharged
from dryers

Analysis, temperature, moisture content, and dust content
of the stack gases

Kilns:

Capacity

Fuel consumption

Draft

Temperatures at front and back ends

Fineness of coal

Percentage of primary and secondary air

Waste heat boilers:

Temperature, heat content, and other characteristics of
gases entering and leaving the boiler system

Efficiency

Dust collectors:

Characteristics of gases entering and leaving the collectors

Quantity and chemical and physical properties of the dust

Efficiency

Much can be accomplished by the application of experimental engineering methods to kiln operation. For the best
results, every item entering and leaving the kiln must be
known and controlled within narrow limits. The data should
include the quantity of each item per barrel of clinker produced. The information on water, and heat input and output
should be as nearly complete as obtainable. The following is
a portion of a preliminary outline for the experimental engineering study of a kiln. It is customary to express some
values in units of weight, and other values in units of volume.
Calculations from volume to weight, or from weight to volume, may be made readily.

Items entering:

1. Slurry
   Quantity,   pounds per barrel of clinker

Solids,	total, per cent by weight
Water,	per cent by weight
Fineness of solids,	per cent passing through 200-mesh sieve

2. Coal

Quantity,	pounds per barrel of clinker
Fineness,	per cent passing through 200-mesh sieve

B.t.u. per pound

3. Air

Total,	pounds and cubic feet per pound of coal
Primary,	per cent of total
Secondary,	per cent of total
Water content,	per cent by weight
Heat content.	B.t.u. per pound

Items leaving:

1. Clinker

Heat content,	B.t.u. per pound

2. Gases

Composition,	per cent by weight
Water content,	per cent by weight
Heat content,	B.t.u. per pound

Basic Principles

The value of instrumentation, whether for control or for testing, depends on a number of factors.

Well-established procedure:

There should be followed well-established procedures such as the test codes of the American Society of Mechanical Engineers. Some of these codes are:[3]

General Instructions
Definitions and Values
Solid Fuels
Gaseous Fuels
Stationary Steam-Generating Units
Coal Pulverizers
Steam Turbines
Centrifugal Pumps
Fans

Evaporating Apparatus
Dust-Separating Apparatus
Electrical Measurements in Power Circuits

The accuracy attainable in any given measurement is dependent upon four things:

a) The method of applying the instrument
b) The accuracy of the instrument itself
c) The accuracy of the observer, and
d) The characteristics of the quantity being measured.

The instruments should be selected after careful consideration has been given to the information desired, test conditions, and the characteristics of all the available instruments of a given type.

> Instruments . . . must be tested to such an extent as may be necessary to determine their condition and their accuracy. Such testing is commonly known as *calibration* of the instrument. . . . In general, the calibration should include determination of the following throughout the range which is to be used:
>
> *a*) Responsiveness and its relation to the sensitivity of the instrument
>
> *b*) Variance and its relation to the sensitivity of the instrument
>
> *c*) Accuracy, or error of indication at different points within the useful range.

It is not necessary to point out that the value of experimental engineering work without calibrated and carefully adjusted instruments is zero. In fact, conditions may arise in which misleading conclusions make the value of such work less than zero.

Records:

Data sheets, $8\frac{1}{2} \times 11$ inches, should be printed or mimeographed, spaces being provided for test number, date, equipment unit, and observer. These sheets should be punched for standard notebooks, but in the field, it is convenient to place them in clipboards. All readings, such as temperatures, pressures, and the like should be recorded. The data sheets containing all original records should be collected at the end of the test, and filed for permanent reference.

Calculation of results:

The last step in any test is to make the necessary calculations, and prepare any graphs that may be desirable in the correlation and study of the results. All calculations should be checked carefully. In checking, it is usually preferable to follow a different procedure than that employed in making the original calculation.

Automatic Control of Kilns

One of the goals of experimental engineering in the cement industry is the automatic control of kilns. It is not difficult to become enthusiastic over the possibilities, highest quality of product, maximum output, minimum fuel, minimum labor, minimum over-all cost. The subject has received much study and discussion.*

In a patent [4] on the subject, issued to Reaser and Tanner in 1942 four objectives are listed. Three forms of the invention are discussed in the specification. The first form is shown in Figure 12:1, and a portion of the inventor's description of this form follows:

An inclined kiln 1 is rotatably supported on suitable bearings 2 and is rotated by a suitable driving means illustrated as an electric motor 3. The motor 3 may be of any suitable type and is preferably provided with a manually operated speed controller so that the speed at which the kiln 1 is rotated may be adjusted to suit varying conditions. However, our invention contemplates that the optimum speed of the kiln 1 once having been arrived at, thereafter its speed will be maintained constant unless there is a material change in operating conditions, such as a change in the character of the materials being fed, or in the character of product desired.

The raw material, either in a dry or in a wet state (known as slurry), is introduced into the kiln at the high end through suitable feeding means diagrammatically shown at 4. The feeding means is preferably adjustable so that material may be fed into the kiln at any desired rate, which may or may not be in proportion to kiln speed. Our invention again contemplates that after the optimum feeding speed has been determined, thereafter it will be maintained constant. The material in the kiln gradually travels to the low end or hood 1A of the kiln from which it is discharged into a cooler 5 provided with a perforated shaking or traveling grate 6 through

---

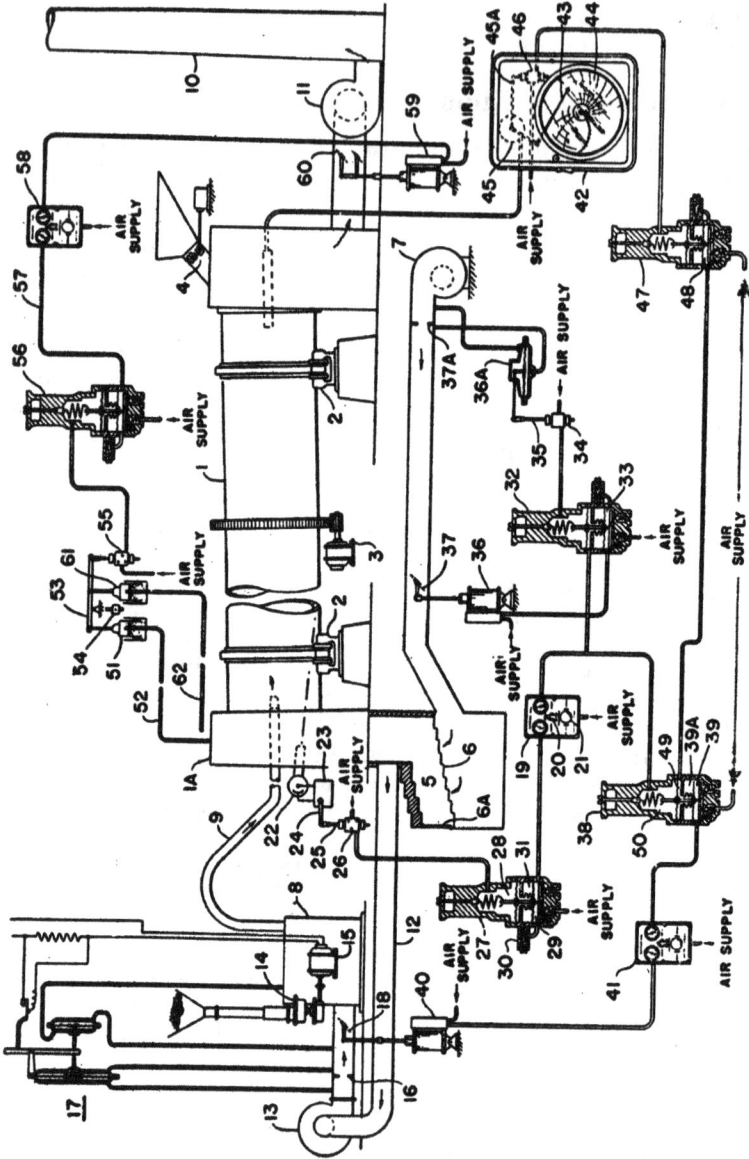

FIG. 12:1. CONTROL APPARATUS FOR KILNS (U.S. Patent 2 298 257)

which cooling air is blown by means of a fan **7**.

This air, after passing through the cooler **5**, provides a convenient source of heated primary and secondary air for combustion, as will be described more in detail hereinafter. The clinker is discharged from the cooler **5** through a swinging gate or door **6A**.

Any suitable fuel may be used to fire the kiln. In the embodiment of our invention that we have chosen to illustrate we have shown the kiln as being fired with pulverized coal from a unit air-swept pulverizer **8**. The pulverized coal is transported from the pulverizer **8** to the kiln **1** through a burner pipe **9**. The products of combustion resulting from the burning of the pulverized coal within the kiln **1** and any excess air present, which for convenience we collectively term "flue gas," are drawn through the kiln **1** by means of an induced draft fan **11** and discharged to the atmosphere through a stack **10**.

The pulverizer **8** is illustrated as being of the air-swept type, through which a stream of air, commonly called primary air, is passed. The primary air may be drawn from the cooler **5** through a duct **12** and forced through the pulverizer **8** by means of a fan **13**. As will be readily understood by those familiar with the art, tempering air may be bled into the duct **12** from the atmosphere to maintain the primary air entering the pulverizer at a predetermined temperature sufficiently high to dry the coal in the pulverizer; but lower than that which would be likely to ignite the coal. The cooler **5** is also shown as being in communication with the hood **1A** and accordingly also provides a source of heated secondary air for combustion.

Raw coal is fed the pulverizer **8** through a feeder **14** driven by a variable speed motor **15**. As known, the quantity of pulverized coal discharged from an air-swept pulverizer is substantially proportional to the rate of flow of primary air therethrough so long as a proportional amount of coal is maintained therein. A measure of the primary air passing through the pulverizer may be obtained by means of a restriction in the primary air duct **12**, such as an orifice **16**, which produces a differential pressure varying in functional relation to the rate of primary air flow. A measure of the quantity of coal in the pulverizer may be obtained by measuring the pressure differential across the whole or a part of the pulverizer. These pressure differentials are transmitted to a controller, generally indicated at **17**, which operates to vary the speed of the feeder motor **15** to maintain the correct quantity of coal in the pulverizer for the existing rate of flow of primary air. The rate at which the pulverized fuel is discharged into the kiln **1** is adjusted by means of a damper **18** controlling the flow of primary air to the pulverizer. Thus, if it is desired to increase the rate of heat liberation in the kiln **1** the rate of flow of primary air is increased proportional to the increase desired, and vice versa.

In the arrangement shown, we establish a pneumatic loading pressure corresponding to the desired rate of heat release within the kiln **1**. Such loading pressure may be established by an operator, who, on the basis of observation of the material in the kiln, from time to time manually adjusts the loading pressure so that the desired heat release is obtained; or automatically from an indication of the temperature at a given location within the kiln.

The operator may select the method by which the loading pressure is to be established by means of a selector valve **19**, which may be of the type illustrated and described in U. S. Patent 2 202 485 to Edward W. Fitch, to which reference may be made for a more detailed description. With the transfer knob **20** in the "automatic" position, a loading pressure automatically established by a temperature-responsive device (hereinafter to be described) is rendered effective. With the transfer knob **20** in the "hand" position, a loading pressure established by manual manipulation of the knob **21** is rendered effective.

We show the temperature-responsive device for establishing a pneumatic loading pressure as a photoelectric device **22**, focused to receive light from the material passing a predetermined spot within the kiln **1**. The correct spot upon which the photoelectric device **22** is focused may be determined experimentally, and it usually corresponds to the point in the kiln at which the so-called clinkering zone begins. If the light, and accordingly the temperature of the material at the spot upon which the device is focused, is above a predetermined amount, it indicates that the heat release is greater than necessary, and that the clinkering zone is beginning too far up in the kiln, which will result in overburned clinker and an unsatisfactory product. Conversely, if the temperature at the spot upon which the photoelectric device is focused is below a predetermined value, it indicates that the heat release is insufficient, and if the condition is permitted to continue underburned clinker will result.

The photoelectric device **22**, through a suitable amplifier shown diagrammatically at **23**, serves to position an arm **24** in accordance with the temperature of the material at the spot on which it is focused. The arm **24** in turn positions the movable valve member **25** of a pneumatic pilot valve **26**, which is illustrated as being of the type described in U. S.

Patent 2 054 464 to Clarence Johnson, to which reference may be made for a more detailed description. This pilot valve serves to establish a pneumatic loading pressure corresponding to the position of the arm 24, and hence corresponding to the temperature of the material on which the device 22 is focused.

The loading pressure established by the pilot valve 26 may be transmitted directly to the selector valve 19, and thence to the devices regulating the rate of fuel and air supply to the kiln 1. However, we usually prefer to have the loading pressure established by the pilot 26 first transmitted to a device, such as a standardizing relay 27, which may be of the type described in U. S. Patent 2 098 914 to H. H. Gorrie.

The purpose of the device 27 is to maintain a sufficient rate of heat release within the kiln 1 so that the temperature of the spot upon which the photoelectric device 22 is focused will remain precisely at the desired value without overshooting or hunting. As shown, the pilot valve 26 is connected to a chamber 28 of the relay 27. Pressure changes in the chamber 28 are immediately reproduced in a relay chamber 29. Following such immediate reproduction of pressure changes a slow continuing change of the pressure within the chamber 29 occurs by virtue of an adjustable bleed passage 30 and chamber 31. The device 27 is initially adjusted so that when the loading pressure established by the pilot 26 corresponds to the desired temperature, it is in equilibrium, and the pressure within the chamber 29 remains constant. Any change in pressure within the chamber 28, however, acts, as stated, to produce an immediate corresponding change in pressure within the chamber 29 followed by a slow continuing change until the pressure within the chamber 28 is restored to that value corresponding to the desired temperature.

Pressures established within the chamber 29 are transmitted to the selector valve 19 and, if the transfer knob 20 is in the "automatic" position, relay to a differential standardizing relay 32. The relay 32, similar to the relay 27 heretofore described, operates to establish a loading pressure in a chamber 33 which may be considered as the resultant of the loading pressures established through the selector valve 19 and one established by a pneumatic pilot valve 34. The movable valve member 35 of the pilot valve 34 is operated by a rate-of-flow meter comprising a differential pressure responsive device 36A and an orifice 37A positioned in the duct leading from the fan 7 to the cooler 5. Accordingly, the pilot valve 34 establishes a pneumatic loading pressure corresponding to the rate of flow of air supplied to the cooler 5 and used for combustion of the fuel within the kiln 1. The relay

32 is so adjusted that when the loading pressure transmitted from the selector valve 19 is equal to or in predetermined proportion to that established by the pilot 34, it is in equilibrium. However, if the loading pressure established by the pilot valve 34 is not equal to or in predetermined proportion to the loading pressure transmitted from the selector valve 19, then the pressure within the chamber 33 will vary. Thus, for example, assuming the loading pressure transmitted from the selector valve 19 and that established by the pilot valve 34 to be equal, if a change occurs in the loading pressure transmitted from the selector valve 19 an immediate and proportionate change in loading pressure will occur in chamber 33 and thereafter a slow continuing change will occur therein until the two loading pressures are again returned to equality or stand in predetermined proportion to each other.

The loading pressure within the chamber 33 is transmitted to a servo-motor 36 operating a damper 37 in the duct leading from the fan 7 to the cooler 5. Thus it will be evident that upon a change in the loading pressure transmitted from the selector valve 19 the rate at which air is supplied the cooler 5, and accordingly utilized in combustion of the fuel within the kiln 1, will continue to vary until a change in the rate of air supply has occurred which is proportional to the change in loading pressure transmitted from the selector valve 19. In other words, for each and every loading pressure transmitted from the selector valve 19 there will be established a predetermined definite rate of combustion air supply, which, by virtue of the fact that the damper 37 is positioned in response to changes in rate of air flow, will be precisely maintained. Briefly, it may be said that the control operates first to set up a desired rate of air supply and thereafter serves to maintain the actual rate of air supply equal to the desired rate.

The loading pressure, either automatically established by means of the photoelectric device 22 or manually established by means of the knob 21 of the selector valve 19, may also be transmitted to a chamber 50 of an averaging relay 38, somewhat similar to the relay 27 but which operates to produce changes in loading pressure within a chamber 39 proportional to changes in the difference between the pressures within the chamber 50 and a chamber 49. The pressure within the chamber 39 is transmitted to a servo-motor 40 operating the damper 18 in the primary air duct 12. Accordingly, it will be seen that upon a change in temperature of the material at the spot within the kiln 1, upon which the photoelectric device 22 is focused, an immediate and proportionate change will occur in the rate at which air is supplied for combustion and in the rate of fuel flow from the pulverizer 8. If desired, a selector valve 41 may be introduced between the chamber 39 and servo-motor 40 so that changes in rate of fuel supply may be made manually rather than automatically.

We effect changes in the rate at which fuel is supplied to the kiln 1 in accordance with changes in the loading pressure established by the pilot valve 26 (in the form of our invention illustrated in the figure), merely to anticipate required changes therein to maintain the desired heat release per unit weight of air supplied for combustion. Essentially, the loading pressure established either automatically by means of the photoelectric device 22 or manually by means of the selector valve 19 is used to establish the desired rate of air supply, and accordingly the desired heat release per unit of time within the kiln 1. However, if a change in either of these loading pressures is made, it is evident that a corresponding change should be made in the rate at which fuel is supplied. Accordingly, such changes in either of these loading pressures may be used to anticipate changes in fuel supply which would later be found necessary. In order, however, to be assured that for each unit weight of air supply a predetermined heat release is obtained, we determine the oxygen content in the flue gases leaving the kiln and adjust the rate of fuel supply to maintain that oxygen content in the flue gas which from the observation and experiment has been found to give optimum combustion conditions.

Arranged to continuously analyze the flue gases leaving the kiln 1 for oxygen we show a recorder 42 which may be of the type forming the subject matter of an application to Clarence Johnson filed in the United States Patent Office on January 16, 1940, Ser. No. 314,189, to which reference may be made for a more complete description. The recorder 42 is arranged to position an index 43 relative to a chart 44 to inscribe thereon a record of the oxygen content of the flue gas leaving the kiln 1. As diagrammatically illustrated, the mechanism 45 for operating the recording pen 43 may be provided with an auxiliary arm 45A for positioning the movable valve member of a pilot valve 46, which will operate to establish a pneumatic loading pressure corresponding to the oxygen content of the flue gas.

The loading pressure established by the pilot valve 46 is transmitted to a standardizing relay 47, which operates to establish in a chamber 48 pressure changes corresponding to changes in the loading pressure established by the pilot 46 and a continuing change corresponding to the amount of departure of the loading pressure established by the pilot **46** from that existing when the oxygen content of the flue gas is at the desired value. The loading pressure within the chamber **48** is transmitted to the chamber **49** of the relay **38**, so that changes therein cause proportionate changes in the chamber **39** which are transmitted through the selector valve **41** to the servo-motor **40**. It will be observed that the relay **38** is not provided with a bleed connection between the chamber **39** and a chamber **39A**, but that the latter is open to the atmosphere so that changes in pressure within the chamber **39** occur only when there is a change in the loading pressure transmitted through the selector valve **19**, or from the relay **47**. By virtue of the standardizing relay **47** when the oxygen content in the flue gas is other than that desired, the loading pressure in the chamber **49** continues to vary in a direction necessary to adjust the rate of fuel supply in order that the oxygen content will be restored to the desired value.

While in the figure we have shown our invention applied to a kiln having a fan **7** for providing forced draft, it may fully as well be applied to a kiln having no forced draft and where the air for combustion is supplied solely by induced draft. The figure indicates that notwithstanding that the fan **7** is eliminated the orifice **37A** will still provide a measure of the air supplied for combustion. It is also evident that the location of the orifice **37A** may be changed without departing from the teachings of our invention; for example, in some installations, it may be preferable to locate the orifice **37A** in the duct leading from the kiln **1** to the induced draft fan **11**. It is further evident that any suitable flow-measuring device other than an orifice may be employed if desired; or that in some instances such an element may be dispensed with entirely and the pressure drop through the kiln used as a measure of the rate of flow of air for combustion. So also a different form of clinker cooler than the specific type we have shown or the elimination thereof entirely will not render our invention any the less applicable to a particular kiln.

Regardless of whether or not the kiln is provided with forced draft and/or a clinker cooler, it is highly important that a predetermined draft be maintained in the hood of the kiln so that air indraft will be held constant. It is usually preferable to maintain a slightly negative draft in the hood so that the air infiltration through inspection openings and clearances will be held at a minimum, while at the same time preventing smoke and flame from being blown into the atmosphere.

In the embodiments of our invention illustrated we maintain a constant hood draft of predetermined amount by

means of a suitable draft gage comprising an oil-sealed bell
51, the interior of which is connected to the interior of the
hood **1A** by means of a pipe **52**. The bell **51** is pivotally connected to a fulcrumed beam **53** to which is secured a pendu-
lum **54** so that for each and every draft within the hood of
the kiln there will be a corresponding position assumed by
the beam **53**. The movable valve member of a pilot valve **55**
is pivotally connected to the beam **53** so that a loading pressure is established corresponding to the draft within the hood
of the kiln. The loading pressure so established is transmitted
to a standardizing relay **56** which is similar to the standardizing relay **27**, and which establishes a control pressure
transmitted through a pipe **57**, a selector valve **58**, and
thence to a servo-motor **59** arranged to operate dampers **60**
within the duct leading from the kiln **1** to the induced draft
fan **11**. The arrangement is such that upon a departure of
the draft within the hood of the kiln from that desired an
immediate and proportional change takes place in the position of the damper **60** in a direction tending to restore the
draft within the hood to the desired value. Thereafter the
standardizing relay **56** operates gradually to position the
dampers **60** until the draft within the hood is restored to
exactly the predetermined desired value. The selector valve
58 provides a convenient means for transferring operation
of the damper **60** from automatic to manual control, as it
will be appreciated that during certain times, for example in
the starting up or shutting down the kiln, it is preferable to
have the hood draft controlled manually rather than automatically.

Usually it is necessary because of the severe atmospheric
conditions existing adjacent a kiln, to locate the draft-
responsive device at a different elevation and at a considerable distance from the kiln, for, as will be appreciated, such
devices are relatively delicate and may not be able to with-
stand the high temperatures and dust-laden air immediately
adjacent the kiln. As the ambient temperature surrounding
the draft-responsive device may therefore be quite different
than that surrounding the kiln, an error, particularly where
the draft responsive device is located at a different elevation
than the kiln, may be introduced into its operation. To pre-
vent such error we provide a second oil-sealed bell **61** pivotally
suspended from the beam **53** the interior of which is connected by a pipe **62** to the atmosphere immediately adjacent
the hood of the kiln. The pipe **62** throughout its entire length
is located adjacent the pipe **52**, so that both pipes will be
maintained at substantially the same temperature and be
subjected to the same atmospheric conditions throughout

their length. Because of this arrangement the draft-responsive device measures the difference in pressure existing between the interior of the hood **1A** and the atmosphere immediately adjacent thereto, rather than the difference in pressure existing between the interior of the hood **1A** and that surrounding the bell **51**, which would be the case if but a single oil-sealed bell were provided. It will be apparent that we have shown an oil-sealed draft-responsive device merely by way of example, it being evident that any suitable differential diaphragm device or other type of gage may be substituted.

Future Developments

It seems likely that experimental engineering in relation to cement manufacture will continue to increase. We may expect improvements in present instruments, new instruments, and advancements in the application of instruments. There are many spoken and written comments about what has been accomplished, what is being accomplished, and what will be accomplished.

The installation of recording instruments means some little responsibility to plant management. Calibrations, adjustments, repairs, and the periodic changing of charts are required if the value of the instruments is to be realized.* In addition, readings are to be recorded and tabulated, calculations are to be made, and graphs are to be drawn—to be followed by the application of the information so obtained in plant operation.

* During an official visit to a plant I observed an elaborate system of instruments. In reply to my comment on the time that must be required in connection with the instruments, the man in charge said the time requirement had been found to be prohibitive and the instruments had been allowed to fall into disuse.

REFERENCES

1. W. B. Heinz and W. W. Starke, *Mech. Eng.*, 66(7):451 (1944).
2. Herman Diederichs and W. C. Andrae, *Experimental Mechanical Engineering* (v. 1, *Engineering Instruments*); John Wiley and Sons, Inc., New York, 1930.
3. *Mechanical Engineer's Catalog* (1963); ASME, 345 East 47th Street, New York (10017).
4. William E. Reaser and Henry C. Tanner, U.S. Patent 2 298 257 (1942).

Chapter 13

THE NATURE OF CEMENT

In the first portion of this chapter, clinker, rather than cement, will be discussed. Whenever it is necessary to shift back and forth between clinker and cement, there should be no confusion, because, as it has been explained in Chapter 9, cement is simply clinker that has been ground with the addition of a small percentage of gypsum. Also we shall exclude subatomic phenomena, and the storage of materials in sealed, or otherwise special, containers, so that it will not be necessary to mention and explain exceptions as the discussion proceeds.

Every material consists of one or more chemical elements, one or more chemical compounds, or some mixture of elements and compounds. From the early days of the industry, the principal elements in clinker were known, as were also some of the physical characteristics of the material. In time, the importance of knowing as much as possible about the constitution of clinker was recognized. When we compare our present-day information on clinker with what was available a century ago, it is surprising that any kind of cement could have been manufactured then.

The following is a list of the principal elements, excepting oxygen, that are present in clinker, and an oxide of each.

Silicon (Si)	Silicon dioxide (SiO_2)
Aluminum (Al)	Aluminum oxide (Al_2O_3)
Iron (Fe)	Ferric oxide (Fe_2O_3)
Calcium (Ca)	Calcium oxide (CaO)
Magnesium (Mg)	Magnesium oxide (MgO)

The free elements can be eliminated at once as components of clinker. Iron, in negligible quantities, occurs in nature in the uncombined condition, but none of the other elements occurs in the free state. All five elements combine with oxygen more or less readily, and if they were present most of them would become oxides. Within a few years after the invention of portland cement, the existence of four of the five elements was known.

TABLE 13:1

SOME ANALYSES OF TYPE ONE PORTLAND CEMENTS
MANUFACTURED IN AMERICA

			Percentage of			
Number	Silicon Dioxide (SiO_2)	Aluminum Oxide (Al_2O_3)	Ferric Oxide (Fe_2O_3)	Calcium Oxide (CaO)	Magnesium Oxide (MgO)	Sulfur Trioxide (SO_3)
1	19.92	7.52	2.28	62.48	3.19	1.51
2	21.14	6.94	2.30	63.24	3.26	1.12
3	19.64	7.52	2.80	62.31	3.04	1.60
4	21.82	8.03	2.51	62.19	2.71	1.02
5	21.94	6.87	2.37	60.25	2.78	1.38
6	22.26	5.36	2.10	63.32	3.81	0.89
7	22.20	6.69	2.27	62.61	3.00	1.32
8	20.32	7.12	2.50	62.94	3.38	1.45
9	24.16	5.10	1.45	62.95	3.12	1.35
10	20.72	7.17	2.85	62.64	1.97	1.42
11	21.84	6.77	5.05	62.66	0.80	1.24
12	20.74	7.06	3.72	62.76	1.78	1.12
13	23.94	5.62	3.20	62.32	1.77	0.90
14	21.31	6.54	2.81	63.01	2.71	1.42
15	25.38	3.34	1.20	62.96	1.20	0.35
16	22.34	7.00	3.30	60.72	1.30	1.05
17	23.12	6.18	2.49	63.47	0.88	1.34
18	19.36	9.18	4.10	63.20	1.16	1.18
19	21.88	7.14	2.85	64.94	trace	0.73
20	21.24	7.85	4.14	63.22	0.28	1.11
21	21.22	7.51	3.83	63.75	0.82	1.58
22	21.93	5.99	2.35	62.92	1.10	1.55
23	22.90	6.80	3.60	63.90	0.70	0.40
24	21.30	6.95	2.00	62.50	1.20	0.98
25	22.04	6.45	3.41	60.92	3.53	1.25
26	22.41	8.12	2.51	62.01	1.68	1.40
27	23.06	8.16	2.88	62.10	1.88	1.57

As important components of clinker, the oxides may be eliminated almost as easily. SiO_2. Al_2O_3, and Fe_2O_3 are inactive substances under ordinary conditions, and they have no cementitious characteristics. CaO and MgO absorb moisture and carbon dioxide from the atmosphere, becoming mixtures of hydroxides and carbonates. In this form, they are said to be air-slaked, and they are valueless as cementitious materials. When CaO alone or mixed with MgO, is converted into the hydroxide by the proper application of water, a cementitious material is formed, but this does not have the hydraulic characteristics of portland cement.

From these statements, it is evident that the usefulness of portland cement is due to compounds other than individual oxides. A chemical compound has a constant composition. As the composition of cement is not constant, it is not a single compound. See Tables 13:1 and 13:2.

TABLE 13:2

SOME MAXIMUM AND MINIMUM PERCENTAGES
IN COMPOSITION OF CEMENTS

Ingredient	Maximum	Minimum
Silicon dioxide (SiO_2)	23.06	19.32
Aluminum oxide (Al_2O_3)	7.59	4.05
Ferric oxide (Fe_2O_3)	5.77	1.45
Calcium oxide (CaO)	66.93	62.01
Magnesium oxide (MgO)	4.91	0.68
Tricalcium silicate	65.9	38.5
Dicalcium silicate	37.2	7.2
Tricalcium aluminate	15.6	2.1
Tetracalcium aluminoferrite	17.5	5.7

Clinker had been manufactured for many years before anyone, apparently, devoted much thought to its composition. By trial and error, procedures for selecting raw materials, and proportioning and processing them had been developed. Notwithstanding the crudity of the early portland cement, and the almost complete ignorance that surrounded it, the product soon became the best in the field of cementitious materials. It was inevitable that, eventually, the constitution of such an important material would be studied, because such information is of great importance in the manufacture and use of any product.

Our information on the constitution of clinker has come from the application of several branches of science, including mathematics, physical chemistry, petrography, and x-rays. The use of the electron microscope has been started.[1] For a long time, the data on clinker were derived almost exclusively from chemical analyses and physical tests. The analyses were used primarily as a guide to obtain cement having the desired physical characteristics, as indicated by the physical tests. It is not surprising that chemical analyses produced only limited information, because, in general, clinker and the raw materials from which it is manufactured contain the same elements, in different proportions. Comparatively little was accomplished, until methods other than those of chemical analyses and physical tests were applied.

Clinker Compounds

Clinker as we know it today consists of a mixture of compounds, or synthetic minerals. The most important of these are tricalcium silicate, dicalcium silicate, tricalcium aluminate, and tetracalcium aluminoferrite. The credit for the discovery of tricalcium silicate is usually given to Le Chatelier.[2] He reported the existence of this compound in some nodules obtained by heating some siliceous limestone, but was unable to synthesize it from calcium oxide and silicon dioxide. In another respect, also, his work was outstanding. It is said that he was the first to employ a petrographic microscope in this field of clinker research.

Extensive researches at the Geophysical Laboratory led to the discovery of two more of the four compounds mentioned. These are dicalcium silicate and tricalcium aluminate. There are three phases of dicalcium silicate, the alpha, beta, and gamma. When this silicate is mentioned in connection with clinker, the beta phase is usually understood, unless one of the others is specified. The results of this work have been published in a number of papers. A very satisfactory account of the whole project is given by Rankin.[3]

For a long time, four compounds were generally accepted as the principal components of portland cement clinker. Of these four, the last one identified, and the only one containing iron, was tetracalcium, aluminoferrite,[4] $4CaO \cdot Al_2O_3 \cdot Fe_2O_3$. Its composition is as follows:

CaO	46.1%
Al_2O_3	21.0
Fe_2O_3	32.9
Density	3.77
Melting point	2570–2588° F

Bogue [5] has reviewed the literature on the constitution of clinker from Roman times to 1925; see also Lea and Desch.[6]

Since 1925, research on the constitution of clinker has included investigations of the compounds of some of the minor elements. The compound $K_2O \cdot 23CaO \cdot 12SiO_2$ has been identified.[7] Then Swayze announced the discovery of a new ternary compound, $6CaO \cdot 2Al_2O_3 \cdot Fe_2O_3$, and a series of solid solutions extending from $6CaO \cdot 2Al_2O_3 \cdot Fe_2O_3$ to $4CaO \cdot Al_2O_3 \cdot Fe_2O_3$.

After the clinker compounds had been identified and synthesized, and their characteristics studied, it became desirable to estimate the percentages present in commercial clinker. This was done by the use of the petrographic microscope, employing three types of specimens,

1) Powdered sample
2) Thin polished section, examined by transmitted light, and
3) Polished and etched specimen, examined by reflected light

In comparing 1 and 2, it may be said that 2 is preferable in accuracy and gives satisfactory results in general, but 1 requires less time for the preparation of the specimen, and the sample is more representative. The polished and etched specimen is a more recent development and is probably preferred by most petrographers. Figure 13:1 shows photomicrographs of cement clinker, in polished sections, etched with distilled water.[8]

In 1921, Colony [9] published a procedure for calculating the compound composition of clinkers and cements from the chemical analyses. The method had many faults and inaccuracies. At the time, there was no method available for determining the uncombined calcium oxide. It would have been possible to calculate erroneously the percentage of tricalcium

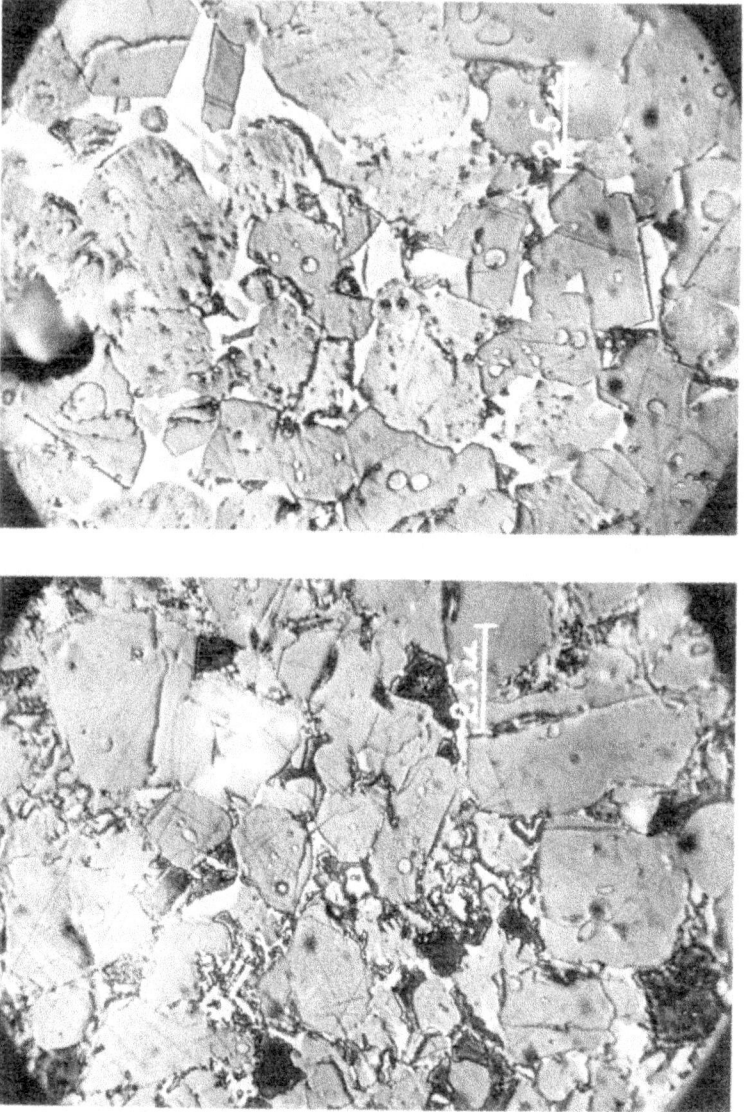

FIG. 13:1. POLISHED SECTIONS OF CEMENT CLINKER, ETCHED WITH DISTILLED WATER

silicate in a sample of ignited raw mix which contained no tricalcium silicate. I pointed out some of these faults [10] and the matter lay dormant for years.

Later Colony's method was revived, and in the face of much opposition, it became widely accepted and found its way into some cement specifications. Since the discovery of tetracalcium aluminoferrite and the development of a method for determining uncombined calcium oxide, the procedure for calculation has been on a firmer basis than formerly. If all the chemical determinations known to affect the results were taken into account, however, the total time required would be so great that the calculation of compound composition could not be used, at least not so extensively as at present. The original method called for the calculation of tricalcium and dicalcium silicates by simultaneous equations, after the calculation of the other compounds had been completed. I avoided the simultaneous equations by calculating all the silica to dicalcium silicate, and then combining the remaining calcium oxide with an equivalent weight of the dicalcium silicate to find the percentage of tricalcium silicate. Bogue [11] published arithmetical and diagrammatical methods for making the calculations, and later a special slide rule [12] was designed for the purpose.

At best, the calculation of compound composition may be considered only an approximation. When we consider that many thousands of tons of clinker minerals are manufactured annually, the need for an accurate procedure is apparent. Possibly some selective solvents may be found, or some chemical-microscopical method may be developed. Many problems, which are apparently much more difficult, have been solved.

Multiple Synthesis

In most cases, the purpose of a synthesis is to obtain a single compound, although some by-products may be formed in the operation. As has been explained, clinker consists of four compounds proportioned within certain limits. Therefore, the clinkering process may be considered a multiple synthesis. This requires a rather complex system of reacting

TABLE 13:3

MULTIPLE SYNTHESIS

Calcium oxide......
Calcium oxide......
Calcium oxide......
Silica............
Tricalcium silicate..
Calcium oxide......
Calcium oxide......
Calcium oxide......
Calcium oxide......
Silica............
Alumina............
Ferric oxide........
Tricalcium silicate..
Dicalcium silicate..
Tricalcium aluminate............
Tetracalcium aluminoferrite............

$3CaO \cdot SiO_2$ $3CaO \cdot SiO_2$ $3CaO \cdot SiO_2$ $3CaO \cdot SiO_2$ $3CaO \cdot SiO_2$ $CaO \cdot SiO_2$ $CaO \cdot SiO_2$ $CaO \cdot SiO_2$ $CaO \cdot SiO_2$

$2CaO \cdot SiO_2$ $2CaO \cdot SiO_2$ $2CaO \cdot SiO_2$ $2CaO \cdot SiO_2$ $2CaO \cdot SiO_2$ $2CaO \cdot SiO_2$ $2CaO \cdot SiO_2$

$3CaO \cdot Al_2O_3$

$4CaO \cdot Al_2O_3 \cdot Fe_2O_3$

TABLE 13:3 (Continued)

	Molecular Weigh	Number of Moles	Total Weight in Pounds	Percentage, Weight	Percentage Moles
REACTING COMPOUNDS					
Calcium oxide	56.08	53	2972.24	67.29	71.622
Silica	60.06	18	1081.08	24.48	23.324
Alumina	101.94	2	203.88	4.62	2.703
Ferric oxide	159.68	1	159.68	3.61	1.351
Totals		74	4416.88	100.00	100.000
PRODUCTS					
Tricalcium silicate	228.30	10	2283.00	51.69	50
Dicalcium silicate	172.22	8	1377.76	31.19	40
Tricalcium aluminate	270.18	1	270.18	6.12	5
Tetracalcium aluminoferrite	485.94	1	485.94	11.00	5
Totals		20	4416.88	100.00	100.00

compounds, which may be illustrated by the example shown in Table 13:3. In this system, 53 moles of calcium oxide, 18 of silica, 2 of alumina, and 1 of ferric oxide, form clinker compounds in the quantities indicated. The minor compounds present in commercial raw materials have been omitted, and it is assumed that all reactions are complete. Moles are expressed in pounds, but any other unit of weight may be employed.

Complete conversion of the reacting compounds to the products requires that if there is an increase or decrease in the weight of one of the reacting compounds, the weights of the others remaining constant, the increase or decrease must be an integer multiple of the molecular weight. For example, if the weight of the calcium oxide were modified by adding or subtracting a quantity other than 56.08 pounds, or an integral multiple of this, the system would not remain in balance, and there would be an excess of one or more of the oxides.

If a change in the weight of the reacting compound is on the basis of its molecular weight, extensive changes may be made in the system without interfering with complete combination. Within limits that are readily apparent, an increase in tricalcium silicate and a decrease in dicalcium silicate would result from increasing the calcium oxide, or decreasing the silica; or vice versa. Corresponding statements could be made concerning alumina and ferric oxide.

Although it is customary to express the relative quantities of the compounds present in clinker on the basis of percentage by weight, molar percentages are more significant. For example, the percentage of tetracalcium aluminoferrite, by weight, in the clinker shown in Table 13:3 is almost twice as great as that of tricalcium aluminate. This is because of the difference in molecular weights. The molar percentages are the same.

Hydration of Clinker Compounds

When a paste is prepared by mixing cement with water, the paste will set or harden within a few hours. The setting is affected by a number of factors, such as:

Percentages of the clinker minerals present
Percentage of gypsum ground with the clinker
Particle size of the solids
Ratio of water to cement
Temperature

Although the space allotted to it here might indicate otherwise, the hydration of cement is a very complex process, and our information concerning it is far from complete.

Assume that each of the clinker minerals separately has been treated with water, and that the reactions and the hydration products have been studied. The following is taken from a paper by Bogue and Lerch: [13]

The products of the reaction of water on tricalcium silicate, under the conditions of these experiments, are crystalline calcium hydroxide and an amorphous hydrated calcium silicate of composition approaching $2CaO \cdot SiO_2 \cdot xH_2O$. The beta dicalcium silicate hydrates to approximately the same amorphous hydrated silicate, but only traces of crystalline calcium hydroxide have been observed.

The compounds containing alumina react rapidly with water to form a fluffy, crystalline, isotropic, hydrated calcium aluminate of a composition $3CaO \cdot Al_2O_3 \cdot 6H_2O$.

The $4CaO \cdot Al_2O_3 \cdot Fe_2O_3$ reacts with water to form an amorphous hydrate, the composition of which has not definitely been established, and crystalline $3CaO \cdot Al_2O_3 \cdot 6H_2O$.

Gypsum reacts with the alumina that enters into solution, with the formation of crystalline calcium sulfoaluminate, $3CaO \cdot Al_2O_3 \cdot 3CaSO_4 \cdot 31H_2O$. This occasions a delay in the development of the crystalline hydrated tricalcium aluminate and so retards the initial set.

The following is from a paper by Brownmiller on "The Microscopic Structure of Hydrated Portland Cement." [14]

Photomicrographs show that a considerable amount of unhydrated cement remains in type 1 and type 2 cements after hydration for 28 days. Type 3 cements show much smaller percentages of unhydrated material even at earlier ages.

Most of the principal constituents of the original clinker can be recognized in the unhydrated fractions. The photographs give no evidence that any major constituent of the cement is selectively or completely hydrated at any age. The rate of hydration depends more specifically on the surface exposed to the action of the water than on the chemical constitution.

The effects of laitance formation are shown by illustrations of the difference in particle size distribution in the laitance as compared to that within the main body of the cement.

Other photographs show the size, amount, and distribution of the $Ca(OH)_2$ which is liberated during the hydration processes. It is estimated that about 15 per cent of $Ca(OH)_2$ has formed at 28 days in the cement examined. The polishing and etching technique described could be applied readily for the preparation of specimens for accurate measurement of the $Ca(OH)_2$ by means of mechanical devices such as a Wentworth micrometer.

The final photographs show some detail of the structure of the hydrates other than the $Ca(OH)_2$. That structure is extremely complicated, but a further development of microscopic technique should be useful in solving some of the riddles which confront cement technicians in attempting to evaluate cements on the basis of performance in concrete.

Bogue [15] has reviewed the literature on hydration, from the third century to the twentieth.

Some Analyses and Ratios

Table 13:1 contains some analyses of cements collected by Meade;[16] in fact, most of the analyses were made by him. In general, these cements are similar to those manufactured now. There is a tendency, however, for present-day cements to be higher in calcium oxide and in sulfur trioxide (SO_3). The range in the principal oxides and clinker compounds of more than one hundred portland cements manufactured in the United States is shown in Table 13:2. Although no definite limits for the composition of cement in general have been established, limits for one or more constituents may be set directly or indirectly by specifications. If a cement consisted only of the compound tricalcium silicate, the calcium oxide content would be 73.69 per cent, and the silica 26.31 per cent. In the same way, if it consisted of dicalcium silicate only, the calcium oxide content would be 65.11 per cent, and the silica content would be 34.89 per cent. As no commercial cement consists of either of these compounds exclusively, nor of a mixture of only the two, the calcium oxide is always less than 73.69 per cent, and the silica less than 34.89 per cent. Cement always contains either alumina or ferric oxide, and practically always both of them. The lower

limit of each of these is zero, but there is no definite upper
limit. In the same way, for each of the other oxides, the lower
limit is zero and there is no definite upper limit.

Other chemical elements in relatively small quantities
sometimes are found in cements. Sodium, potassium, man-
ganese, titanium, and phosphorus may be mentioned. Almost
any element may be present in a fractional percentage without affecting the quality of the product.

As a guide in proportioning raw materials for the manufacture of cement, and for comparing and studying various
cement analyses, several ratios were employed for many
years, and are used, to some extent, at present. Usually,
each oxide is expressed as a percentage. The following are
some examples:

$$\frac{SiO_2}{Al_2O_3} \tag{1}$$

$$\frac{SiO_2}{Al_2O_3 + Fe_2O_3} \tag{2}$$

$$\frac{CaO}{SiO_2 + Al_2O_3 + Fe_2O_3} \tag{3}$$

$$\frac{CaO + MgO}{SiO_2 + Al_2O_3 + Fe_2O_3} \tag{4}$$

$$\frac{SiO_2 + Al_2O_3 + Fe_2O_3}{CaO} \tag{5}$$

$$\frac{SiO_2 + Al_2O_3 + Fe_2O_3}{CaO + MgO} \tag{6}$$

The usefulness of these ratios has decreased because it has
become customary to calculate the compound composition of
clinker and of cement from the analyses, and particularly because compound compositions are mentioned in some specifications. Ratios in which the numerator contains more than
one term should not be used, unless the relation of these
terms remains constant. The same, of course, is true of the
denominator. The reason for this is that the molecular
weights of the oxides are not the same. In each case, the

percentage of the oxide should be divided by the molecular weight, thus expressing the oxide in terms of moles.

This may be illustrated by substituting two sets of values in ratio 4:

$a)$
$$\frac{63 + 1}{23 + 6 + 3} = 2$$

$b)$
$$\frac{62 + 2}{21 + 7.5 + 3.5} = 2$$

As two very different cements have the same ratio, it is evident that the ratio has little significance. Many other examples could be given. The values may be expressed as moles by dividing each percentage by the corresponding molecular weight. The molecular weight of each oxide is: CaO, 56.08; MgO, 40.32; Al_2O_3, 101.94; Fe_2O_3, 159.68; and SiO_2, 60.06. The ratio a then becomes

$$\frac{\dfrac{63}{56.08} + \dfrac{1}{40.32}}{\dfrac{23}{60.06} + \dfrac{6}{101.94} + \dfrac{3}{159.68}} = \frac{1.12 + 0.02}{0.38 + 0.06 + 0.02} = 2.48$$

When the corresponding values are substituted in b the result is 2.61. That is, the two cements, different in composition, do not have the same ratio. In some cases, combining weights as well as molecular weights must be taken into account.

The principal clinker compounds have been mentioned. Some abbreviations and formulas by which they are identified are:

Compound		Abbreviation or Formula		
	1	2	3	4
Tricalcium silicate	C_3S	3CS	$3CaO \cdot SiO_2$	Ca_3SiO_5
Dicalcium silicate	C_2S	2CS	$2CaO \cdot SiO_2$	Ca_2SiO_4
Tricalcium aluminate	C_3A	3CA	$3CaO \cdot Al_2O_3$	$Ca_3Al_2O_6$
Tetracalcium aluminoferrite	C_4AF	4CAF	$4CaO \cdot Al_2O_3 \cdot Fe_2O_3$	$Ca_4Al_2Fe_2O_{10}$

In column 1 there is the abbreviation most commonly used. It is simple and convenient, but it has one serious disadvantage. The form suggests a chemical formula and it is therefore misleading. For example, C_4AF is the chemical formula for a compound of carbon, argon, and fluorine.[17] Such a compound could not exist. In column 2 there is an abbreviation just as convenient as that in column 1, but it does not suggest a chemical formula. In column 3 there is the chemical formula ordinarily employed. It is correct, but is subject to some misinterpretation. In the formula $3CaO \cdot SiO_2$, for example, there is a suggestion that CaO and SiO_2 retain their identity to some extent, which is not the case. In column 4 there is the most satisfactory formula.

Winchell [18] lists the optic constants of each of the four compounds.

Specifications

Cements are manufactured and sold in conformity with specifications. These specifications are written by federal, state, and municipal governments, by technical societies, and by other organizations. Unless otherwise explained, all specifications to which reference is made in this book are those of the American Society for Testing and Materials (ASTM). According to an early specification of the ASTM (C 9–30),

> Portland cement is the product obtained by finely pulverizing clinker produced by calcining to incipient fusion an intimate and properly proportioned mixture of argillaceous and calcareous materials, with no additions subsequent to calcination excepting water and calcined or uncalcined gypsum.

This was the accepted definition of portland cement for many years, and probably most men now in the industry memorized it, or tried to memorize it, as the first step in their study of portland cement. In my opinion, it is the best definition for portland cement that has been suggested, although it has some faults.*

---

* The cement-manufacturing process is almost complete when the clinker has been produced. There remains only grinding, with the addition of relatively small quantities of one or more materials, only one of which is essential. The statement concerning the production of clinker, in the old definition, has been succeeded by the phrase, "consisting essentially of hydraulic calcium silicates . . ." which is not adequate.

Criticisms of the definition increased with the years, until it was replaced by the following: (ASTM C 150–41)

Portland cement is the product obtained by pulverizing clinker consisting essentially of hydraulic calcium silicates, to which no additions have been made subsequent to calcination other than water and/or untreated calcium sulfate, except that other materials may be added provided that they have been shown to be acceptable, in the amounts indicated, by tests carried out or reviewed by Committee C 1.

"What the hell is cement?" This most famous question of the industry was asked Robert W. Lesley,[19] in 1874. He had agreed to sell 200 barrels of "Cumberland Cement" on commission. This first consignment looked greater to him than the many millions of barrels he sold subsequently during his career. Mr. Lesley was to become the first president of the Portland Cement Association, and one of its most prominent and most beloved members. In his later years, he was known affectionately as the dean of the cement industry. It is not surprising that the question was a difficult one. Now, after more than ninety years, we still cannot answer it in all its detail.

The latest ASTM Specification for portland cement follows:

Standard Specification for
PORTLAND CEMENT[1]

ASTM Designation: C 150 – 64

ADOPTED, 1941; REVISED, 1942, 1944, 1946, 1947, 1949, 1952, 1953, 1955, 1956, 1959, 1960, 1961, 1962, 1963, 1964.[2]

This Standard of the American Society for Testing and Materials is issued under the fixed designation C 150; the final number indicates the year of original adoption as standard or, in the case of revision, the year of last revision.

Scope

1. This specification covers five types of portland cement, as follows:

Type I.—For use in general concrete construction when the special properties specified for types II, III, IV, and V are not required.

Type II.—For use in general concrete construction exposed to moderate sulfate action, or where moderate heat of hydration is required.

Type III.—For use when high early strength is required.

Type IV.—For use when a low heat of hydration is required (Note 1).

Type V.—For use when high sulfate resistance is required (Note 1).

NOTE 1.—Attention is called to the fact that cements conforming to the requirements for type IV and type V are not usually carried in stock. In advance of specifying their use, purchasers or their representatives should determine whether these types of cement are, or can be made available.

Basis of Purchase

2. The purchaser should specify the type or types desired. When no type is specified, the requirements of type I shall govern.

Definition

3. *Portland Cement*, for the purpose of this specification, is the product obtained by pulverizing clinker consisting essentially of hydraulic calcium silicates.

Additions

4. (*a*) The cement covered by this specification shall contain no addition except as provided for below.

(*b*) Water or untreated calcium sulfate, or both, may be added in amounts such that the limits shown in Table I

---

[1] Under the standardization procedure of the Society, this specification is under the jurisdiction of the ASTM Committee C-1 on Cement.

[2] Latest revision accepted August 31, 1964, by action of the Society at the Annual Meeting and confirming letter ballot.

Prior to adoption as standard, this specification was published as tentative from 1940 to 1941, being revised in 1941.

This specification replaces the former Standard Specifications for Portland Cement (C 9 - 38), and High-Early-Strength Portland Cement (C 74 - 39), which were accordingly discontinued in September, 1941.

TABLE I.—CHEMICAL REQUIREMENTS.

	Type I	Type II	Type III	Type IV^a	Type V^a
Silicon dioxide (SiO$_2$), min, per cent.................	...	21.0	...	...	
Aluminum oxide (Al$_2$O$_3$), max, per cent.............	...	6.0	...	...	b
Ferric oxide (Fe$_2$O$_3$), max, per cent.................	...	6.0	...	6.5	b
Magnesium oxide (MgO), max, per cent.............	5.0	5.0	5.0	5.0	4.0
Sulfur trioxide (SO$_3$), max, per cent:					
When 3CaO·Al$_2$O$_3$ is 8 per cent or less.............	2.5	2.5	3.0	2.3	2.3
When 3CaO·Al$_2$O$_3$ is more than 8 per cent..........	3.0	...	4.0	...	...
Loss on ignition, max, per cent.....................	3.0	3.0	3.0	2.5	3.0
Insoluble residue, max, per cent....................	0.75	0.75	0.75	0.75	0.75
Tricalcium silicate (3CaO·SiO$_2$),^{c} max, per cent.......	...	...	...	35	...
Dicalcium silicate (2CaO·SiO$_2$),^{c} min, per cent........	...	...	...	40	...
Tricalcium aluminate (3CaO·Al$_2$O$_3$),^{c} max, per cent....	...	8	15^d	7	5
Sum of tricalcium silicate and tricalcium aluminate, max, per cent...................................	...	58^e	...	...	...
Tetracalcium aluminoferrite plus twice the tricalcium aluminate^c (4 CaO·Al$_2$O$_3$·Fe$_2$O$_3$ + 2(3 CaO·Al$_2$O$_3$), or solid solution (4 CaO·Al$_2$O$_3$·Fe$_2$O$_3$ + 2 CaO· Fe$_2$O$_3$), as applicable, max, per cent.............	...	...	...	...	20.0

^{a} See Note 1.

^{b} The tricalcium aluminate shall not exceed 5 per cent, and the tetracalcium aluminoferrite (4CaO·Al$_2$O$_3$·Fe$_2$O$_3$) plus twice the amount of tricalcium aluminate shall not exceed 20 per cent.

^{c} The expressing of chemical limitations by means of calculated assumed compounds does not necessarily mean that the oxides are actually or entirely present as such compounds.

When the ratio of percentages of aluminum oxide to ferric oxide is 0.64 or more, the percentages of tricalcium silicate, dicalcium silicate, tricalcium aluminate and tetracalcium aluminoferrite shall be calculated from the chemical analysis as follows:

Tricalcium silicate = (4.071 × per cent CaO) − (7.600 × per cent SiO$_2$) − (6.718 × per cent Al$_2$O$_3$) − (1.430 × per cent Fe$_2$O$_3$) − (2.852 × per cent SO$_3$)

Dicalcium silicate = (2.867 × per cent SiO$_2$) − (0.7544 × per cent C$_3$S)

Tricalcium aluminate = (2.650 × per cent Al$_2$O$_3$) − (1.692 × per cent Fe$_2$O$_3$)

Tetracalcium aluminoferrite = 3.043 × per cent Fe$_2$O$_3$

When the alumina-ferric oxide ratio is less than 0.64, a calcium aluminoferrite solid solution (expressed as ss (C$_4$AF + C$_2$F)) is formed. Contents of this solid solution and of tricalcium silicate shall be calculated by the following formulas:

ss (C$_4$AF + C$_2$F) = (2.100 × per cent Al$_2$O$_3$) + (1.702 × per cent Fe$_2$O$_3$)

Tricalcium silicate = (4.071 × per cent CaO) − (7.600 × per cent SiO$_2$) − (4.479 × per cent Al$_2$O$_3$) − (2.859 × per cent Fe$_2$O$_3$) − (2.852 × per cent SO$_3$).

No tricalcium aluminate will be present in cements of this composition. Dicalcium silicate shall be calculated as previously shown.

In the calculation of C$_3$A, the values of Al$_2$O$_3$ and Fe$_2$O$_3$ determined to the nearest 0.01 per cent shall be used. In the calculation of other compounds the oxides determined to the nearest 0.1 per cent shall be used.

Values for C$_3$A and for the sum of C$_4$AF + 2C$_3$A shall be reported to the nearest 0.1 per cent. Values for other compounds shall be reported to the nearest 1 per cent.

^{d} When moderate sulfate resistance is required for type III cement, tricalcium aluminate may be limited to 8 per cent. When high sulfate resistance is required, the tricalcium aluminate may be limited to 5 per cent.

^{e} This limit applies when moderate heat of hydration is required and tests for heat of hydration are not requested.

TABLE II—PHYSICAL REQUIREMENTS.

	Type I	Type II	Type III	Type IV[a]	Type V[a]
Fineness, specific surface, sq cm per g (alternate methods):[b]					
Turbidimeter test:					
Average value, min.	1600	1600	...	1600	1600
Minimum value, any one sample.	1500	1500	...	1500	1500
Air permeability test:					
Average value, min.	2800	2800	...	2800	2800
Minimum value, any one sample.	2600	2600	...	2600	2600
Soundness:					
Autoclave expansion, max, per cent.	0.80	0.80	0.80	0.80	0.80
Time of setting (alternate methods):[c]					
Gillmore test:					
Initial set, min, not less than.	60	60	60	60	60
Final set, hr, not more than.	10	10	10	10	10
Vicat test (Method C 191):					
Set, min, not less than.	45	45	45	45	45
Air content of mortar, prepared and tested in accordance with Method C 185, max, per cent by volume, less than.	12.0	12.0	12.0	12.0	12.0
Compressive strength, psi:[d] The compressive strength of mortar cubes, composed of 1 part cement and 2.75 parts graded standard sand, by weight, prepared and tested in accordance with Method C 109, shall be equal to or higher than the values specified for the ages indicated below:					
1 day in moist air.	...	...	1700	...	...
1 day in moist air, 2 days in water.	1200	1000	3000	...	...
1 day in moist air, 6 days in water.	2100	1800	...	800	1500
1 day in moist air, 27 days in water.	3500	3500	...	2000	3000
Tensile strength, psi:[d] The tensile strength of mortar briquets composed of 1 part cement and 3 parts standard sand, by weight, prepared and tested in accordance with Method C 190, shall be equal to or higher than the values specified for the ages indicated below:					
1 day in moist air.	...	...	275	...	...
1 day in moist air, 2 days in water.	150	125	375	...	...
1 day in moist air, 6 days in water.	275	250	...	175	250
1 day in moist air, 27 days in water.	350	325	d	300	325
Heat of hydration:[e]					
7 days, max, cal per g.	...	70	...	...	...
28 days, max, cal per g.	...	80	...	...	...
False set, final penetration, min, per cent[f].	50	50	50	50	50

[a] See Note 1.

[b] Either of the two alternate fineness methods may be used at the option of the testing laboratory. However, in case of dispute, or when the sample fails to meet the requirements of the Blaine meter, the Wagner turbidimeter shall be used, and the requirements in Table II for this method shall govern.

[c] The purchaser should specify the type of setting time test required. In case he does not so specify, or in case of dispute, the requirement of the Vicat test only shall govern.

[d] The purchaser should specify the type of strength test required. In case he does not so specify, the requirements of the compressive strength test only shall govern. The strength at any age shall be higher than the strength at the next preceding age. Unless otherwise specified, the compressive and tensile strength tests for types I and II cement will be made only at 3 and 7 days. If, at the option of the purchaser, a 7-day test is required on type III cement, the strength at 7 days shall be higher than at 3 days.

[e] These requirements apply only when specifically requested; when the heat of hydration requirements are specified, the strength requirements for type II shall be 80 per cent of the values listed in Table II.

[f] This requirement applies only when specifically requested.

for sulfur trioxide and loss on ignition shall not be exceeded.

(c) At the option of the manufacturer, processing additions may be used in the manufacture of the cement, provided such materials in the amounts used have been shown to be not harmful in accordance with the Specifications for Processing Additions for Use in the Manufacture of Portland Cement (ASTM Designation: C 465).[3]

Manufacturer's Statement

5. At the request of the purchaser, the manufacturer shall state in writing the nature, amount, and identity of any processing addition that may have been used and also, if requested, shall supply test data showing compliance of such processing addition with Specification C 465.

Chemical Requirements

6. Portland cement of each of the five types shown in Section 1 shall conform to the respective chemical requirements prescribed in Table I.

NOTE 2.—Cement containing not more than 0.60 per cent alkalies calculated as the percentage of Na_2O plus 0.658 times the percentage of K_2O, may be specified when the cement is to be used in concrete with aggregates that may be deleteriously reactive. Reference should be made to the Specifications for Concrete Aggregates (ASTM Designation: C 33)[4] for suitable criteria of deleterious reactivity.

Physical Requirements

7. Portland cement of each of the five types shown in Section 1 shall conform to the respective physical requirements prescribed in Table II.

Packaging and Marking

8. When cement is delivered in packages, the name and brand of the manufacturer and the type under this specifi-

---

[3] 1964 Book of ASTM Standards, Part 9.
[4] Appears in this publication.

cation shall be plainly indicated thereon, except that in the case of type I the type need not be indicated. Similar information shall be provided in the shipping advices accompanying the shipment of packaged or bulk cement. A bag shall contain 94 lb net. A barrel shall consist of 376 lb net. All packages shall be in good condition at the time of inspection.

Storage

9. The cement shall be stored in such a manner as to permit easy access for proper inspection and identification of each shipment, and in a suitable weather-tight building that will protect the cement from dampness and minimize warehouse set.

Inspection

10. Every facility shall be provided the purchaser for careful sampling and inspection, either at the mill or at the site of the work, as may be specified by the purchaser. The following periods from time of sampling shall be allowed for completion of testing:

1-day test	6 days
3-day test	8 days
7-day test	12 days
28-day test	33 days

Rejection

11. (a) The cement may be rejected if it fails to meet any of the requirements of this specification.

(b) Cement remaining in bulk storage at the mill, prior to shipment, for more than 6 months, or cement in bags in local storage in the hands of a vendor for more than 3 months, after completion of tests, may be retested before use and may be rejected if it fails to conform to any of the requirements of this specification.

(c) Packages varying more than 5 per cent from the specified weight may be rejected; and if the average weight of packages in any shipment, as shown by weighing 50 packages taken at random, is less than that specified, the entire shipment may be rejected.

(d) Cement failing to meet the test for soundness in the autoclave may be accepted if it passes a retest, using a new sample, at any time within 28 days thereafter. The provisional acceptance of the cement at the mill shall not deprive the purchaser of the right of rejection on a retest of soundness at the time of delivery of the cement to the purchaser.

Methods of Testing

12. The cement shall be sampled and the properties enumerated in this specification shall be determined in accordance with the following methods:

(a) *Sampling.*—Methods of Sampling Hydraulic Cement (ASTM Designation: C 183).[5]

(b) *Chemical Analysis.*—Standard Methods of Chemical Analysis of Portland Cement (ASTM Designation: C 114 – 63).[3]

(c) *Fineness by Turbidimeter.*—Method of Test for Fineness of Portland Cement by the Turbidimeter (ASTM Designation: C 115).[3]

(d) *Fineness by Air Permeability.*—Method of Test for Fineness of Portland Cement by Air Permeability Apparatus (ASTM Designation: C 204).[3]

(e) *Autoclave Expansion.*—Method of Test for Autoclave Expansion of Portland Cement (ASTM Designation: C 151).[5]

(f) *Time of Setting by Gillmore Needles.*—Method of Test for Time of Setting of Hydraulic Cement by Gillmore Needles (ASTM Designation: C 266).[3]

(g) *Time of Setting by Vicat Needle.*—Method of Test for Time of Setting of Hydraulic Cement by Vicat Needle (ASTM Designation: C 191).[3]

(h) *Air Content of Mortar.*—Method of Test for Air Content of Hydraulic Cement Mortar (ASTM Designation: C 185).[3]

(i) *Compressive Strength.*—Method of Test for Compressive Strength of Hydraulic Cement Mortars (Using 2-In. Cube Specimens) (ASTM Designation: C 109).[5]

(j) *Tensile Strength.*—Method of Test for Tensile Strength of Hydraulic Cement Mortars (ASTM Designation: C 190).[3]

(k) *False Set.*—Method of Test for False Set of Portland Cement (Paste Method) (ASTM Designation: C 451).[5]

---

[5] Will be published in Part 9 of the **1965** Book of ASTM Standards. Currently available as a separate reprint.

REFERENCES

1. C. M. Sliepcevich, L. Gildart, and D. L. Katz, *Ind. Eng. Chem.*, 35(11):1178 (1943).
2. H. Le Chatelier, *Recherches experimentales zur la constitution des mortiers hydraulics*, Dunod, Paris, 1904.
   ——, *Experimental Researches on the Constitution of Hydraulic Mortars*, translated by Joseph Lathrop Mack, McGraw-Hill Publishing Co., Inc., New York, 1905.
3. G. A. Rankin, *J. Franklin Inst.*, 181:747 (1916).
4. W. C. Hansen, L. T. Brownmiller, and R. H. Bogue, *J. Am. Chem. Soc.*, 50:396 (1928).
5. R. H. Bogue, A Digest of the Literature on the Constitution of Portland Cement Clinker, *Concrete*, 29(1):14; (2):27; (3):31; (4):31; (5):31; (6):37 (1926); 30(1):33; (2):36 (1927).
6. F. M. Lea, and C. H. Desch, *The Chemistry of Cement and Concrete;* Edward Arnold and Co., London, 1935.
7. W. C. Taylor, *J. Research* (National Bureau of Standards), 29:437 (1942).
8. L. S. Brown, Portland Cement Association Laboratories, 2(3): 23–34 (1959).
9. R. J. Colony, *Eng. News-Record*, 86:637 (1921), reproduced from *Trans. Am. Inst. Mining Met. Engrs.*
10. J. C. Witt, *Eng. News-Record*, 87:650 (1921).
11. R. H. Bogue, *Ind. Eng. Chem., Anal. Ed.*, 1(4):192 (1929).
12. L. A. Dahl, *Rock Products*, 32(23):50 (1929).
13. R. H. Bogue and Wm. Lerch, *Ind. Eng. Chem.*, 26:837 (1934).
14. L. T. Brownmiller, *J. Am. Concrete Inst.*, 14:193 (1943).
15. R. H. Bogue, Digest of Literature on Nature of Setting and Hardening Processes in Portland Cement, *Rock Products*, 31(10):69; (12):62; (14):61; (18):65; (20):65 (1928).
16. R. K. Meade, *Portland Cement*, Chemical Publishing Co., Inc., Easton, Pa., 1930.
17. *Concrete, Cement Mill Section*, 43(5):41 (1935).
18. A. N. Winchell, *The Microscopic Characters of Artificial Minerals;* John Wiley and Sons, Inc., New York, 1931.
19. R. W. Lesley, *History of the Portland Cement Industry in the United States;* International Trade Press, Inc., Chicago, 1924.
20. M. A. Swayze, *Am. J. Sci.*, 244:1, 65 (1946).

For additional references see *Bibliographies*, Chapter 24.

Chapter 14

ANALYSES AND TESTS

To one who has followed this book from the beginning, it should not be surprising that many chemical analyses and physical tests are required for the operation of a cement plant; nor that, in addition, chemical analyses, physical tests, and other examinations and measurements in considerable variety are suggested by economy and by good practice in general. For simplicity, the general term *tests* may be applied to all laboratory operations. It will be recalled that proportioning two or more raw materials on the basis of their chemical composition is one of the basic principles of portland cement manufacture. In a study of the test schedule at one cement plant, it was found that a test was made on an average of every few minutes, day and night.

For several reasons, some test methods are presented. First, they are essential in the manufacture and use of cement; second, many workers in the cement and concrete industries are interested particularly in this phase of the work; and, third, some information on test procedures is essential for a general understanding of cement. This third reason applies especially to personnel of departments other than laboratories.

Classification of Tests

The following outline lists the tests under three headings:
1. *Type of test:*
   a) Chemical analyses
   b) Physical tests
   c) Other examinations and measurements

2. *Materials tested:*
   a) Raw materials
   b) Raw mixes
   c) Fuels
   d) Clinker
   e) Gypsum
   f) Cement
   g) Miscellaneous supplies

3. *Tests required: For*
   a) Materials received
   b) Manufacturing processes
   c) Products sold
   d) Research and development

Samples

As all tests are made on samples, the importance of sampling is obvious. There are two types of sample, the individual, and the general. The first is a portion of a substance which is tested independently of any other sample. A general sample consists of a mixture of individual samples. Usually, the individual sample is taken manually. The general sample may be taken manually, but usually it is taken mechanically. The importance of automatic sampling devices is receiving more and more recognition. It should be pointed out, however, that although the general sample is usually preferable, many of the irregularities in composition that may be present in a material do not appear in a general sample. Consequently, at times individual samples are required.

It is not likely that an absolutely representative sample can be obtained. To take a sample as nearly representative as practicable is a complex procedure, involving many variables. Some of these are: the quantity of an individual sample, sometimes called an *increment*, in relation to the total quantity of material to be samples; particle size distribution in the individual sample; the number of individual samples for one general sample; and the subsequent treatment of the general sample. The study of these problems becomes rather mathematical.

Coal is a good example of a substance, the sampling of

which has received much study. Committee D-5 of the American Society for Testing and Materials has been working on this for years. See A.S.T.M. Designation: D 492–48.

Laboratory Buildings

Before the discussion of laboratory procedures, with which this chapter is concerned primarily, some comments on laboratory buildings are in order. The many characteristics of laboratory buildings could be discussed at great length, but only the items that affect the quality and quantity of the work will be mentioned here.

Location:

For obtaining samples, inspecting operating units, and the like, a convenient location for the laboratory building is near the center of the plant. The building should not, however, be too far from the office and other departments with which the laboratory must be in frequent contact. The building should not be so near the operating buildings that vibration or noises interfere with the laboratory equipment or processes. Dust and high temperatures are not included as factors limiting the location of the building, because every laboratory should be air-conditioned.

At some plants, a compromise is effected by placing the principal laboratory in a location convenient for administration purposes, and the distribution of reports and correspondence, and then installing one or more branch laboratories in locations convenient to the plant buildings. The writer does not favor this arrangement. The department laboratory is likely to receive less supervision that if it were in the main laboratory building, and costly errors may result. Facilities for communication and the transportation of samples should be considered.

Type of building:

There is no standard type of laboratory building. Frequently, there is a basement and one or two floors above grade. If two floors, the physical laboratory may be on the first floor and the chemical laboratory on the second. The best arrangement is a one-story building without basement.

At most plants there is land available which provides the maximum in light, drainage, and safety. A long list of other advantages could be mentioned.

The laboratory is not the most hazardous department in the plant, but neither is it the least hazardous. One essential is sufficient room. Many of the industrial laboratories I have seen have been too crowded. When a laboratory building is occupied for the first time, there should be some floor space available for expansion, in most cases not less than 25 per cent of the total. New equipment must be added from time to time. Supplies, samples, records, correspondence, books and periodicals tend to accumulate. Some accumulation is legitimate, regardless of how much care is devoted to systematic arrangement, and the disposal of material no longer needed.

Laboratory equipment should be housed in a separate building designed for the purpose. Too often it is located in a portion of a building previously erected for some other purpose. In this case, the equipment is adjusted to the building. The reverse of this should be the relation between building and equipment.

One item that cannot be overemphasized is the distribution of electric lines, piping, and drainage, and their accessibility for maintenance and extension. This refers particularly to the chemical laboratory. In many laboratory rooms, especially those built some years ago, the distribution of facilities is very limited. For example, water and drainage are available in only one location, electric current in another, gas in another, and so on. When such an arrangement exists, the frequent services of electricians and pipe fitters are required, and there are delays and inconveniences.

There are some items that could be listed in connection with either buildings or equipment. One of these is the laboratory hood.[1] Formerly some defects were nearly always present, but better design is now rather prevalent.

Equipment:

The precautions that should be taken in the selection of the equipment most suitable for a given laboratory are well known to experienced workers, and no one else should attempt selecting it. Other factors being equal, preference should be given to equipment units that are well established and have been rendering good service in the industry. Careful consideration should be given, however, to the new units that are announced from time to time. One of these may be very desirable in increasing the accuracy and speed of some laboratory operation.

For keeping up to date on all these matters, the attending of chemical expositions, such as those at New York and Chicago, is well worth while. New ideas in laboratory procedures, calculations, apparatus, publications, and furniture may be found applicable directly or indirectly to one's own problems. A new apparatus unit may increase accuracy, decrease required time, or both.

CHEMICAL METHODS: GENERAL

The general requirements for chemical analyses are only outlined here. No attempt will be made to present a complete set of chemical methods. Because of differences in raw materials, plant equipment, and manufacturing processes, each laboratory has its individual problems. For the best results, the man in charge decides the accuracy needed for each operation, and then selects the method by which such results may be obtained most satisfactorily. No one book contains all the required information.

In this chapter, some methods are given for raw materials, and for cement. Those for cement are applicable to clinker, also. In fact all these methods have much in common, many features having been based on the work of Hillebrand,[2] on the analysis of silicate and carbonate rocks. Bulletins 148, 176, 305, 422, and 700 have been issued by the United States Geological Survey. The following is taken from Bulletin 700:

> Prior to the publication of Bulletin 148 of the United States Geological Survey, the literature relating to the analysis of silicates, though extensive, was so widely scattered that in no single work was there to be found, in the light of modern methods, a satisfactory exposition of the procedures to be followed or the precautions to be observed, especially in the search for some of the rarer constituents or those which, with-

out being rare, had come to be recognized as occurring persistently in small amounts.

In Bulletin 305 there was added a section on carbonate rocks, for the reasons that the chief carbonate rocks form a most important element in the earth's crust, and that the knowledge of their composition is of moment to the geologist as well as to the cement worker who now makes enormous use of them. The methods applied to their analyses differ in but unimportant respects from those used with the more siliceous rocks, for they are to a great extent themselves siliceous, and contain essentially the same constituents, and therefore, the same principles apply to both.

The gravimetric method for calcium oxide is basically more accurate than the volumetric method, in the hands of a capable, experienced analyst. The volumetric method, however, may give better results in the hands of a less skillful analyst.

It is preferable to standardize the potassium permanganate solution against Iceland spar, rather than against sodium oxalate. The calcium carbonate content of Iceland spar of high purity is very nearly one hundred per cent, but this should be determined by analysis in advance. One-half-gram portions of the spar should be dissolved in hydrochloric acid. From this point, the procedure for analysis should be the same as that for the cement, starting with the filtrate from the silicon dioxide, and ending with the potassium permanganate titration, except that the aluminum oxide and ferric oxide determinations, in the case of the spar, need not be completed.

A solution for volumetric analysis should be standardized with the same ion that is to be determined in the analysis. In this case the potassium permanganate should be standardized against calcium rather than against oxalate, or against iron, as sometimes has been suggested. Calcium provides a direct standardization, the others are indirect.

The relative merits of gravimetric and volumetric methods, and the standardization of potassium permanganate solution are some of the highly controversial points in analytical procedures.

Analysis of Raw Materials and Raw Mixes

The methods of analysis given here are intended primarily for argillaceous limestones or cement rocks, containing less than 85 per cent of calcium carbonate.[3] Usually, the procedure for limestones that contain more than 85 per cent of calcium carbonate may be simplified by omitting the fusion of the sample. The numbers in parentheses refer to the list of reagents, given further on in the text.

Raw materials other than limestones:

The raw materials other than limestones that may be analyzed by these methods include clays, shales, slates, traprocks, and sandstones. Modifications, either necessary or desirable, may suggest themselves to the analyst.

Raw mixes:

The methods for the complete analysis of raw mixes are much the same as those for raw materials. For the routine checking of raw mixes in connection with plant control, several methods are available, including the acid-alkali titration, potassium permanganate titration, and the determination of carbon dioxide. The nature of the raw materials must be taken into account in the selection of a method.

If either raw material present contains a calcium compound insoluble in hydrochloric acid, the raw mix must be fused. In some cases, individual samples of raw mix are titrated, and a 24-hour average sample analyzed by fusion. A correction factor may be calculated and applied to each titration result. Before such a procedure is adopted, the results obtained by various methods should be studied and correlated.[4]

Preparation of Sample

Dry the sample at 105 to 110° C and grind it to pass through a 100-mesh sieve.

Determination of loss on ignition:

Weigh exactly 1 gram of the sample into a weighed platinum crucible and ignite it for 30 minutes over a blast lamp or in an electric muffle at 1050° C. When a blast lamp is used, place a well-fitting cover on the crucible. Cool in a desiccator and weigh. Multiply the loss in weight by 100 to obtain the percentage of loss on ignition. If the sample contains any carbonaceous matter, make a preliminary heating for 30 minutes with the open crucible in an inclined position over a Bunsen burner or in the front part of an electric muffle with the door partly open at 950 to 1000° C.

Decomposition of the sample:

Weigh exactly 0.5 gram of the sample into a platinum crucible and mix thoroughly with 0.5 gram of fusion mixture (No. 43) * when the sample contains from 65 to 85 per cent $CaCO_3$. Use 1 gram of fusion mixture when the sample contains from 50 to 65 per cent $CaCO_3$; use 2 grams of fusion mixture when the sample contains less than 50 per cent $CaCO_3$.

Cover the crucible with a lid and heat for 10 minutes over a Bunsen burner or in the cooler part of an electric muffle. Then heat for 15 minutes over a blast lamp or in the muffle at 1050° C. Cool and place the crucible on its side in a 210-milliliter porcelain casserole or a 250-milliliter beaker. Add 15 to 30 milliliters of hydrochloric acid (No. 1), cover immediately with a watch glass and heat until solution is complete. Remove the crucible with a stirring rod and wash it thoroughly with hot water. Heat the crucible for a few minutes to bring out any iron staining. If there is any evidence of iron staining, heat the crucible with 5 milliliters of hydrochloric acid until the stain is removed, as checked with a second heating. Add 5 drops of nitric acid (No. 7) and cover loosely with the watch glass. Evaporate to dryness at a temperature just below the boiling point. Continue the heating on a hot plate at a temperature not exceeding 120° C until the acid fumes are completely driven off.

Determination of Silica

Remove the crucible from the heat, cool it, and add about 15 milliliters of hydrochloric acid (No. 1). Cover with a watch glass and boil for a few minutes. Rinse the under side of the watch glass and the sides of the vessel with a stream of hot water. Then filter through an 11-centimeter filter paper, Grade B, into a 400-milliliter beaker. Scrub the vessel thoroughly with a rubber policeman and rinse the silica with

---

* Numbers in parentheses refer to list of reagents that follows this section on analyses.

hot water to the filter. Wash the filter once with hot hydrochloric acid (No. 1), then 5 times with hot water.

Transfer the filtrate from the beaker back to the vessel in which the first evaporation was made and rinse the beaker carefully with hot water. Evaporate again to dryness at a temperature just below the boiling point. Continue heating on a hot plate at a temperature not exceeding 120° C for at least 1 hour after the acid fumes are driven off.

Remove from the heat, cool, and add about 15 milliliters of hydrochloric acid (No. 1). If for any reason the vessel must be left standing cold for some time before adding the acid, the dried residue will pick up moisture from the air and it should be put back on the hot plate for at least 1 hour before adding the acid. Cover and boil for a few minutes. Rinse the under side of the watch glass and the sides of the vessel with a stream of hot water. Then filter through another 11-centimeter filter paper, Grade B, into the 400-milliliter beaker. Scrub the vessel with a rubber policeman and rinse the remaining silica with hot water to the filter. Wash the filter once with hot hydrochloric acid (No. 1), then 5 times with hot water.

Transfer the 2 filter papers containing the first and the second silica precipitates to a weighed platinum crucible. Dry and char the papers with gentle heat. Burn off the carbon and ignite the crucible for at least 3 hours in an electric muffle at 1050 to 1100° C or cover the crucible and heat it for 30 minutes over a strong blast lamp. Cool in a desiccator and weigh as SiO_2. Multiply the weight of the SiO_2 by 200 to obtain the percentage of SiO_2.

After weighing, add to the residue in the crucible 5 drops of sulfuric acid (No. 4) and about 10 milliliters of hydrofluoric acid (No. 6). Evaporate under the hood to dryness at a gentle temperature, then increase the temperature slowly until the sulfuric acid fumes are driven off and ignite for about 15 minutes in the muffle at 1000° C or over a blast lamp. Cool in a desiccator and weigh the remaining residue in the crucible.

Subtract the weight of the residue from the original silica weight and add it to the R_2O_3 weight obtained later. Use the same platinum crucible for the ignition of the R_2O_3 precipitate. Multiply the weight of the SiO_2 by 200 to obtain the percentage of SiO_2.

Perchloric acid method:

Weigh exactly 0.5 gram of the sample into a large platinum crucible and mix well with 5 grams of fusion mixture (No. 43). Cover the crucible with a lid and heat over a Bunsen burner or in the cooler part of an electric muffle until the fusion is quiet. Then twirl the crucible gently to mix the fusion and heat for 15 minutes over a blast lamp or in the muffle at 1000° C. Again twirl the crucible and heat for 5 minutes more. Twirl the crucible to coat the fusion on the sides of the crucible and after cooling roll the crucible between the fingers to loosen the fusion. Place the crucible on its side into a 210-milliliter casserole or a 250-milliliter beaker and add about 10 milliliters of hot water and 10 milliliters of perchloric acid (No. 3). Boil until the fusion is dissolved, then remove and wash the clean crucible with as little hot water as possible. Grind the sample with the flattened end of a glass rod to break up larger particles if necessary. Then add 30 milliliters of perchloric acid (No. 3) and 5 to 6 drops of nitric acid (No. 7). Cover with a watch glass and boil until copious fumes of perchloric acid are driven off. Continue heating for 5 minutes, then cool and add 15 milliliters of hydrochloric acid (No. 1). Heat to boiling, then rinse the under side of the watch glass and the sides of the vessel with about 50 milliliters of water, then boil again for 3 minutes. Filter through an 11-centimeter filter paper, Grade B, into a 400-milliliter beaker. Scrub the vessel with a rubber policeman and rinse the silica with a stream of hot water into the filter. Wash the filter once with hydrochloric acid (No. 1) and 8 times with hot water.

Transfer the filter paper into a weighed platinum crucible. Dry and char it with gentle heat. Burn off the carbon and ignite the crucible for at least 3 hours in an electric muffle at 1050 to 1100° C or cover the crucible and heat it for 30 minutes over a strong blast lamp. Cool in a desiccator and weigh as SiO_2. Multiply the weight of the SiO_2 by 200 to obtain the percentage of SiO_2.

Determination of Alumina

Add to the filtrate from the silica determination a few drops of methyl red solution (No. 45) and, while stirring, ammonium hydroxide (No. 15) until the solution is distinctly alkaline. Add hydrochloric acid (No. 1) until the solution is slightly acid, cover with a watch glass and heat to boiling. Then add slowly ammonium hydroxide (No. 15) until the red color just changes to a light yellow. Boil the solution for 1 to 2 minutes or until the odor of ammonia is very faint. Avoid too-long boiling, as this renders the precipitate difficult to filter. (Therefore, do not add too large an excess of ammonium hydroxide so that not more than a few minutes of boiling are necessary to reduce the amount of ammonia to the proper point.) Rinse the under side of the watch glass, and filter at once through an 11-centimeter filter paper, Grade B, into a 400-milliliter beaker. Rinse the beaker once and wash the filter 3 times with hot ammonium nitrate solution (No. 28).

With a fine stream of water, wash the precipitate from the filter back into the beaker in which the precipitation was made without injuring the filter paper. Dissolve the precipitate in about 10 milliliters of hydrochloric acid (No. 1) with gentle heating. Repeat the precipitation with ammonium hydroxide (No. 15), boiling as before. Filter through the same filter paper into the beaker which contains the filtrate from the first precipitation. Scrub the beaker and rinse the remainder of the precipitate with a stream of hot ammonium nitrate solution (No. 28) in the filter. Wash the filter 5 times with the same hot solution.

Transfer the filter paper to a weighed platinum crucible, dry and char with gentle heat. Burn off the carbon and ignite the crucible for 15 minutes in an electric muffle at 1050 to 1100° C or over a strong blast lamp. In order to avoid loss of material during the blasting of the open crucible, mount the crucible through a hole in an asbestos hood. Cool in a desiccator, and weigh quickly. Multiply the weight of the R_2O_3 by 200 to obtain the percentage of R_2O_3. Subtract from the percentage of R_2O_3 the percentage of Fe_2O_3 to obtain the percentage of Al_2O_3.

Determination of Ferric Oxide

Permanganate method:

Weigh exactly 2 grams (or exactly 5 grams, if the sample is very low in Fe_2O_3) of the sample into a platinum crucible and mix thoroughly with 2 grams of C.P. calcium carbonate when the sample contains from 70 to 85 per cent $CaCO_3$. Use 3 grams of calcium carbonate when the sample contains from 60 to 70 per cent $CaCO_3$, use 4 grams of calcium carbonate when the sample contains from 50 to 60 per cent $CaCO_3$, and use 6 grams of calcium carbonate when the sample contains less than 50 per cent $CaCO_3$. (Use 2.5 times the specified amounts of calcium carbonate when a 5-gram sample is used.)

Cover the crucible with a lid and ignite it for 30 to 40 minutes at the highest possible temperature over a strong blast lamp or in an electric muffle at 1100° C. After cooling transfer the sample completely by careful brushing into a 400-milliliter beaker and add slowly about 30 to 50 milliliters of hot water.

Measure out the proper amount of hydrochloric acid (No. 1) to dissolve the sample. Use 25, 38 or 50 milliliters of acid when 5, 10 or 15 grams of material is clinkered. Interpolate the milliliters of acid for intermediate amounts of clinkered material. Too much acid causes indistinct and fading endpoints. Clean the crucible with a portion of the acid. Add this portion and the remainder of the acid to the sample in the beaker, cover with a watch glass and boil until the solution is complete as indicated by the absence of dark-colored particles. Use a glass rod with a flattened end to break up larger particles if necessary.

Add to the boiling solution drop by drop, stannous chloride solution (No. 42) until the color of the solution has changed to colorless, then add 3 drops of stannous chloride solution in excess. Cool the solution by placing the covered beaker into a pan with cold water not allowing the solution to stand exposed to the air any longer than necessary. When cool, add 20 milliliters of mercuric chloride solution (No. 36) and stir until a silky precipitate has formed. Then add 40 milli-

liters of guard solution (No. 34), dilute with water to about
300 milliliters and titrate immediately with standard potassium permanganate solution (No. 21) to the first appearance
of a faint pink color. Multiply the number of milliliters used
by the Fe_2O_3 equivalent of the permanganate solution and
by 50 (by 20, if a 5-gram sample is used) to obtain the percentage of Fe_2O_3.

> NOTE: The excess stannous chloride is necessary to prevent a possible
> reoxidation of the ferrous chloride during cooling. If not enough stannous
> chloride has been used in reducing the iron, no precipitate will form on
> addition of the mercuric chloride. If too much stannous chloride has been
> added, the precipitate will appear heavy and granular instead of silky,
> or may even show gray metallic mercury. In either case repeat the determination. Run a blank determination occasionally and subtract its titration value from titrations of samples.

Dichromate method:

Follow the directions given in the *Permanganate Method*
up to the point where the silky precipitate is formed. Then
add 15 milliliters of phosphoric-sulfuric acid mixture (No.
13) and 4 drops of diphenylamine solution (No. 47), dilute
with water to about 150 milliliters and titrate with standard
potassium dichromate solution (No. 24) until the green color
of the solution changes to a blue-green. Then add with
vigorous stirring the dichromate solution dropwise until the
color changes to an intense violet-blue. Multiply the number
of milliliters used by the Fe_2O_3 equivalent of the dichromate
solution and by 50 (by 20, if a 5-gram sample is used) to
obtain the percentage of Fe_2O_3.

Sodium sulfite method:

Weigh exactly 2.0 grams (or 5.0 grams, if the sample is
very low in Fe_2O_3) of the sample into a platinum dish, add
5 to 8 milliliters of sulfuric acid (No. 5) and 20 to 50 milliliters of hydrofluoric acid (No. 6), depending on the amount
of silica present in the sample, and evaporate slowly to dryness until the sulfuric acid fumes are completely driven off.
Fuse the residue in the dish with 10 to 20 grams of potassium
bisulfate until a clear fusion is obtained. Heat very slowly
at first to avoid spattering. Cool and break up the cake on a
sheet of clean paper. Transfer it to a 1-liter flask and rinse
the platinum dish with hot water in the flask. Dilute with

hot water to about 400 milliliters and boil until the solution is clear.

Add to the boiling solution ammonium hydroxide (No. 16) until the precipitation is complete, avoiding much excess. Boil until the excess ammonia is practically removed and then add 1 to 3 grams of fresh sodium sulfite, depending on the amount of iron present in the sample. Shake the flask until the sodium sulfite is dissolved. While keeping the solution just about a boiling temperature without active boiling, add very slowly sulfuric acid (No. 5), about 1 milliliter at a time at first, with intermittent shaking, until about 50 milliliters have been added. The solution should now be clear and colorless. Place a number of glass beads in the flask and boil the solution vigorously to drive off the excess SO_2. After 20 minutes' boiling, test the issuing steam for SO_2 by inserting a stopper with a bent glass tube dipping into 50 milliliters of water containing 5 milliliters of sulfuric acid (No. 5) and 1 drop of standard potassium permanganate solution (No. 21). If the solution is not decolorized at the end of 1 minute, the SO_2 can be assumed to have been driven off.

Cool the flask quickly under a stream of tap water, dilute the contents with cold water to about 600 milliliters, and titrate immediately with standard potassium permanganate solution (No. 21) to the first appearance of a faint pink color. Make a blank determination in the same manner and subtract the number of milliliters used by the blank from the milliliters used by the sample. Multiply the difference by the Fe_2O_3 equivalent of the permanganate solution and by 50 (by 20, if a 5-gram sample was used) to obtain the percentage of Fe_2O_3.

NOTE: The slow addition of sulfuric acid is very important, because the reduction of iron with SO_2 takes best in a very slightly acid solution, and plenty of time should be given for the reduction before adding a large excess of acid. Use as little excess of sodium sulfite as possible to avoid long boiling in removing the excess SO_2.

Determination of Calcium Oxide

Heat the filtrate from the R_2O_3 precipitation, obtained in the determination of alumina, covered with a watch glass, to boiling. Add while boiling 5 milliliters of ammonium hydroxide (No. 16), and slowly 25 milliliters of ammonium oxalate solution (No. 29) which has been heated to boiling. Continue boiling for about 3 minutes, then remove the beaker from the heat, rinse the under side of the watch glass into the beaker and let the precipitate settle.

Filter through a 12.5-centimeter filter paper, Grade C, which contains a little filter pulp, into a 600-milliliter beaker. Rinse the beaker 5 times thoroughly with hot water to remove all soluble oxalates turning the washings into the filter and allowing to drain between washings. Then wash the filter twice with hot water and remove the filtrate. Place another beaker beneath the funnel and wash the filter 6 times with hot water.

Remove the filter paper from the funnel and drop it with the precipitate into the beaker in which the precipitation was made. Then add about 300 milliliters of hot water and 35 milliliters of sulfuric acid (No. 5) and stir to dissolve all calcium oxalate. Titrate the hot solution immediately with standard potassium permanganate solution (No. 20) running from a 100 milliliter buret, preferably one with zero overflow, until a faint pink color is produced which remains for several minutes. Read the buret, add the buret correction and deduct the filter paper-correction. Multiply the number of milliliters used by the CaO equivalent of the solution and by 200 to obtain the percentage of CaO. Multiply the number of milliliters used by the $CaCO_3$ equivalent and by 200 if the result in terms of $CaCO_3$ is desired.

Determination of magnesia:

Cool the filtrate from the CaO determination, which should not be more than 350 milliliters and pour about 20 milliliters of sodium ammonium phosphate solution (No. 40) through an 11-centimeter filter paper, Grade C, into the solution. Add through the same filter paper about 50 milliliters of ammonium hydroxide (No. 15) in small portions while stirring the solution vigorously. Continue stirring until a precipitate starts to form. Cover the beaker with a watch glass and allow the precipitate to settle in a cool place for at least four hours or over night.

Filter through a 12.5-centimeter filter paper, Grade B, into a 600-milliliter beaker. Scrub the beaker with a rubber policeman, then rinse the remaining precipitate completely into the filter with a stream of ammoniacal ammonium nitrate solution (No. 26). Rinse the beaker twice and wash the filter paper eight times with the same solution.

Transfer the filter paper to a weighed platinum or porcelain crucible and dry carefully with gentle heat. Then increase the temperature to char the paper completely. In placing the paper in a platinum crucible, take care that the paper is loosely folded so as to have free access to air during ignition or else serious injury to the platinum may result. Burn off the carbon and then ignite the crucible for 10 minutes over a Bunsen burner or in an electric muffle at 1000° C. Cool in a desiccator and weigh as $Mg_2P_2O_7$. Multiply the weight of the $Mg_2P_2O_7$ by 72.42 to obtain the percentage of MgO.

Reagents for Analyses of Raw Materials

In the original methods, directions for the preparation of all reagents and for the standardization of solutions are given in detail. Here the composition of the reagents is indicated, and procedures for standardization have been omitted. Any statements concerning the contents of one liter of a solution is on the basis of the chemicals used in preparing the solution, without taking into account reactions that may have taken place among the components of the solution.

1. Hydrochloric acid (20 per cent)
2. Hydrochloric acid (5 per cent)
3. Perchloric acid, concentrated
4. Sulphuric acid, concentrated
5. Sulfuric acid (30 per cent)
6. Hydrofluoric acid (40 per cent)
7. Nitric acid, concentrated
8. Nitric acid (20 per cent)
9. Nitric acid (3 per cent)
10. Oxalic acid (5.5 per cent)
11. Acetic acid, concentrated
12. Phosphoric acid, concentrated
13. Phosphoric-sulfuric acid mixture:
    One liter contains 150 milliliters of No. 12 and 150 milliliters of No. 4.

14. Acid mixture:
   To 484 milliliters of water, add 115 milliliters of No. 4, 100 milliliters of No. 7, and 300 milliliters of No. 1.

15. Ammonium hydroxide, concentrated

16. Ammonium hydroxide (13 per cent NH_3)

17. Sodium hydroxide (30 per cent)

18. Standard nitric acid:
   One liter contains 22.5 milliliters of No. 7.

19. Standard sodium hydroxide:
   One liter contains 48 milliliters of No. 17.

20. Standard potassium permanganate:
   One liter contains 105 milliliters of No. 38.

21. Standard potassium permanganate:
   One liter contains 500 milliliters of No. 20.

22. Standard sodium arsenite:
   One liter contains 0.9008 gram of arsenious oxide, dissolved in dilute sodium carbonate solution.

23. Standard sodium arsenite:
   One liter contains 65 milliliters of No. 41.

24. Standard potassium dichromate:
   One liter contains 3.071 grams of the solute.

25. Standard iodine:
   One liter contains 2 grams of iodine and 5 grams of potassium iodide.

26. Ammoniacal ammonium nitrate:
   One liter contains 109 milliliters of No. 15, and 27 milliliters of No. 7.

27. Ammonium acetate:
   One liter contains 320 milliliters of No. 11, neutralized with No. 15; 20 grams of citric acid, then No. 15 to volume.

28. Ammonium nitrate:
   One liter contains 20 milliliters of No. 7, neutralized (plus slight excess) with No. 16.

29. Ammonium oxalate (4 per cent)

30. Barium chloride (10 per cent):
   Dissolve 200 grams of C.P. barium chloride ($BaCl_2 \cdot 2H_2O$) in about 500 milliliters of warm water. Filter and dilute with water to 2 liters. This solution contains approximately 10 per cent $BaCl_2$.

31. Bromine water

32. Cadmium chloride:
   One liter contains 4 grams, dissolved in 100 milliliters of No. 15.

33. Ferric sulfate:
      One liter contains 100 grams of ferric sulfate, 50 milli-
        liters of sulfuric acid, and 100 milliliters of phos-
        phoric acid.

34. Guard solution:
      One liter contains 224 milliliters of No. 35, 128 milli-
        liters of No. 4, 140 milliliters of No. 12, and a
        sufficient quantity of No. 20 to produce a faint
        permanent pink color.

35. Manganous sulfate (30 per cent)

36. Mercuric chloride (10 per cent)

37. Potassium nitrite (5 per cent)

38. Potassium permanganate (3.6 per cent)

39. Silver nitrate (0.2 per cent)

40. Sodium ammonium phosphate (14 per cent)

41. Sodium arsenite (95 per cent in dilute sodium carbonate):
      One liter contains 5 grams of arsenious oxide and 15
        grams of sodium carbonate.

42. Stannous chloride:
      One liter contains 100 grams of stannous chloride and
        100 milliliters of hydrochloric acid.

43. Fusion mixture:
      Sodium carbonate plus 2 per cent of potassium nitrate.

44. Methyl orange (0.1 per cent):
      Dissolve 1.0 gram of methyl orange in about 300
        milliliters of hot water and allow to stand, stirring
        from time to time until completely dissolved. Filter
        if necessary and dilute with water to 1 liter.

45. Methyl red (0.2 per cent in alcohol)

46. Phenolphthalein (0.6 per cent in 50 per cent alcohol)

47. Diphenylamine (1 gram dissolved in 100 milliliters of
      No. 4).

Analysis of Cements

The methods for the chemical analysis of portland cements
selected for this book are those of the American Society for
Testing and Materials; Designation: C 114–63. Issued in the
same booklet is Information on analytical balances and
weights.

Limestone, clay, and raw mix have the same constituents

as cements, and in some cases the same chemical methods may be used. In other cases, however, cement methods are not applicable. In fact, it is often necessary to study the results obtained by various methods before selecting the methods that are most suitable for a given set of raw materials. Much depends on whether or not all the calcium is present in compounds soluble in acid. Because of these facts, some methods for raw materials and raw mixes have been included.

The American Society for Testing and Materials is greatly important in the cement industry, not only in establishing working standards, but also in tending to guide some developments. To make this clear, and for convenience, the first edition of the book contains a rather long series of ASTM specifications. It was planned to include several in this second edition, however, since the Society no longer encourages the reprinting of standards, the number has been greatly reduced, and reference numbers have been substituted. Partial adjustment is provided in that all standards are now published annually, rather than triennially as formerly.

In the last decade or two there have been extensive modifications in laboratory test procedures. These modifications have been brought about by many factors, including desire for accuracy and additional information.* Two disadvantages of the transition should be pointed out. Long records of former test results become less valuable, and an operator must wait longer to be assured that a product meets specifications.

In some manufactured products, the quality may be determined largely by careful chemical analyses, with possibly a few physical tests. The development of the cement industry has not reached this point, and, because of the complexity of this situation, may never reach it.

Other methods of testing are summarized at the end of Specification C 150–64, in the preceding chapter.

---

* For additional information I prefer to continue testing for soundness over boiling water (C 189–44) and for sieve fineness (C 184–44). C 189–44 is no longer listed in the current book of standards.

REFERENCES

1. J. C. Witt, *Ind. Eng. Chem.*, 12:598 (1920).
2. W. F. Hillebrand, *U.S. Geol. Survey, Bull.* 700; Washington, D.C. (1919).
3. *Standard Chemical Methods for Cement Plants;* Universal Atlas Cement Co., Chicago, 1937. An unpublished manuscript. Reproduced here by permission.
4. J. C. Witt, *Philippine J. Sci.*, 15(1):107 (1919).

Chapter 15

UNIFORMITY

My recognition of the prime importance of uniformity developed only after years of experience in cement manufacture. For a long time uniformity was simply one of the factors that eventually should receive consideration in any project.

This recognition resulted from the solution of a number of problems that came my way, and the correlation of the information so obtained. These incidents led to a study of the characteristics of uniformity itself, and as a result it became the first factor to receive attention in any investigation. Stated in another way, uniformity has become the starting point for the development of processes—and is no longer to be considered incidental.

The development of the cement industry has brought about ever-increasing sizes of mills, kilns, and other production units; and, consequently, corresponding decreases in the number of each type of unit for a given output. These changes have been desirable in that they increased output and decreased production costs. Unavoidably, however, they have tended to decrease the uniformity of intermediates and final product.

Several of the projects in which uniformity was found to be paramount will be mentioned here.

1. At a one-kiln, dry-process plant there was no dust precipitator, and no mechanical equipment for returning to the process the dust that accumulated at the base of the kiln stack. It so happened that it was necessary to shut down the kiln at rather frequent intervals because of the poor quality

of the lining. During the shut-down periods, the dust was removed from the stack and stored. When the plant was again in operation the dust was dumped into the screw conveyor from the raw mill. In this way, the accumulation of dust was returned to the system in a day or two.

The uniformity of the clinker was never excellent, but from time to time it was very undesirable. At first the cause of such variation was not known, but eventually it was traced to the return dust. Analyses revealed that the average composition of the dust, on the ignited basis, differed greatly from the average composition of the raw mix.

Corrective measures did not suggest themselves at first. Economically, the dust could not be rejected, and there was no favorable location for disposal. The opening into the screw conveyor was closed, and a door was installed in the housing of the elevator transporting the raw mix to the kiln feed silo. The men were then directed to shovel the dust into the elevator. In this way, the dust entered the raw mix stream so gradually that the composition was not greatly affected. The small increase in labor cost was well justified.

2. At a multi-kiln, dry-process plant there were dust collectors, and screw conveyors to return the collected dust to the system. At intervals, the variations in the composition of the clinker, with no changes in raw materials, plant equipment, nor general operating procedure, greatly concerned the superintendent.

In this consultation assignment, no solution of the problem was at once apparent. Finally it was learned that at times a shut-down of the dust conveyors was required because of the high temperature of the dust. The dust then accumulated for several days, was returned to the system rapidly, and kiln feed of irregular composition resulted. Previous experience suggested the cause, and the solution of the problem followed. After the next shut-down, the dust was returned to the system much less rapidly. Later, improved conveyors were installed and the difficulty disappeared.

3. In another problem, the difficulty lay with the analysis and the setting time of the cement. For the product to meet the specifications of the day, the sulfur trioxide could not be

too high, and the initial set could not be too fast. The sulfur trioxide range was very short. For months shipment after shipment was rejected on the basis of set, and it seemed that the two requirements were simply incompatible. Considerable variations in the sulfur trioxide were observed. To maintain values always below the specified percentage, it was necessary to allow for these variations—which made the situation still more difficult.

At the time, the procedure was to proportion the clinker and gypsum by volume. Neither the clinker nor the gypsum was analyzed regularly in advance. It was found desirable to change the whole procedure for proportioning the two materials. Each was analyzed in advance, and proportioning by weight was adopted. Complaints and rejections ceased, and the problem was solved. There was a small increase in production cost, but the situation certainly justified this.

4. At a large dry-process plant, wide variations in clinker composition had been observed for years. Raw materials, equipment, and laboratory control had been checked repeatedly but offered no suggestions for improvement. There were three eight-hour laboratory shifts. Eventually it was observed that the greatest variations occurred at night. It was decided to follow through the scale changes resulting from application of the analytical data by the night men. The procedure was found to be faulty. When this was corrected, the uniformity of the clinker was greatly improved— with no increase in production cost.

It has been explained that at the beginning of the cement industry information on related science and engineering was limited. Consequently the development of the industry has been much less rapid than it might otherwise have been. If the concept of cement should present itself for the first time today, however, how many years would be required for the development of the product as we now know it? Even with today's science and engineering information available, the assembling of the many steps in technology required to produce a material of construction with all its demanded qualifications would constitute a major accomplishment.

Three principles have resulted from my various encounters with uniformity:

1. In a program for a research project, the objective of which is to improve quality, increase output, or decrease the production costs of an existing product, the first step is to investigate the uniformity of raw materials, intermediates and finished product. If in any case the uniformity is not satisfactory, procedures for improvement should be studied carefully. If good uniformity seems unlikely, a satisfactory outcome of the project cannot be expected.

2. If the project involves the development of a new product, essentially the same procedure is to be followed, but more difficult problems probably will arise.

3. Assume that there is a series of research reports to be considered for commercial application at a given plant. The reports may be divided into three groups: First, those that require only existing uniformity; second, existing uniformity is not sufficient but increased uniformity seems feasible; third, reports that belong neither to the first nor to the second group. Such reports may have great intrinsic value but no commercial application can be expected soon.

Although every effort should be made to increase uniformity, absolute uniformity is not to be expected. For example in Figures 13:1 and 13:2, considerable variation of composition in length of 25 microns, or approximately one ten-thousandth-inch may be observed.

It is evident that the demands of uniformity are many and difficult. To make acceptable products, the cement manufacturer and the concrete producer must exert care at every step. But the benefits of uniformity are many and desirable, including lower costs, and higher quality.

The concrete producer can expect to receive cement with quality well above specifications. But this extra quality is in no sense guaranteed, and it may decrease at any time. To be safe, he should always base his design calculations on the specified values.

Chapter 16

THE NONAQUEOUS LIQUID PROCESS

The first portion of this chapter is from one of my papers.[1]

HEAT INPUT, THEORETICAL HEAT, THERMAL EFFICIENCY, ADDITIONAL HEAT

Although this is primarily a practical paper, a few words on fundamental theory are considered desirable. All values are typical examples, on the basis of experience, observations and calculations. In no sense are they to be considered averages. Although the statements are in relation to cement manufacture, many are applicable to manufacture in general.

Let us consider the well-known equation

$$\frac{\text{Output}}{\text{Input}} = \text{thermal efficiency}$$

For heat calculations, this may be written

$$\frac{\text{Theoretical heat}}{\text{Heat input}} = \text{thermal efficiency}$$

or transposing

$$\text{Heat input} = \frac{\text{theoretical heat}}{\text{thermal efficiency}}$$

Heat input is one of the principal factors in the cost of the production of cement. In the United States, it is generally expressed in terms of Btu per barrel of clinker. In comparing the cost of the fuels available in a given district, it is custo-

213

mary to calculate the pounds of coal, the gallons of oil, and the cubic feet of gas per barrel of product.

Obviously, heat input is vitally important to the cement manufacturer. It is important also to fuel producers, plant-equipment manufacturers, cement users, and to many other groups, the incomes of which are affected directly or indirectly by the cement industry.

Heat input may be expressed by another equation:

Heat input = theoretical heat + additional heat

Here, additional heat is the heat in excess of the theoretical heat required to produce one barrel (376 pounds) of clinker of a specified composition, from an equivalent quantity and composition of dry raw mix.

For the raw mix and clinker as stated, the theoretical heat, as calculated thermodynamically, is a constant, not subject to any modification whatever. If clinker of another composition is specified, adjustments in the quantity and composition of the raw mix must be made. If a catalyst or flux is added to the raw mix to facilitate the synthesis of the clinker minerals, the catalyst or flux becomes a component of the raw mix, and corresponding adjustments must be made. The theoretical heat is then a constant for that raw mix and the resulting clinker.

The thermal efficiency is always less than 1. Therefore, the heat input is always greater than the theoretical heat, and there is always some additional heat. This additional heat results from the combined effect of all factors influencing clinker formation, in excess of the theoretical heat. The effect of each factor is positive. None can be negative. For convenience in discussion, the term *additional heat*, rather than thermal efficiency, will be employed in general.

In Figure 16:1, the rectangles represent quantities of theoretical or additional heat. They are not drawn to scale but they represent roughly comparable heat quantities along the x-axis. It will be noted that the y-axis length remains constant. Let us assume that each of the five rectangles A to E represents the theoretical heat required for the production of a barrel of clinker, and that all the clinker produced has the same composition.

A represents theoretical heat in general. *B* to *E* show the theoretical heat for each of four types of kilns to be discussed. Each of these five rectangles represents roughly 500,000 B.t.u.

The Vertical Kiln

When portland cement was patented in 1824, there was only one type of kiln—the vertical one. In comparison with modern kilns, our information on the vertical kiln is very limited. Fundamentally the quantity of additional heat required for this type of kiln was small. In the beginning, however, there were numerous factors which brought about relatively high additional heat. For example, the raw mix was not so finely ground as it is today, and its composition was more irregular. It was necessary to sort the clinker and reject a portion of it. As experience in the industry increased, however, there was gradual improvement in technique and a lower additional heat resulted.

Rectangle *F* indicates the rather small quantity of additional heat for the vertical kiln, in comparison with rotary kilns.

The Dry-process Rotary Kiln

The invention of the rotary kiln about 1866, together with the development of the technique of pulverized coal, was one of the great advances in the cement industry, because it brought about so many improvements. Clinker production became a continuous operation, labor requirements were decreased, and the percentage of rejected clinker became much lower—when it was necessary to reject any at all. Although at first there was considerable opposition to the rotary kiln, as is to be expected when any really new equipment or process is introduced, its merit gradually won recognition, and the vertical kiln gradually disappeared. When we consider the billions of barrels of clinker that have been produced and the great industrial importance of concrete, the rotary kiln may be considered one of the world's great inventions.

The success of the rotary kiln has been accomplished by its many advantages, notwithstanding one important handi-

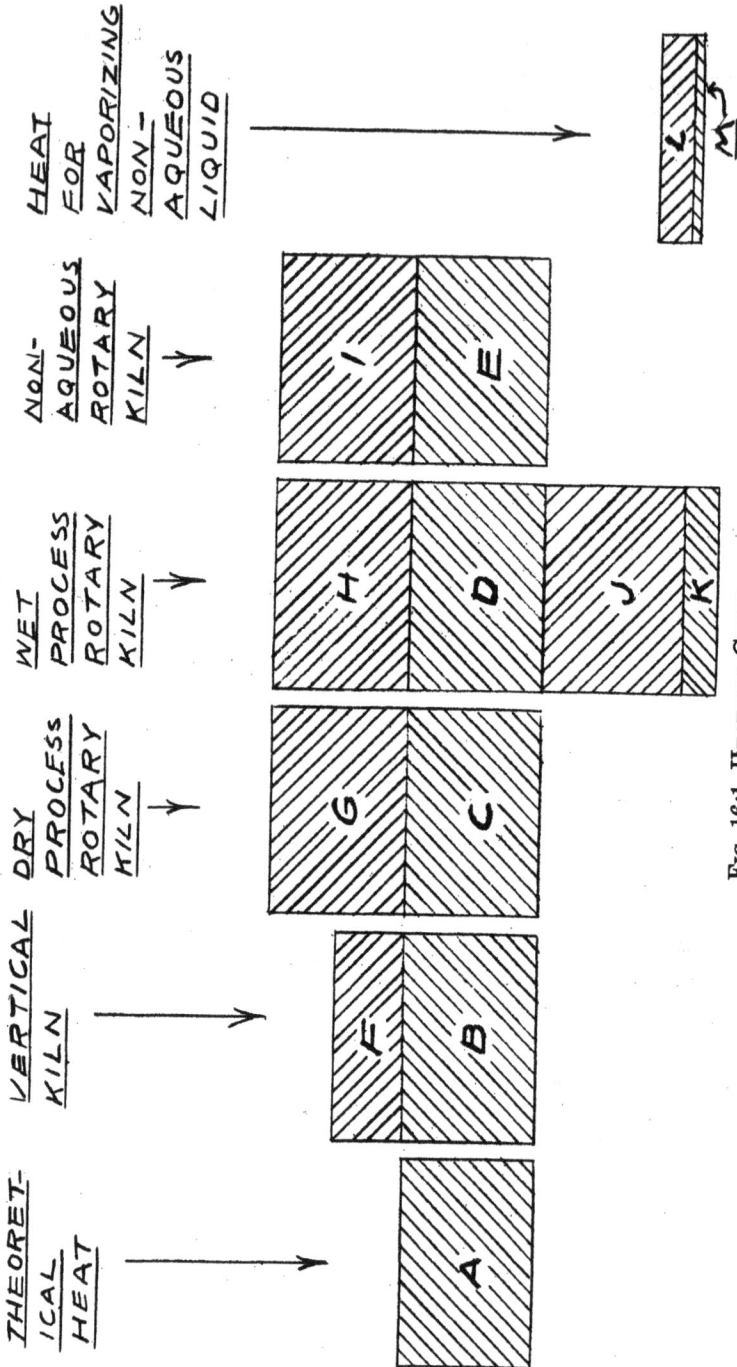

FIG. 16:1. HEAT FOR CLINKERING

cap. Soon after the kiln was commercialized, it became evident that the heat input was much higher than that of the vertical kiln. As the theoretical heat was not affected, the condition was due entirely to an increase in the additional heat, represented by rectangle G. At first, the additional heat probably greatly exceeded the theoretical in many cases. Even today, kilns are in operation for which the additional heat approximates the theoretical, plus or minus a few hundred B.t.u. In the nearly-a-century that has elapsed, this additional heat has required millions of tons of coal, or the B.t.u. equivalent of other fuels.

The high fuel consumption is caused by such factors as heat transfer and radiation, which are fundamentally related to the characteristics of the kiln itself. Paradoxically, however, this handicap of high fuel-consumption has had one important advantage. An analysis of the situation reveals that almost every important advance in kiln design and operation, and in the development of accessory equipment, has been brought about directly or indirectly, or at least encouraged by, a desire for fuel conservation. Here are some examples:

Increase in the size of kilns, kiln insulation, preheating kiln feed, preheating combustion air, recovery of radiated heat, waste heat boilers. Also such developments as increased fineness and uniformity of raw mix, and increased fineness of pulverized coal, have been activated in part, at least, by fuel conservation.

The Wet-Process Rotary Kiln

Just as the heat input of the dry-process rotary kiln is double the theoretical heat for clinkering dry raw materials, so the heat input of the wet-process rotary kiln is triple the theoretical heat for clinkering dry raw materials. And just as the dry-process rotary won its place in the industry, notwithstanding a large heat handicap, so has the wet-process rotary, with a much larger heat handicap, been installed in most cement plants that have been constructed in the United States in the last few decades. Here again the many advantages have prevailed.

The dry-process kiln receives 600 lb of solids for the production of one barrel of clinker. The wet-process kiln receives the same quantity of solids plus 400 lb of water—the suspension of solids in water being called *slurry*. The theoretical heat for vaporizing this water (plus the heat for superheating the steam to the exit temperature of the kiln gases) is approximately 500,000 B.t.u. Now because the thermal efficiency of this operation is not perfect, some additional heat is required.

The heat input for the wet-process kiln is illustrated in Figure 16:1. D is the theoretical heat of clinker formation from dry raw materials, the value being equal to A, B, C, and E.

J is the theoretical heat for vaporizing the water. $D + J$ is the theoretical heat of clinker formation from slurry. H (corresponding to G) is the additional heat of clinker formation from dry raw materials. K is the additional heat for vaporizing the water. $H + K$ is the total additional heat for clinker formation from slurry. Therefore, the total heat input for the wet rotary is $H + D + J + K$.

Most of the comments on G apply to H also. As the popularity of the wet process has increased in the United States, however, the use of waste-heat boilers has tended to decrease. In a dry-process plant, all the power for operation may be generated by waste-heat boilers. In a wet-process plant, usually the total power requirement cannot be so generated, but stand-by electric energy must be purchased, or auxiliary prime movers must be installed. It is not claimed that the decrease in waste-heat-boiler installation is entirely due to the wet process, but this certainly is an important factor.

Research Preceding Development of Nonaqueous Process

Preceding the development of the nonaqueous process for manufacture of portland cement there were years of research on cement manufacture, carried on by my associates and myself. This involved physical, chemical, and petrographic research in the laboratories, experimental engineering, and pilot-plant operation. Most of this research was related directly or indirectly to fuel conservation. Numerous reports were prepared but few papers were published.

In formulating industrial research programs, it has been found desirable to prepare outlines, taking into account what has been accomplished in a given field and showing what may be called the probability of obtaining positive results along various lines of attack. Such an outline on kiln heat input is shown in Table 16:1 with reference to Figure 16:1. It serves to correlate many of the statements made previously. As has been stated, for raw mix and clinker of given composition the theoretical heat, A to E, is a constant.

Theoretical heat J can be diminished only by reducing the water employed in grinding the raw materials, or by removing a portion of the water from the slurry. By dissolving in the slurry water one of a group of substances called in the industry fluidizers, the quantity of water may be reduced slightly. Of course, any substance dissolved in water reduces the vapor pressure, and consequently raises the boiling point. Also any substance undesirable in clinker is to be avoided.

About 1925, the first filters for removing a portion of the water from slurry were put into operation. For some years, the increase in the number of installations was rapid, but at present, filters are not generally employed in wet-process plants. Several operating difficulties appeared. The filter cloth is subject to deterioration. The filter cake is sticky and often is difficult to handle. Collected dust cannot be added to water slurry because it facilitates the setting of slurry solids. Because of colloidal conditions, some water slurries cannot be filtered.

Additional heat G, H, and I has presented the greatest opportunities for fuel conservation. It seems that almost every conceivable lead has been investigated by someone, and much of the technical development of the industry has come about from such investigations. New leads do not appear often, but new methods of attack appear from time to time. The most important developments have been in connection with G. Most of them are applicable to H, and all of them are applicable to I. Table 16:1 shows only a few of the many examples that could be listed.

The procedures for decreasing additional heat G, H and I are divided into two groups—direct and indirect. In the direct group, the heat input is reduced. In the second group, the heat entering the kiln is not affected, but some of the heat is recovered and the total heat input is credited with this quantity of heat, thus reducing the net heat input. Additional heat K has some factors in common with G, H, and I. Such an outline as Table 16:1 is helpful in studying and evaluating a recent equipment unit or process as it is announced. It may be something really new, or it may be another application of a principle that was applied years ago in the industry.

TABLE 16:1

HEAT INPUT OF CEMENT KILNS

Classification	Location *	Procedures for decreasing the quantity of heat
Theoretical	A, B, C, D, E	Change in the composition of raw materials, or clinker, or both.
Theoretical	J	Reduction in the quantity of water entering the kiln. Examples: Filters, fluidizers.
Additional	F, G, H, I	Direct: More finely ground raw materials and coal, kiln design, kiln insulation, instrumentation. Indirect: Recovery and application of waste heat.
Additional	K	In general, the same procedures as for G, H, I.

* This refers to Figure 16:1.

Development of Nonaqueous Process

Both the dry and wet processes have numerous advantages and disadvantages. The principal advantage of the dry process is the low heat input (in comparison with the wet process), and the principal advantage of the wet process is the composition uniformity of the kiln feed. Incidentally, the effect of this uniformity is not restricted to improvement in the uniformity of the finished product. It tends to decrease the costs of raw materials and fuel and has other advantages that could be listed.

After all the years the two processes have been in operation, and all the research to increase the uniformity of the dry process, and to decrease the heat input of the wet process, it has become evident that fundamental difficulties limit the accomplishments that can be expected.

Some years ago I started an investigation, the object of which was to find a procedure for combining the advantages of the wet and the dry processes. Study was based on such items as raw materials, fuels, plant-equipment units, and conventional manufacturing processes.

After several months, the investigation was almost abandoned—but not completely so. There was just no lead in sight. During this preliminary period, the utilization of water was taken for granted, and no particular attention was given to the nature of the liquid. Later, however, it was decided to make a critical study of water itself. Soon a fundamentally important and interesting fact was revealed. The advantages of the wet process are due to the presence of a mobile liquid— water. This facilitates grinding, blending, pumping, and storing. Each of the disadvantages of the wet process can be traced to one or more of the physical and chemical properties of the liquid.

This concept provided a firm basis for the investigation. Evidently the next step was to search for some liquid other than water, in the presence of which raw materials could be ground. A specification for such a liquid was formulated as follows:

1. To remain liquid over a long range of temperature.
2. No chemical reactions with water or with the compounds present in cement raw materials, even at rather high temperatures.
3. No oxidation or reduction.
4. No deliquescence.
5. Boiling point approximately 300–600 F.
6. Low latent heat of vaporization.
7. Reasonable cost and availability.

A review of the properties of many nonaqueous liquids revealed that each of them could be eliminated at once because of failure to meet one or more clauses of the specification.

As the search for a suitable liquid continued, it began to appear unlikely that any liquid on the market would meet the specification. Probably this would mean a long costly research program with no assurance of success, and consequent delays in the development of the process. It was pleasing therefore to find a group of petroleum distillates each of which meets the specification as satisfactorily as if it had been synthesized for this one use. Of course the way is clear for research in the synthesis of a still more satisfactory liquid or for the blending of existing liquids. The liquid most used so far is known as Fuel Oil Number One.

The first patent on the nonaqueous process was U.S. 2 611 714 issued in September, 1952. A second patent, U.S. 2 801 932, was issued in August, 1957 and the work under way probably will result in applications for additional patents.

Some Advantages of the Nonaqueous Process

Until a suitable liquid had been located, the process was just an idea. Following that event, the development has been rapid. Sinclair studied the petroleum technology, and several other organizations have cooperated in the development of the process by consultations, calculations, test runs, and the like.

The preparation of nonaqueous slurry and aqueous slurry is practically identical in all respects—except the nature of the liquid. The raw materials, the proportion of solids and liquid, and the grinding equipment are all the same. But the characteristics of the two suspensions differ almost unbelievably. So far, all these differences in characteristics have meant advantages for the process, and there are indications that additional advantages will appear. Some disadvantages may appear, but the backlog of fuel conservation in this process is so great that any disadvantages probably will have little effect on the over-all economy of the process.

Nonaqueous slurry does not freeze, does not require continuous stirring, is not affected by the addition of collected dusts. One-fourth of the liquid may be removed by decantation, and one-half by filtration, leaving only one-fourth to be vaporized.

Filtration is optional, but is desirable economically. The undesirable features mentioned with reference to the filtration of water slurry just disappear. Nonaqueous slurry filtration requires a smaller filter area, proceeds more rapidly; and approximately half as much liquid remains in the filter cake. After the removal of the liquid, the solids enter the rotary as preheated raw mix, resulting in increased clinker production and decreased heat input. The kiln is actually a dry-process rotary kiln with all the advantages of such a unit. As none of the heat carried by the combustion gases has been removed for preheating the raw mix, waste-heat boilers may be installed advantageously. The raw mix flows more readily than ordinary raw mix because it is lubricated by a few tenths of one per cent of the nonaqueous liquid.

Referring to Figure 16:1 and Table 16:1, the nonaqueous process eliminates J and K. The heat requirements $I + E$ are identical with $G + C$ of the dry rotary kiln.

Before the slurry solids reach the kiln they are subjected to theoretical heat L and additional heat M for removal of the remainder of the slurry liquid. The quantity of L is in round numbers one-fifth of the quantity of J, which means a saving of 400,000 B.t.u. per barrel of clinker. Some of heat L is carried by the raw mix entering the kiln, and some is released when the vaporized liquid is condensed. The sum can amount to a large portion of L, but this has not been taken into account in the saving of 400,000 B.t.u.

Heats A, B, C, D, E, J, L may be calculated thermodynamically. F need not be taken into account. G, H, I have been calculated in connection with many kiln and boiler tests. The value of K is small, and it may be assumed within rather narrow limits on the basis of boiler tests. This accounts for all but one. Arrangements are under way for determining M, which may be expected to be small.

No one with whom I have discussed the process has challenged the fundamental principles. One or two engineers have pointed out that some of the heat content of combustion gases consumed in the evaporation of slurry water would otherwise be lost, and the wet rotary should not be debited with all of it. This may be true but to demonstrate would be difficult. Such a statement applies to K rather than to J. In calculating fuel economy I have compared J and L. Probably the comparison of $J + K$ with $L + M$ would indicate still greater economy.

So far in explaining the nonaqueous process, it is assumed that the raw materials are dry at the time the liquid is added. There are several procedures by which this drying may be accomplished. For example, a small fraction of the heat in the kiln gases may be utilized. Calculations indicate that the heat required for drying is not relatively important.

Recent tests indicate that it is not necessary to dry all the raw materials in advance. A mixture of water and nonaqueous liquid may be removed by vaporization. The presence of water lowers the boiling point of the nonaqueous liquid. Mixtures of the two types of liquids in any proportion separate quickly.

Technology has many examples of the use of nonaqueous liquids. When a liquid is needed for any purpose, water first suggests itself because of its cheapness and availability. In some cases it is dismissed at once because of one or more of its characteristics. In other cases, water can be used but some other liquid has been proved to be much more satisfactory, as in the manufacture of cement. Probably other processes will be developed. A book on nonaqueous liquids has been published.[2]

The following is taken from another paper.[3]

Nonaqueous slurry does not freeze and it does not require stirring. The liquid may be separated in a variety of ways. Most of it can be removed mechanically, the remainder being either vaporized, or burned with heat recovery. The saving in heat input is considerable. The nonaqueous process combines the uniformity of product characteristic of the wet process with the fuel economy of the dry process.

Preparation of Cement from Clinker

The nonaqueous process has two divisions: (1) the preparation of kiln feed, and (2) the preparation of cement from clinker.

Cement may be prepared from clinker by either the dry or the nonaqueous procedure. In the dry procedure, clinker discharged from the kiln is delivered to the consumer with no additional processing except grinding, with the addition

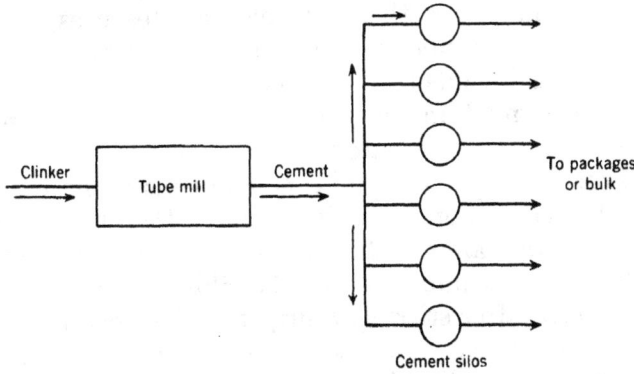

FIG. 16:2. CLINKER TO CEMENT
Dry Grinding

of a small percentage of retarder. There is no provision for adjustment or correction. As indicated in Figure 16:2, cement discharged from the tube mills is stored in a series of silos, from which it is shipped in sacks or in bulk. Not much blending is possible, and such blending as results is fortuitous rather than systematic.

Clinker cannot be ground in the presence of water. Dry grinding is accompanied by the annoyance of coatings that adhere to the surface of the grinding media; the problem of disposing of the "fringe" cement produced when there is a change from one type of cement to another; and the formation of small lumps of cement in the silos.

Grinding cement from clinker by the nonaqueous process

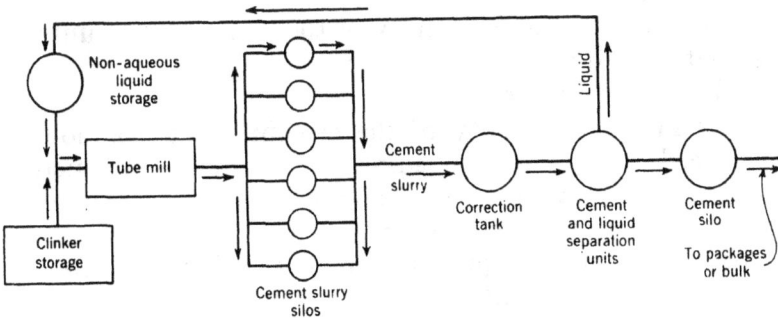

FIG. 16:3. CLINKER TO CEMENT
Nonaqueous Liquid Process

(Fig. 16:3) eliminates these troublesome features, and has the advantage, previously mentioned, that large quantities of finely ground solids can be blended easily in a liquid. The operation has much in common with the wet process for the preparation of the raw material fed to kilns.

In Figure 16:3, clinker and a nonaqueous liquid enter the tube mill. The equipment units between the clinker storage and the tube mill are not shown. Cement slurry is discharged from the mill and pumped into the silos. The slurry in any silo is uniform after stirring. Slurry may be drawn from any silo, in any quantity, and transferred to the correction tank. After mixing and checking, the slurry passes through the liquid recovery units. Here the liquid and solid phases are separated. The liquid is returned to storage, and the cement is conveyed to the silo for finished cement, from which it is shipped to the consumer in packages or in bulk.

Uniformity and Versatility

Several decades ago the multiple type of cement manufacture started to displace the single type. Today five basic types are recognized by ASTM specifications, and there is no reason to believe that the number of types will not increase. When it is realized that all five types of cement contain the same four oxides and the same four synthetic minerals, the importance of uniformity for each type cannot be overemphasized. Without some degree of uniformity, no one type can be maintained separate from the others in chemical and physical properties.

Regardless of the uniformity of the clinker, more uniform cement can be produced by the nonaqueous procedure than by the dry procedure.

The improved quality of the cements that can now be produced permits corresponding improvements in concrete. The designer now has a wider range of cementitious materials to choose from than ever before.

Although the nonaqueous process was developed for the production of portland cement in general, it now appears to be particularly adapted to special cements, such as white, high-alumina, oil-well, blended, and masonry—with all of which I have had experience. The advantage of the new method is especially marked when one raw material is much more expensive, difficult to obtain, or relatively less in quantity than the others.

From time to time cement users have requested some modification of existing cements by grinding with them small percentages of various types of admixtures, such as pigments, for example. In many cases such requests have been considered economically impracticable. The new process changes this situation materially.

Although the foundation has been laid for improvements in the manufacture of portland cement as here described, no one can say what the eventual development will be. Great advances in concrete technology have been made in the past few decades. The new processes for the manufacture of cement should aid in making even greater strides forward in the decades to come.

It should be pointed out that although Figure 16:3 shows one arrangement for the application of the nonaqueous process to the preparation of cement from clinker, numerous others could be chosen. It seems likely that many modifications will be developed.

Recently it has been reported that clinker was ground in the presence of water, and the cement slurry thus produced was transferred directly to a concrete mixer. This procedure should improve economy, uniformity, and the quality of the product.

The development of the nonaqueous process did not end with the research by Sinclair. The work has been continued by individuals and organizations. Each step of the process has been checked by pilot-scale tests. The accumulation of information is too large to present here.

In Chapter 15 it was stated that the study of uniformity has led to its being recognized as one of the prime factors of cement production, and that it may well serve as a starting point for research programs.

The nonaqueous process is an example. The objective was to find a procedure for obtaining uniformity as satisfactory as that of the wet process, with a fuel consumption comparable to that of the dry process. The appearance of other advantages was unexpected.

An interesting feature of the nonaqueous process is that it presents so many advantages over the wet process which has been in operation for so many years. With long operating experience, it seems likely that resulting information will advance it still further.

REFERENCES

1. J. C. Witt, paper presented at the Diamond Jubilee Semi-Annual Meeting of the American Society of Mechanical Engineers, 1955; Editorial Staff Report, *Chemical Engineering*, 207 (Nov. 1955).
2. L. F. Audrieth and J. Kleinberg, *Nonaqueous Solvents;* John Wiley and Sons, New York, 1953.
3. J. C. Witt, Cement and concrete of the future, *Civil Engineering*, 26:50 (1956).

Chapter 17

THE COUNTER-CYCLONE CLINKERER

In preceding chapters it has been explained that when portland cement was invented the vertical kiln was the only clinker-producing unit available, that the development of the rotary kiln was an important advancement in portland cement technology, but that as experience with a rotary kiln and quality demands of the product increased, many disadvantages of the unit became apparent.

Stated simply, the kiln receives raw materials, fuel, and air, and discharges clinker but the situation is far from simple. In Chapter 7 eight functions of the kiln are listed and further study might result in additions to the list. Other factors remaining equal, the difficulty of designing a manufacturing unit varies directly with the number of its functions. Multiplicity of functions tends to require compromises in design and operation, and may be considered one of the causes of undesirable features, several of which will be mentioned here. The next two paragraphs are from one of my papers.[1]

First, the exceptionally low thermal efficiency, making it one of the most wasteful of fuel-burning mechanisms. Then, as the dimensions of the kiln and the complexity of the kiln system increased, in the interests of decreased fuel consumption and increased clinker output, other undesirable items appeared—such as the time required for clinker formation, inflexibility, and capital investment. The increase in the number of types of cements on the market has made inflexibility particularly troublesome.

Some years ago, plant operation, experimental engineering, and research convinced me that although improvements of the rotary kiln might further decrease its disadvantages, an entirely new unit was required for the continued rapid development of the industry. Until rather recently no other expression of such opinion came to my attention, but within the last year there have been several independent opinions.

My first patent for *Apparatus for Producing Cement, Clinkers and the Like* (U.S. 2 489 211), was issued in November, 1949. Then a well-known research foundation was selected for further development of the *Counter-cyclone Clinkerer*, under my direction. The consultation and pilot plant work extended over a period of two years. Many reports were issued, all of which indicated the soundness of the new type of clinker-producing unit. The specifications and claims of my second patent. U.S. 2 634 116, issued in April, 1953, are as follows:

U.S. PATENT 2 634 116

METHOD OF AND APPARATUS FOR CLINKERING CEMENT RAW MATERIALS AND THE LIKE

(April 7, 1953)

This invention relates to the production of hydraulic cement, and has for its principal object the provision of an improved method of and apparatus for clinkering finely divided Portland cement-making materials.

The invention has for a further object the overcoming of many of the deficiencies of the methods and apparatus heretofore proposed and/or employed, whereby to provide a commercially practical mode of continuously clinkering cement raw mix and similar materials in stationary shafts or kilns of relatively short length as compared to that of the rotary kilns now most commonly used for the purpose, with material savings in the costs of manufacture, installation, maintenance and operation, and by which a high percentage of substantially uniform completely clinkered product may be obtained by a single passage of the materials through the apparatus. The clinkering chamber may be a stationary vertical or inclined shaft, or even a stationary horizontal kiln, of appropriate cross sectional area and shape, and a length not exceeding 100 feet.

In its broadest aspect the invention comprises the establishment and maintenance of a relatively high velocity whirling or cyclonically moving mass of burning fuel and hot gases in each of a multiplicity of longitudinally contiguous zones throughout the length of an elongated chamber, the burning masses in adjacent zones preferably whirling in opposite directions about the chamber axis and intermingling to produce a high degree of turbulence and a more or less uniform clinkering temperature in all parts of the chamber; the continual movement of the components of the burning masses longitudinally of the chamber to an exit port at one end thereof; the continuous feeding of the finely divided raw materials at relatively high velocity into the other end of the chamber for immediate high velocity contact of the individual raw mix particles with the burning fuel and their movement serially through the several zones of the chamber to the exit port, said particles being continuously subjected to said substantially uniform clinkering temperature from beginning to end of their traverse through the chamber, whereby a high-percentage of them become completely clinkered by the time they reach said port; the continuous discharge of the products of combustion and their burden of clinkered materials through said exit port of the chamber; and the separation of the clinkered materials from the combustion products.

More specifically, and with particular reference to the exemplification shown in the accompanying drawing, the invention comprises the introduction of dry pulverized raw materials into one end of a stationary shaft or kiln at a velocity on the order of 75 feet or more per second, and in a whirling or cyclonic manner relative to the shaft axis; the introduction into the shaft, at a multiplicity of points spaced throughout substantially its entire length, of high velocity cyclonically moving masses of burning fuel, such as gas, oil or powdered coal, the masses in longitudinally adjacent zones of the chamber whirling in opposite directions and intermingling to produce a high degree of turbulence and substantially uniform clinkering temperatures throughout all portions of the chamber; discharging the gaseous products of combustion and the clinkered materials carried thereby in a stream from the end of the chamber opposite that at which the raw materials are introduced; and effecting an abrupt change in the direction of movement of the discharged gases, whereby the inertia of the clinkered solids will discharge them from said stream.

The length of the chamber is such, with regard to the temperatures and velocities employed, as to provide a time of contact between the raw materials and the mass of flame and hot gases sufficient to insure complete clinkering of

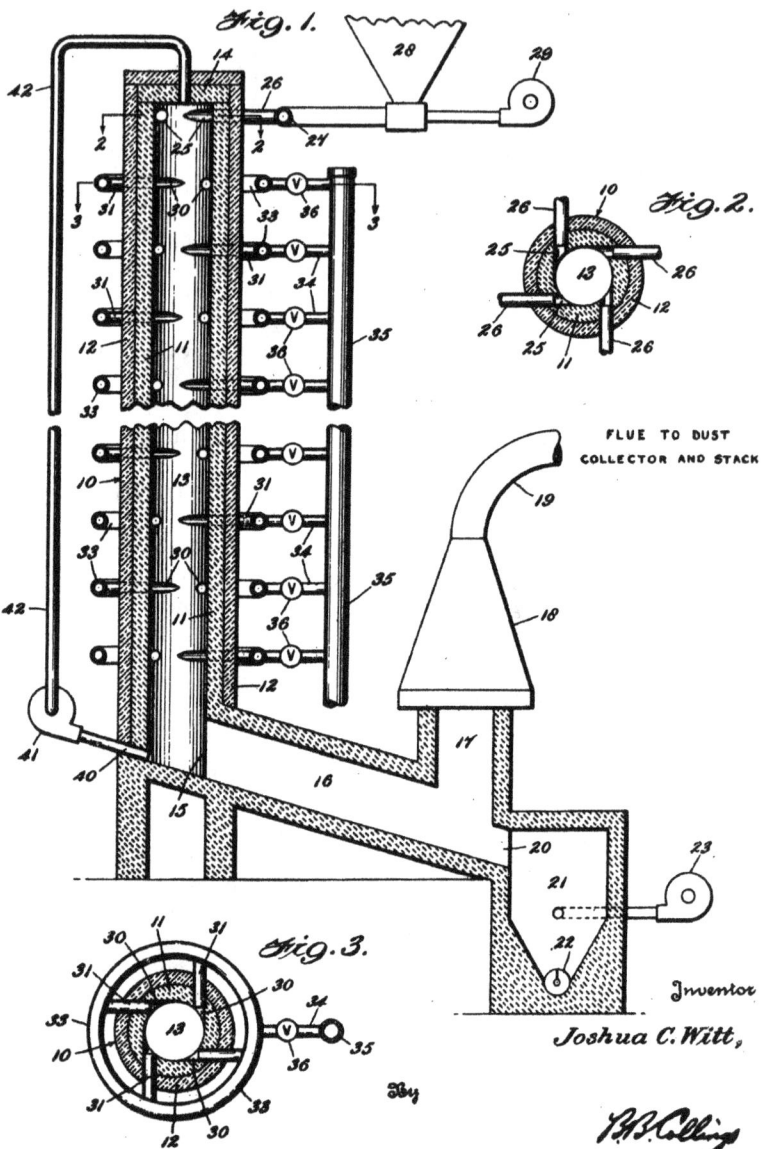

Fig. 1.

Fig. 2.

FLUE TO DUST
COLLECTOR AND STACK

Fig. 3.

Inventor

Joshua C. Witt,

By

P.B.Collins

ATTORNEY

substantially all of said materials, but in no case need such length exceed 100 feet and in most instances it may be substantially less.

The invention further comprises the steps and combinations of steps constituting the method, and the combinations and arrangements of parts constituting the apparatus, more fully described below and particularly pointed out in the claims.

In the accompanying drawing forming a part of this specification there is illustrated somewhat diagrammatically the essential elements of one form of apparatus suitable for carrying out the present method of clinkering cement raw mix. In the said drawing, in which like reference characters designate like parts in all the views:

Figure 1 is a vertical sectional view, partly in elevation, of the apparatus; and

Figs. 2 and 3 are cross sectional views, taken on the planes indicated, respectively by the lines 2—2 and 3—3 in Fig. 1, looking down.

Referring to said drawing in detail, 10 designates a vertical shaft, the walls of which comprise a refractory lining 11 surrounded by a layer of heat insulating material 12, providing an elongated clinkering chamber 13 of circular cross section. The chamber is closed at its top by a cap 14, and at its bottom it has an open port 15 communicating with a somewhat downwardly inclined laterally extending discharge passage 16 having an abruptly angular off-take 17 communicating with a hood 18 and a flue 19 which extends to an appropriate dust collector and thence to a stack which will provide a draft downwardly through the chamber 13. If desired, a fan or blower may be interposed in the flue 19 to assist in maintaining the draft, but since this as well as the dust collector and stack are common expedients in the art, such elements have not been shown in the drawing.

At or near the junction of the passages 16 and 17 the passage wall is provided with a port 20 through which the clinker may be discharged into a collecting and cooling chamber 21, as will appear more fully below. The clinker is removed from said chamber by a screw or other suitable conveyer 22, and cooling air may be forced into and through this chamber by a fan or blower 23.

Adjacent the top of the shaft 10 its wall is provided with a set of raw mix induction ports comprising tangentially disposed passages 25 which communicate by means of pipes 26 with a manifold 27 to which the pulverized raw materials may be supplied by any suitable feeding mechanism, here conventionally indicated as comprising a hopper 28 for containing the materials, and a blower or pump 29 for supplying a current of air to entrain and carry said materials at relatively high velocity through manifold 27, pipes 26 and induction ports 25 into chamber 13.

Below the raw mix induction ports 25 the wall of the shaft 10, throughout substantially its entire height, is pierced with a multiplicity of fuel induction ports arranged in vertically spaced sets, each of which comprises a plurality of tangentially disposed passages 30 discharging into a horizontal zone of the chamber 13, which zones are longitudinally contiguous from one end of the chamber to the other. As here shown, the uppermost set of passages 30 and each alternate set therebelow discharges into the chamber 13 in a direction opposite to that in which the raw mix induction passages 25 discharge, while the remaining sets of fuel passages discharge in the same direction as said raw mix passages. While this arrangement insures maximum turbulence in the burning fuel mass and temperature uniformity throughout the length of the chamber, in some instances at least some adjacent sets of fuel passages 30 may discharge in the same direction.

Each passage 30 houses a nozzle or burner 31 and the nozzles of each set are connected to a manifold 33, the several manifolds in turn being connected by pipes 34 to a header 35 connected to a fuel supply. Each of the pipes 34 may be provided with a valve 36, whereby to control each set of burners. The arrangement here shown is a more or less conventional illustration of a fuel supplying apparatus in which gas is the fuel employed, and of course the specific details thereof will be varied as necessary if oil or powdered coal is to be used.

In preparing clinker in the apparatus just described, fuel issuing from the several sets of burners 31 is ignited, producing a multiplicity of counter-cyclonically moving high velocity masses of flame and hot gases in a series of longitudinally contiguous zones from top to bottom of the chamber 13. After a suitable preheating period, the pulverized raw materials are introduced at high velocity through the tangential ports 25 and immediately meet the high velocity oppositely whirling mass of flame and hot gases from the uppermost set of burners 31. A state of extreme turbulence is created by such meeting which tends to insure individual contact of the material particles with the hot gases and this is substantially maintained throughout the clinkering operation due to the alternating cyclonic movements of the several masses of burning fuel which the materials encounter as they pass serially through the superposed zones of the chamber. Thus, the said particles may attain clinkering temperature comparatively rapidly and be completely clinkered by the time they reach the exit port **15**, which traverse ordinarily is completed in less than one minute.

The alternating whirling movements of the fuel masses in the several zones also produce a scouring action on the raw mix and fuel particles which tends (*a*) to remove the film of air adsorbed by each raw mix particle when it enters the chamber; (*b*) to remove the layer of carbon dioxide which forms around each raw mix particle, due to calcination of carbonates; and (*c*) to remove carbon dioxide from the burning fuel particles and substitute therefor oxygen from the atmosphere.

As each set of burner nozzles **31** is continuously discharging a certain volume of flame and hot gases into its particular zone of the chamber **13**, and since said chamber is provided with but one exit port **15** through which the products of combustion may escape, it follows that irrespective of the draft created by the stack and/or fan associated with the flue **19**, continuous movement of the flame and hot gases in the chamber will be maintained longitudinally thereof toward the port **15**. As the raw materials are introduced adjacent the closed end **14** of the chamber and likewise may exit only through the port **15**, their travel through the chamber is in the same direction as and concomitant with that of the flame and hot gases.

As the gaseous products of combustion and the clinkered materials carried thereby pass through the discharge duct **16** the draft from the stack will draw most of the gases upwardly through the off-take **17**, thus imparting a sharp change of direction to their movement. However, the inertia of the clinkered solids will cause most of them to continue through the port **20** and into the collecting and cooling chamber **21**, from which they may be removed by the conveyer **22** for further processing by conventional procedures. Any solids drawn upwardly through the off-take **17**, hood **18** and flue **19** will be separated from the gases by the dust collector, and may be recovered therefrom.

To avoid undesirable deposit and piling up of the clinkered solids at the bottom of the shaft **10**, due to the change in direction of movement of the gases as they pass through port **15** and into discharge passage **16**, one or more nozzles **40** may be located in the shaft wall in line with the bottom of said passage, through which nozzles a blast of air or hot gases may be projected by a fan or blower **41**. In the example shown in the drawing, a duct **42** supplies hot gases from the upper part of the chamber **13** to the intake of such blower.

While the apparatus here shown is intended to operate on the "down-draft" principle, the method is equally susceptible of being carried out in an "up-draft" apparatus, i.e., one in which the exit port **15** is located at the top of the shaft and the raw mix feed ports **25** are adjacent the bottom thereof; and as previously stated, the shaft need not necessarily be vertical.

The time necessary to effect clinkering by this process will depend somewhat upon the temperatures employed (which may vary within the usual limits) and the height of the shaft in turn will be dependent upon the time element. However, when operating at temperatures on the order of 2800° F to 3000° F, satisfactory clinker may be produced as a result of contact between the raw mix particles and burning fuel for less than one minute, and such contact may be attained in an apparatus of the character above described in which the height of the shaft ranges from 30 to 100 feet.

While velocities on the order of 75 feet per second have been mentioned above in connection with the introduction of the raw mix and fuel into the clinkering chamber, this is intended merely as illustrative of what is meant by "relatively high velocity" as herein employed, and not as a specific limitation thereto.

What is claimed is:

1. In the manufacture of Portland cement and the like, the method of clinkering finely divided raw materials which comprises maintaining a turbulent mass of burning fuel throughout all portions of an elongated chamber by cyclonically introducing said fuel at relatively high velocity into each of a series of longitudinally contiguous zones of the chamber, the cyclonic motion of the fuel in some of said zones being in a direction opposite to that of the fuel in other of the zones; discharging the products of combustion from one end of the chamber whereby to maintain a longitudinal flow of flame and hot gases therein; introducing the finely divided raw materials into the other end of the chamber for high velocity contact with the burning mass and clinkering traverse therewith serially through the several zones of the chamber; and separating the clinkered materials from the discharged combustion products.

2. In the manufacture of Portland cement and the like, the method of clinkering finely divided raw materials which comprises maintaining a turbulent mass of burning fuel throughout all portions of an elongated chamber by cyclon-

ically introducing said fuel at relatively high velocity into each of a series of longitudinally contiguous zones of the chamber, said series extending from end to end of the chamber, and the cyclonic motion of the fuel in alternate zones being in a direction opposite to that of the fuel in the remaining zones; discharging the products of combustion from one end of the chamber to induce a longitudinal flow of flame and hot gases therein; pressurally introducing the finely divided raw materials into the other end of the chamber for immediate high velocity contact of the material particles with the burning fuel mass, the longitudinal movement of the latter carrying the clinkering particles serially through the several zones of the chamber; and separating the clinkered materials from the discharged combustion products.

3. In the manufacture of Portland cement and the like, the method of clinkering finely divided raw materials which comprises continuously cyclonically introducing burning fuel at high velocity into each of a series of longitudinally contiguous zones throughout the length of an elongated chamber, the cyclonic motion of the fuel in some of the zones being in a direction opposite to that of the fuel in other of the zones, whereby to maintain a turbulent mass of burning fuel and substantially uniform clinkering temperatures in all portions of the chamber; discharging the products of combustion from one end only of the chamber, whereby to maintain a longitudinal flow of flame and hot gases therein toward said end; introducing the finely divided raw materials into the other end of the chamber for immediate high velocity contact of the discrete particles with the burning fuel mass and movement concomitantly therewith serially through the several zones of the chamber, whereby said particles will be clinkered and discharged from the exit end of the chamber suspended in the stream of combustion products; and effecting an abrupt change in the direction of movement of the discharged combustion products whereby the inertia of the clinkered solids may separate them from such products.

4. In apparatus for clinkering finely divided cement raw materials and the like, the combination of means providing an elongated clinkering chamber closed at one end and having a discharge port at its other end; a multiplicity of sets of fuel feeding devices spaced longitudinally of the chamber throughout its length, each set comprising a plurality of nozzles discharging tangentially into one of a series of longitudinally contiguous zones of the chamber, some of the sets discharging in one direction and some in an opposite direction, whereby to provide a turbulent mass

of burning fuel and substantially uniform clinkering temperatures in all parts of the chamber; means for cyclonically introducing the finely divided raw materials into the chamber adjacent its closed end for high velocity contact with the burning mass and travel serially through the several zones of the chamber to said discharge port; and a discharge conduit leading from said port.

5. In apparatus for clinkering finely divided cement raw materials and the like, the combination of means providing an elongated clinkering chamber closed at one end and having a discharge port at its other end; high velocity fuel feeding devices spaced longitudinally along the chamber throughout its length and discharging cyclonically into a series of longitudinally contiguous zones of the chamber, the devices feeding adjacent zones being arranged to discharge in opposite directions whereby to provide a turbulent mass of burning fuel and substantially uniform clinkering temperatures in all parts of the chamber, the components of which mass move longitudinally of the chamber from its closed end to said discharge port; means for pressurally introducing the finely divided raw materials into the chamber adjacent its closed end for high velocity contact of the material particles with the burning mass and their traverse concomitantly therewith serially through the several zones of the chamber to the discharge port; and a discharge conduit leading from said port, for conducting the products of combustion and their burden of clinkered materials therefrom.

The direct and continuous application of high-temperature heat results in the synthesis of clinker minerals in several minutes, rather than in several hours, as with the rotary kiln. Many other advantages can be listed, such as the small quantity of raw materials in process, no large moving parts, and low investment per barrel of clinker.

The particle size of the clinker produced in the Clinkerer is small, probably 20–30 mesh. There is no crushing required, and only finishing grinding. Cooling is simple and rapid. The savings in equipment, power, and heat recovery at this point are readily apparent. The particle size is most favorable to blending, and so we come again to a consideration of *uniformity.*

Operating problems may be expected, and changes in design suggested, but the laboratory and pilot investigations strongly indicate that the Clinkerer provides a satisfactory and economical procedure for the production of clinker.

The nonaqueous liquid process (see Chapter 16) and the Counter-cyclone Clinkerer were developed independently. Patents had been issued on each project before it was recognized that they could be operated together to advantage.

REFERENCES

1. J. C. Witt, The passing of the rotary kiln, *Rock Products*, 61(11):64 (1958).

Chapter 18

RESEARCH IN SCIENCE
AND IN ENGINEERING

In this volume it has been stated, or at least suggested, many times that research has been one of the major factors in the development of the cement and concrete industries of today. The situation calls for a chapter to emphasize its really great fundamental importance. This simply could not be overemphasized.

One might believe that because the accomplishments with cement have been of such magnitude, the relative need for research would gradually diminish. But the reverse is true. If one happened to have an idea for improvement in the trial-and-error days of development in the industry, its value could easily be proved or disproved, on the basis of a few simple tests at most. The word *simple* is justified because there were no other types of testing methods in that period. The mathematics that had been applied to cement was far from complex. There was no literature to be consulted.

It is not necessary to point out how greatly the situation has changed. Today the preparation of an adequate program for a research project, with all the searching of the literature, might require more time and work than would be necessary to carry out the project itself. The ever-increasing volume of library work desirable in connection with modern research is mentioned in Chapter 24.

The search of the literature should be most thorough. After the original reference to a piece of equipment or to a procedure, any following reference may contain some inaccuracies, or may be influenced to some extent by interpretation. So the last reference may differ in some respects from the original. It could happen that two papers on the same subject, each having a list of references, could disagree rather widely.

It is well to go back to the original paper when this is practicable. In work on one project I found a series of interlocking references on the same laboratory procedure. One reference mentioned two or three previous papers, and so on. The situation aroused my curiosity. The trail led back and back, predating American journals. Finally the original paper was located in a British science journal.

One of the most encouraging trends in cement manufacture is the increasing application of scientific research. In 1962, published papers on new and modernized plants contain such expressions as *x-ray spectograph, centralized control, computer control, closed circuit TV, maximum automation, ultimate in quality control*. Regardless of the objective of improvements at a given plant, it may be taken for granted that uniformity will be involved. The road to advancement in cement is through uniformity. There are no detours.

An excellent reason for research is competition. I could think of no greater favor to the competition than a moratorium on research in the cement industry.

Competing industries, great and small, are becoming more and more active and emphatic. Recently a manufacturer listed many of the points of excellence of concrete, and claimed that his product is superior in each of them. There is always a chance that some day a cementitious material will be developed that will displace portland cement just as portland has displaced its many predecessors. Let us hope that this will not happen within the next century or two. Although cement has entered a surprising number of markets, there is still a probability of expansion in these markets, and the acquisition of many additional markets.

In any discussion on research a distinction must be made between basic and applied research. There is much disagreement concerning these terms and some exact definitions are greatly needed.

In any program on cement, from raw materials deposits to

concrete in place, extensive research is essential. For convenience, I have placed this research into five divisions, A through E, as shown in Table 18:1.[1]

A, B, and C apply to cement; D and E to concrete. There can be an exchange of information between any one division and any other division. Such exchanges may contribute information to some research projects, and may suggest other projects.

TABLE 18:1

RESEARCH ON CEMENT AND CONCRETE

Division of Research	Application to
A: Basic	Cement
B: Industrial, General	Cement
C: Industrial, Specific	Cement
D: Industrial	Concrete
E: Basic	Concrete

A: This *basic* research includes science in general, particularly chemistry, physics, and petrography. This development was not related in any way to the manufacture of cement. But cement manufacture could not have reached its present status without the information supplied by this research. Atomic weights, analytical methods, microscope technique, and many other topics have been essential.

B: This is *industrial* research in the cement field. Each fact discovered may become useful sometime, but the technique may be so far out of line from present-day manufacturing procedures that application may be far in the future.

C: This is *industrial* research planned for a specific purpose, taking into account how any acquired suitable information could be utilized. In this division, the lag between research and its application is greatly shortened, and application without delay is to be expected.

D: There has been much extensive research in concrete design and in the selection of concrete materials, proportioning, mixing, placing, and curing. The objectives are always strength, durability, beauty, and minimum unit cost.

E: This division of basic research for concrete corresponds to A for cement. It involves primarily physics and engineering. The information needed for plain, reinforced, and prestressed concrete, and for concrete products, is unlimited.

It often happens that the research man devotes so much of his thought and effort to his projects that he pays little attention to the application of his results; and the plant man is so devoted to production and costs that he thinks little about research. It would be helpful for both men to be reminded about the great importance of the commercial application of research.

This paper was nearly complete when an excellent symposium [2] on the planning of research came to my attention. On the basis of this symposium, my division *B* may be considered *unplanned* industrial research; and my division *C*, *planned* research. Many important factors are mentioned in the symposium.

Although everyone recognizes the undesirability of the lag between research and its application, there is another situation that at times may be even worse. This is in the tendency to over-apply research. An example of this came to my attention recently. Some laboratory research results were applied, without justification, to plant operation.

Since the preparation of the paper mentioned here, additional thought has been given to uniformity. Attention is again directed to the principles at the end of Chapter 15.

REFERENCES

1. J. C. Witt, Research in Science and in Engineering, and Application to Industry. Presented at the Fifty-fourth Annual Meeting of the Illinois State Academy of Science (1961).
2. Staff Report. *1961 Chemical and Engineering News*, 39(2):78–84; 39(3):106–20 (1961).

Chapter 19

CONCRETE MATERIALS

Starting with cements that pass ASTM specifications, skill and experience are the principal essentials for the production of excellent concrete. The next step is the selection of the additional concrete materials. To the operator there is available extensive experience of men in the industry, as reflected in ASTM standards. From the 1964 Standards (the latest available at this writing) I have selected eleven:

C 33–63: Concrete Aggregates
C 125–58: Terms Relating to Concrete and Concrete Aggregates
D 448–54: Standard Sizes of Coarse Aggregate for Highway Construction
C 29–60: Unit Weight of Aggregate
C 70–47: Surface Moisture in Fine Aggregate
C 127–59: Specific Gravity and Absorption of Coarse Aggregate
C 128–59: Specific Gravity and Absorption of Fine Aggregate
C 131–55: Abrasion of Coarse Aggregate by Use of the Los Angeles Machine
C 136–63: Sieve or Screen Analysis of Fine and Coarse Aggregate
D 75–59: Sampling Stone, Slag, Gravel, Sand, and Stone Block for Use as Highway Material
D 289–55: Abrasion of Graded Coarse Aggregate by Use of the Deval Machine

The preceding chapters were devoted largely to the manufacture of portland cement. Primarily in the chapters that follow cement is considered a component of concrete.

The cement industry and the concrete industry do not have many points in common. Materials, equipment, processes, personnel, and objectives are largely different. The cement manufacturer must know something about the use of cement, at least, the general principles of concrete construction. Otherwise, he would not be able to make products that give satisfactory results in the field, nor to answer questions, and supply the necessary technical service to the user.

It is not absolutely necessary that the cement user know anything about the manufacture of cement, but it is highly desirable. Even superficial information on the subject should assist him in (1) choosing the best type of cement for a given project, (2) deciding some details of design, (3) selecting the best curing procedures, (4) realizing that cement specifications should be based on performance, rather than on cement manufacturing processes, and such information should assist him in many other ways. These facts suggested that a few chapters on concrete, in a book primarily devoted to cement manufacture, should be of some interest to both manufacturers and users.

Although the cement and concrete industries do not have many points in common, some procedures of the two industries are parallel. First, raw materials are selected; second, the raw materials are proportioned; and third, the mixture of raw materials is processed to make the desired product. Although most raw materials are natural substances, some are manufactured. Cement is a manufactured raw material for concrete, and some concrete aggregates pass through manufacturing processes, in addition to the usual operations of crushing and screening.

Classification of Concrete Materials

For convenience in discussing concrete materials, we shall divide them into two groups, primary and secondary, as has been done with cement raw materials.

Primary Concrete Materials:
    a) Cement
    b) Water

c) Fine aggregates
d) Coarse aggregates

Secondary Concrete Materials:

a) Admixtures (integral)
b) Materials for surface application.

Primary and secondary materials:

Primary concrete materials are those essential for the manufacture of concrete as we know it today. Secondary materials are those that may be used in addition to primary materials for the purpose of modifying the concrete in some way, such as waterproofing, hardening, or decorating.

Cement:

The manufacture, testing, and selection of cement have been discussed.

Water:

Without water there would be no concrete. It is the only substance that combines with clinker compounds to produce the material known as *hydrated cement*. As a material in concrete, water has received very little study, aside from the water-cement ratio, which has to do more with the relative quantity of water rather than with the characteristics of water itself.

To be on the safe side, it is good practice to use the purest water obtainable, but good concrete has been produced with impure waters.[1] The reason for this is that the quantity of materials carried by the mixing water is a small percentage of the total weight of the concrete. These statements apply to the mixing water only—not to water with which concrete comes into contact after placement. According to the Joint Committee,[2]

> Practical experience as well as results of laboratory investigations indicate, in general, that any water which can be safely used for drinking purposes will be satisfactory for use in concrete.

A water unsuitable for drinking may be entirely satisfactory for concrete, and *vice versa*.

Aggregates:

Starting with the next section, the remainder of this chapter will be devoted to aggregates, but a few general statements will be made here.

The separation of fine and coarse aggregates is on the basis of particle size. Fine aggregates pass a number four sieve (4760-micron) and coarse aggregates are retained on this sieve. In this book, it is considered that concrete is a mixture of cement, water, and one or more aggregates. On this basis, concrete includes mortars which consist of cement, water, and a fine aggregate. Referring to the list of concrete materials, at the beginning of this section, we have the following:

$$\text{Cement paste} = a + b$$
$$\text{Concrete} = a + b + c \qquad \text{or}$$
$$a + b + c + d$$

It would be possible to make a concrete of the composition, $a + b + d$, but it is seldom if ever done, and in general it is not to be recommended.

The fine aggregate for masonry mortar differs from that ordinarily used for concrete. The American Society for Testing and Materials has a specification for it (Designation: C 144–52T).

On the basis of density, there are two groups of aggregates, heavy and lightweight. Heavy aggregates are the usual types, such as sand, gravel, crushed stone, and slag. Ordinarily, when there is no statement concerning weight or density, heavy aggregates are understood. Many varieties of lightweight aggregates are used. Some types are manufactured from clay, some from blast-furnace slag, and some from other materials.*

Aggregates

In selecting an aggregate for concrete, general experience with concrete construction, and experience with the aggregates available for a given project, are of prime importance.

---

* Unless otherwise noted, the term *aggregates* in this volume refers to materials of ordinary weight, such as sand, gravel, crushed stone, and the like.

Some of the characteristics of aggregates to be taken into account are:

> Maximum particle size
> Minimum particle size
> Particle size distribution (or grading)
> Soundness
> Presence of impurities
> Presence of deleterious substances
> Resistance to abrasion
> Toughness
> Density
> Chemical composition

The principal aggregates are sand and gravel, crushed stone, and blast-furnace slag. *Crushed stone* and *slag* usually refer to coarse aggregates; however, fine aggregates are made from these materials. There is an association of producers of each of these classes of aggregates:

> National Sand and Gravel Association
> National Crushed Stone Association
> National Slag Association

Each Association studies its products and distributes information in its publications.

In 1960, the total sand and gravel [3] used in construction amounted to 683,020,000 short tons, values at $654,861,000. Incidentally, this is about three times the tonnage reported in 1943. It seems that there is justification for the headline that once appeared in a trade journal

AMERICA IS BUILT WITH AGGREGATE.

A few uses have been found for heavy-weight aggregates, but the production and use of lightweight aggregates have developed into a large industry. A map published by *Pit and Quarry* shows the location of 249 plants in the United States for the production of 13 types of lightweight aggregates (including natural and manufactured materials). Nearly every state has at least one plant; California has thirty-three.

There are three ASTM specifications for lightweight aggregates:

> C 331–59 T: For concrete masonry units
> C 332–61:   For insulating concrete
> C 330–60 T: For structural concrete.

REFERENCES

1. D. A. Abrams, *Proc. Am. Concrete Inst.*, 20:442 (1924).
2. *Report of the Joint Committee on Standard Specifications for Concrete and Reinforced Concrete*, Portland Cement Association, Chicago, 1940.
3. *Minerals Yearbook*, U.S. Bureau of Mines, Washington, D. C., 1961.

Chapter 20

PROPORTIONING MATERIALS FOR CONCRETE

Let us assume that a concrete structure has been designed, the total volume of concrete has been estimated, and the concrete materials have been selected. The next step is designing the mix, or calculating the quantity of each material required for one cubic yard of concrete.

The importance of proportioning concrete materials is being recognized more and more. Although good concrete can not be made with poor materials, it is probable that mediocre concrete results more frequently from faulty proportioning than from faulty materials. It may be said, however, that although there is rarely any difficulty in obtaining good cement and mixing water, first-choice aggregates may not be produced in a given district, and the cost of shipping aggregates to the district may be prohibitive.

A fine aggregate may be used in the form in which it is taken from a deposit, or it may be processed in various ways, such as washing, screening, and crushing. Two or more fine aggregates may be combined to obtain a desired particle size distribution. These statements apply also to coarse aggregates. In discussing proportioning, it is assumed that only one fine aggregate and one coarse aggregate are used. If any preliminary processing or blending were necessary, it is assumed that this was completed before the aggregates were delivered to the job. The proportioning procedures explained in this chapter are not limited to two aggregates. They are applicable to any number of concrete materials.

For many years there was very little development in proportioning procedures. It seems to have been the general belief that any fine or any coarse aggregate could be employed, and that proportioning by the crudest procedures was satisfactory. Unfortunately, some concrete is manufactured on this basis to this day.

One of the most important improvements in concrete technology has resulted from controlling the relative quantities of water and cement. The binding medium of concrete is hydrated cement. This is a product resulting from chemical reactions between the components of cement and water. In many chemical reactions, the required quantity of each of the reacting substances can be calculated, and the ratios of the substances can be established. Some difficulties arise in the case of cement and water. First, the use of hydrated cement was well established before the nature of the reactions was known. Second, the workability of the concrete must be given preference even over the chemical requirements in proportioning the cement and water. Also many other factors, such as the nature and quantities of aggregates, influence the quantity of water required.

The quantitative relation of water to cement in concrete is an example of how much time may elapse between the development of information and the commercial application of this information. In this case, the period was well over a century. The specification of a patent for a hydraulic lime [1] granted to James Parker on June 28, 1796 contains these statements:

> To compose the cement in the best and most advantageous manner, I take two measures of water and five measures of the powder thus described; then I add the powder to the water, or the water to the powder . . . the cement is then made and will set or become indurated either ten or twenty minutes after the operation has ceased in or out of water.
> But although I have described what I consider as the best proportions for the conversion of the cement, it is expressly to be understood that these and all other proportions are to be included within the meaning and purpose of the specification, but that no other proportions will produce so strong a cement in so short a time as those I have here pointed out; and also that I occasionally burn, and grind, and mix the powder before described with lime and stones, clay, sand or

calcined earth in such proportions as may be necessary and useful for the purpose that the cement is intended to be applied to, always observing the less water is used the better, and the sooner the mortar or cement is used after being made the stronger and the more durable it will be.

In present-day terminology, this is equivalent to a water-cement ratio of approximately three gallons of water per sack of cement.

In 1876–7, Mann [2] determined the strength of cement when mixed with various percentages of water. Portions of cement, weighing 32 ounces each, were mixed with 5, 6, 7, 8, 9, and 10 ounces of water, respectively. The seven-day strengths decreased from 412 to 182 pounds, as the ratio of cement to water, by weight, increased from 0.156 to 0.312. The mix containing 5 ounces of water was so dry that it could not be troweled. The mix containing 10 ounces was so wet that it could be troweled.

In 1919, Abrams [3] reported the results of an extensive investigation on concrete. In determining the compressive strengths of large numbers of concrete specimens, it was observed that a relation exists between strength and the ratio of cement to water, and that this relation is independent of many other variables. This led to the formulation of the equation

$$S = \frac{A}{B^z}$$

Where, S is the compressive strength of concrete and X is the ratio of the volume of water to the volume of cement in the batch. A and B are constants whose values depend on the quality of cement used, the age of the concrete, curing conditions, etc.

At one time, some were inclined to believe that if the water-cement ratio is correct, concrete must be satisfactory; and that if concrete has any undesirable characteristics whatever, the water-cement ratio must be at fault. Subsequent studies have shown that although the water-cement ratio is a most important principle, other variables also must be taken into account. [4]

The unit of quantity of concrete is the cubic yard. It is evident that a unit volume of concrete is the sum of the absolute volumes of its components, and it follows that absolute volume is the correct basis for calculating quantity of concrete materials. When the quantities have been calculated, and the weight for unit volume of each material is known, it is more accurate, and is usually more convenient, to proportion the materials by weight, not taking into account any entrained air.

In connection with the proportioning of concrete materials, many expressions have come into use, which indicate limitations of one or more of the materials. Some of these limitations are interrelated. The following are examples:

1. Water-cement ratio. This may be expressed as the ratio by weight or by volume. Customarily, it is expressed in terms of gallons of water per sack of cement.
2. Cement factor. The number of sacks of cement per cubic yard of concrete
3. Maximum particle size of the coarse aggregate
4. Ratio of the volume of fine aggregate to the volume of coarse aggregate
5. Ratio of the volume of fine aggregate to the total volume of aggregates
6. Fraction of a cubic yard occupied by the absolute volume of the coarse aggregate.

In the design of a concrete mix, there must be taken into account four characteristics of the finished concrete: (a) strength, (b) durability, (c) economy, and (d) appearance. There has been much discussion concerning the order in which these characteristics should be considered. It is evident that strength should come first, because without sufficient strength concrete could not be put into service at all. Durability is the next in importance, because concrete must be durable if it is to continue in service. Ordinarily, the best economy, consistent with the degree of appearance desired, comes next, but in some types of construction, economy is secondary to appearance.

When the nature of the mix has been decided, taking into account such ratios as those just mentioned, the quantity of each material required for one cubic yard of concrete is calculated. The principles on which this calculation is based are simple, but of some large number of procedures that are employed, some are unnecessarily long and tedious.

Proportioning Calculations

In 1941, the following paper on *The Proportioning of Concrete Materials* was published.[5]

Methods in large variety are employed in calculating concrete materials quantities. Although no one principle seems to be recognized universally, the water-cement ratio [3] and absolute volumes are probably in more common use than any of the others.[6]

The procedure described here, which for convenience may be called the *unit layer method*, did not result from investigations in the concrete field. It was suggested by some research on the proportioning of materials,[7] and the mixing of solids, in connection with manufacturing processes. In fact, it may be considered a special case of proportioning. It may be said at this point that the method was developed for practical purposes without a preliminary review of the literature. After it was decided to present the material for publication, some of the best known papers [8-15] were reviewed, primarily to locate any duplications that might exist. No attempt has been made to prepare a complete bibliography.

It is customary to express water volumes in U. S. gallons; cement volumes, in sacks or in cubic feet; aggregate volumes, in cubic feet or in cubic yards; and concrete volumes in cubic yards. A cubical tank having a capacity of one cubic yard can be filled with these materials in any proportion. As the base is a constant, the depth of each layer determines the volume of the material. In this way the quantity of each material may be expressed as a fraction of a linear yard, third degree quantities being replaced in effect by first degree quantities. Further simplification has been brought about by selecting, as a unit, a layer 1 yard square and 0.01 yard thick. This is equivalent to 0.01 cubic yard, or to 0.27 cubic foot, or to a cube having an edge of approximately 0.22 yard. In this way, all units are converted into a common unit. Although this common unit is in the English system, it can be handled with nearly the same facility as metric units.

The first step in the solution of a problem is to express all values in terms of unit layers. This greatly simplifies the work and in no way affects any other detail of whatever calculating procedure may be preferred. The answers are in terms of unit layers but they can be converted to any desired units of volume or of weight. The sum of the unit layers in one cubic yard of concrete is 100. This greatly facilitates the checking of results.

Cements and aggregates are considered to consist of unit layers of *voids* and unit layers of *solids*. Percentages of voids are expressed as decimal voids. For example, if an aggregate contains 30 per cent of voids, the decimal voids are 0.30. In making the calculations, it is assumed that the unit layers of voids and of solids may be separated. In this way, if an aggregate, as mentioned above, is placed in a cubic yard tank to a depth of 10 unit layers, the tank will contain three unit layers of voids and seven unit layers of solids.

It is assumed that the volume of cement paste is equal to the volume of cement solids plus the volume of water. There are factors that tend to cause the volume of the cement paste to be either greater or less than the sum of the two components, such as solubility, formation of colloids, absorption, and adsorption. In this work, however, the accuracy may be considered sufficient for the purpose without taking these factors into account. If greater accuracy were desired, a series of measurements could be made, and correction factors calculated. The volume of cement solids is calculated on the basis of a specific gravity of 3.1. Here again greater accuracy may be obtained by determining the specific gravity of the cement to be used.

The following symbols refer to quantities expressed in unit layers:

W = Water
E_v = Voids of cement
E_s = Solids of cement
E = Cement = $E_v + E_s$
F_v = Voids of fine aggregate
F_s = Solids of fine aggregate
F = Fine aggregate = $F_v + F_s$
C_v = Voids of coarse aggregate
C_s = Solids of coarse aggregate
C = Coarse aggregate = $C_v + C_s$
P = Paste = $E_s + W$

The following additional symbols are employed:

U = water-cement ratio in gallons of water per sack of cement

Q = ratio of volume of cement to volume of fine aggregate. $Q = E/F$

T = ratio of volume of cement to volume of coarse aggregate. $T = E/C$

R = ratio of volume of fine aggregate to the sum of the volumes of the aggregates. $R = \dfrac{F}{F + C}$

M = ratio of volume of coarse aggregate to volume of fine aggregate. $M = C/F$

A = decimal voids in fine aggregate

B = decimal voids in coarse aggregate.

Some unit layer equivalents are given in Table 20:1.

Application of Unit Layer Method

To illustrate the use of the unit layer method, four cases will be considered. In the first of these, a mix that would not be workable has been chosen for discussion. This has been done to explain further the basis of the method and to provide data to serve as an example in discussing a procedure for the adjustment of quantities, at the close of the paper. In all formulas and problems, a quantity of material expressed in unit layers refers to the total quantity of that material present in a cubic yard. For simplicity, it is assumed in these cases, that the aggregates are dry and non-absorbent, that there is no entrained air, and that there are no losses of materials. Corrections for these will be discussed later.

CASE 1. The water-cement ratio and the decimal voids in the fine and in the coarse aggregates are stated. The voids of the coarse aggregate are to be filled with the fine aggregate. The voids of the fine aggregate are to be filled with cement paste. The general equation is

$$F_s + C_s + P = 100$$

also, $C = 100,$ $F = BC,$ $P = 100AB$

Example 1: The water-cement ratio is 6 gallons per sack. The fine aggregate has 0.45 decimal voids, and the coarse aggregate has 0.40 decimal voids. Then $F = 0.40 \times 100 = 40$; and $P = 18$ (referring to Table 20:3).

Table 20:1

SOME UNIT LAYER EQUIVALENTS

	A	B	C	D	E
	Material	Ordinary Unit Sack (1 cu ft)	Equivalent Values in Unit Layers	Reciprocal	Contained in One Cubic Yard
1	Cement		3.70	0.2703	27
2	Solids		1.80	0.5556	56
3	Voids		1.90	0.5263	53
4	Total		3.70	0.2703	27
5	Water	Gallon	0.494	2.0243	202
6	Cement Paste	Cubic foot	3.70	0.2703	27
	One sack of cement plus water, in gallons				
7	1		2.29	0.4367	44
8	2		2.79	0.3584	36
9	3		3.29	0.3040	30
10	4		3.78	0.2646	26
11	5		4.27	0.2342	23
12	6		4.76	0.2101	21
13	7		5.26	0.1901	19
14	8		5.75	0.1739	17
15	Aggregate (Solids)	Cubic foot	3.70	0.2703	27
	Volume of solids, for several values for decimal voids				
16	0.25		2.78	0.3597	36
17	0.30		2.59	0.3861	39
18	0.35		2.41	0.4149	41
19	0.40		2.22	0.4505	45
20	0.45		2.04	0.4902	49
21	0.50		1.85	0.5405	54
22	0.25		0.92	1.0870	108
23	0.30		1.11	0.9009	90
24	0.35		1.29	0.7752	78
25	0.40		1.48	0.6757	68
26	0.45		1.66	0.6024	60
27	0.50		1.85	0.5405	54

When the values for A and B are not found in Table 20:3, it is possible that the use of the nearest values in the table will give sufficiently accurate results. If not, interpolation may be employed, or the necessary multiplication may be made.

Referring to Table 20:1, it will be observed that when $P = 18$ for a water-cement ratio of 6 gallons per sack, the cement factor lies between 3 and 4. Interpolation may be employed, or the factor may be found by multiplying 18 by 0.2101, shown in Table 20:1, line 12, column D. The cement factor is found to be 3.8 sacks, and the water is $3.8 \times 6 = 22.8$ gallons.

Collecting the results that have been obtained, the materials required for a cubic yard of concrete are:

Cement	3.8 sacks
Water	22.8 gallons
Fine aggregate	0.4 yard
Coarse aggregate	1.0 yard

Referring again to the general equation, $F_s = 22$, $C_s = 60$, $P = 18$, and the sum is 100. The equivalent weights may be calculated easily, or tables or graphs for the purpose may be prepared.

TABLE 20:2

VOLUME OF CEMENT PASTE, EXPRESSED IN UNIT LAYERS

Sacks of Cement	Gallons of Water per Sack of Cement										
	4.0	4.5	5.0	5.5	6.0	6.5	7.0	7.5	8.0	8.5	9.0
1.0	3.8	4.0	4.3	4.5	4.8	5.0	5.3	5.5	5.8	6.0	6.3
2.0	7.6	8.1	8.6	9.0	9.5	10.1	10.5	11.0	11.5	12.0	12.5
3.0	11.4	12.1	12.8	13.6	14.3	15.1	15.8	16.6	17.3	18.0	18.8
4.0	15.0	16.1	17.1	18.1	19.1	20.1	21.0	22.1	23.0	24.0	25.0
4.5	17.0	18.2	19.4	20.0	21.5	22.7	23.7	24.8	26.0	27.2	28.2
5.0	18.9	20.2	21.4	22.6	23.8	25.0	26.5	27.6	28.8	30.1	31.3
5.5	20.8	22.2	23.6	24.9	26.3	27.7	28.9	30.4	31.7	33.1	34.5
6.0	22.8	24.3	25.7	27.2	28.6	30.2	31.5	33.2	34.5	36.1	37.5
6.5	24.7	26.3	27.9	29.4	31.1	32.7	34.2	36.0	37.5	39.2	40.7
7.0	26.6	28.3	30.0	31.8	33.4	35.3	36.8	38.7	40.3	42.2	43.9
7.5	28.5	30.4	32.3	34.0	35.8	37.7	39.5	41.5	43.2	45.3	47.0
8.0	30.4	32.4	34.3	36.3	38.2	40.4	42.0	44.3	46.0	48.2	50.1
8.5	32.3	34.3	36.5	38.5	40.5	42.8	44.6	47.0	49.0	51.2	53.2
9.0	34.0	36.3	38.5	40.8	43.1	45.3	47.3	49.9	51.9	54.2	56.2

CASE 2. The water-cement ratio, the cement factor, and the decimal voids of the aggregates are stated. The fine aggregate is to be sufficient to fill the voids of the coarse aggregate. The same general equation is applicable. Also

$$F = BC \qquad \text{and} \qquad F_s = F(1 - A)$$

It may be shown that

$$C = (100 - P) \frac{1}{1 - AB}$$

Example 2: The water-cement ratio is 6 gallons per sack; the cement factor 6 sacks; and the decimal voids, of the fine and coarse aggregates, 0.45 and 0.40 respectively. From Table 20:2, $P = 28.6$. In the formula $C = (100 - P)\dfrac{1}{1 - AB}$, $P = 28.6$, and $\dfrac{1}{1 - AB} = 1.22$, from Table 20:3. Substituting, $C = (100 - 28.6) \times 1.22 = 87.1$, $F = 0.40 \times 87.1 = 34.8$.

In other words, P is taken from Table 20:2, C is found by subtracting P from 100, and multiplying the result by 1.22, from Table 20:3; and F is 0.40 times C. The mixture consists of the following:

Cement	6	sacks
Water	36	gallons
Fine aggregate	0.348	cubic yard
Coarse aggregate	0.871	cubic yard

The correctness of these figures may be shown by substituting the values of F_s, C_s and P in the general equation, obtaining $19 + 52 + 29 = 100$.

TABLE 20:3

VALUES FOR 100 AB (UPPER LINE) AND FOR

$$\frac{1}{1 - AB} \text{ (LOWER LINE)}$$

Decimal Voids in Fine Aggregate (A)	Decimal Voids in Coarse Aggregate (B)					
	0.25	0.30	0.35	0.40	0.45	0.50
0.25	6.25	7.50	8.75	10.00	11.25	12.50
	1.07	1.08	1.10	1.11	1.13	1.14
0.30	7.50	9.00	10.50	12.00	13.50	15.00
	1.08	1.10	1.12	1.14	1.16	1.18
0.35	8.75	10.50	12.25	14.00	15.75	17.50
	1.10	1.12	1.14	1.16	1.19	1.21
0.40	10.00	12.00	14.00	16.00	18.00	20.00
	1.11	1.14	1.16	1.19	1.22	1.25
0.45	11.25	13.50	15.75	18.00	20.25	22.50
	1.13	1.16	1.19	1.22	1.25	1.29
0.50	12.50	15.00	17.50	20.00	22.50	25.00
	1.14	1.18	1.21	1.25	1.29	1.33

CASE 3. The water-cement ratio and the proportions of cement, fine aggregate and coarse aggregate are stated. The formulas are:

$$E = \frac{100QT}{T - AT + Q - BQ + 0.494QT + 0.136QTU}$$

$$F = E/Q \qquad C = E/T$$

Example 3: The water-cement ratio is 5.5 gallons per sack. The mix is 1:2:4. Then $Q = 0.50$, $T = 0.25$, and $U = 5.5$. It can be shown that in this problem the voids of both aggregates must be known or assumed. Let $A = 0.45$ and $B = 0.40$.

Substituting these values in the equation for E,

$$E = \frac{100}{4.742} = 21.1 \text{ unit layers. Also,}$$

$$F = 42.2 \text{ unit layers}$$
$$C = 84.4 \text{ unit layers}$$

Therefore the materials required are 5.7 sacks of cement, 31.4 gallons of water, 0.422 cubic yard of fine aggregate, and 0.844 cubic yard of coarse aggregate. Substituting the values of F_s, C_s, and P in the general equation, the result is $23 + 51 + 26 = 100$.

CASE 4. Sometimes the ratio of the fine aggregate to the sum of the fine and coarse aggregates is specified. If this ratio, the water-cement ratio, and the cement factor are specified,

$$C = \frac{(100 - P)(1 - R)}{R(1 - A) + (1 - B)(1 - R)}$$

$$F = \frac{RC}{1 - R}$$

Example 4: The water-cement ratio is 5.5 gallons; the cement factor 5.5 sacks; ratio of fine, to fine plus coarse aggregates, 0.35; and the decimal voids for the fine and coarse aggregates are 0.45 and 0.40, respectively. From Table 20:2, $P = 24.9$. Solving for C and F, $C = 84.1$, $F = 45.3$. Substituting in the general equation, $25 + 50 + 25 = 100$.

Great Variety of Possible Cases

It is evident that cases in almost unlimited variety could be presented. It is not the intention to recommend any particular procedure for designing concrete mixtures. A study of the cases that have been discussed, however, indicates that the following has many advantages:

1. Specify the cement-water ratio, in the interest of strength.
2. Specify the cement factor, in the interest of durability.
3. Specify a ratio of coarse aggregate to fine aggregate,
   $C/F = M$.

This type of problem may be solved very easily. The volume of paste is taken from Table 20:2.

$$F = \frac{100 - P}{(1 - A) + M(1 - B)}$$
$$C = FM$$

In the statement of example 4, suppose that $M = 2$ is used in place of R,

$$\text{then } F = \frac{100 - 24.9}{0.55 + 2(0.6)} = 42.9$$
$$C = 85.8$$

TABLE 20:4

SOME VALUES FOR $(1 - A) + M(1 - B)$, AND

$$\frac{1}{(1 - A) + M(1 - B)}\quad *$$

$(1 - A) + M(1 - B)$	$\dfrac{1}{(1 - A) + M(1 - B)}$
1.20	0.83
1.25	0.80
1.30	0.77
1.35	0.74
1.40	0.71
1.45	0.69
1.50	0.67
1.55	0.65
1.60	0.63
1.65	0.61
1.70	0.59
1.75	0.57
1.80	0.56
1.85	0.54
1.90	0.53
1.95	0.51
2.00	0.50
2.05	0.49
2.10	0.48
2.15	0.47
2.20	0.45
2.25	0.44

* This is applicable when A and B are not less than 0.25 nor greater than 0.50; and M is not less than 1.4 nor greater than 2.0.

The results may be proved by the usual procedure.
It is not difficult to prepare a table for values of

$$\frac{1}{(1 - A) + M(1 - B)}$$

when the range in values of each of the quantities A, B, and
M is not too great. When the values of A and B are not less
than 0.25, nor greater than 0.50; and the value of M is not
less than 1.4, nor greater than 2.0, the minimum and maximum values of $(1 - A) + M(1 - B)$ are 1.20 and 2.25,
respectively. The values of $(1 - A) + M(1 - B)$, at intervals of 0.05, from 1.20 to 2.25, and the corresponding values

for $\dfrac{1}{(1 - A) + M(1 - B)}$, have been calculated. These are
shown in Table 20:4.

This provides a short cut for solving for F. When the
values of $(1 - A) + M(1 - B)$ have been found (and usually this can be done mentally), the nearest value in the first
column of Table 20:4, and the value in the same line of the
second column, are selected. The latter multiplied by
$(100 - P)$ equals F. The value of F, obtained in this way,
is sufficiently accurate, in most cases.

The Effect of Moisture, Absorption, and Entrained Air

The factors of moisture, absorption, and entrained air
must be taken into account, although the first two do not
affect the relative quantities of paste and fine and coarse
aggregates, after they have been blended.

Moisture:

Allowance must be made for the moisture content of an
aggregate, if the aggregate is to be weighed. If the proportioning is by volume, the moisture may be neglected, unless
it is sufficient to cause bulking, or a decrease in the weight
per unit volume due to increased friction between the particles. In any event, the surface water of an aggregate must
be deducted from the total mixing water.

To illustrate, suppose the same in example 2 is saturated
with water, and has a surface water content of 5 per cent by
weight, which causes a bulking of 20 per cent. Assuming that
the dry sand weighs 100 pounds per cubic foot, the free water
amounts to 5.6 gallons. Therefore the volume of mixing water

is 36.0 − 5.6 = 30.4 gallons. One cubic foot of the aggregate (on the dry basis) has a volume of 1.2 cubic feet, when wet. Therefore for one cubic yard of concrete the fine aggregate required is 0.348 × 1.2 = 0.418 cubic yard. If the required aggregate weighs 940 pounds on the saturated, surface-dry basis, the weight on the wet basis would be 1.05 × 940 = 987 pounds.

Absorption:

If an aggregate is not saturated with water at the time the concrete materials are proportioned, the absorption must be determined, and the corresponding water added to the mixing water. If in example 2 the fine aggregate is dry and has an absorption of one per cent by weight, the mixing water should be increased by slightly more than one gallon. Very porous aggregates may absorb some cement paste. This would affect the relative quantities of paste and aggregates. Preliminary tests would be required in advance of any calculations.

Entrained Air:

In this paper it is considered that entrained air replaces a portion of the paste. If in example 2, a cubic yard of concrete contains 5 per cent of air (equivalent to 5 unit layers), the general equation becomes—

$$F_s + C_s + P + \text{Air} = 100,$$
$$or \quad 19 + 52 + 24 + 5 = 100$$

Adjustment of Quantities

Although it has been explained that the unit layer method is a procedure for calculating the quantities of material, it must not be inferred that a concrete mixture prepared on the basis of such calculations is necessarily workable. If, as shown by slump tests, or otherwise, a mixture is not workable an adjustment calculation should be made, followed by another trial batch, and so on.

Concrete made in accordance with Example 1 would not be workable. The ratio of fine to coarse aggregate, or the ratio of paste to fine aggregate, or both, should be increased. Suppose that the paste be increased by 10 unit layers, and the fine aggregate solids by a like number. Then the quantities of materials will be

$$F_s = 32, \qquad C_s = 40, \qquad P = 28 \text{ unit layers}$$

If the mixture were not workable, another increase in F_s, or P, or both, should be made.

In preparing this paper, consistent accuracy has been kept in mind. It is believed that, in general, the procedure is sufficiently accurate for the purpose for which it is intended. In some instances, procedures by which greater accuracy could be obtained have been pointed out, and other procedures will suggest themselves.

A later paper [16] on Proportioning Concrete Materials by the Unit Layer Method included the following:

In the first paper, the unit layer method for proportioning concrete materials was explained. Tables, formulas, and examples were presented. In this method, all units of volume such as sacks (of cement), gallons, cubic feet, and cubic yards are reduced to a common unit, called the unit layer. This is a solid one yard square and 0.01 yard thick.[17]

In this way the volume of each material may be expressed as a fraction of a linear yard, third degree quantities being replaced, in effect, by first degree quantities. It is considered that cement and aggregates may be expressed as unit layers of solids plus unit layers of voids. A unit layer refers to solids, rather than to voids, unless there is a statement to the contrary. Further work with the unit layer method has resulted in the preparation of a chart on which the graph for each material, together with its related scale, is shown in a different color.

The chart is a calculating device, and its scope, in the field for which it was designed, is unlimited. Tables contain specific values and therefore their scope is limited, regardless of their size or number. The chart may be used with facility for the preparation of concrete materials tables.[18,19,20]

REFERENCES

1. G. R. Redgrave and C. Spackman, *Calcareous Cements: Their Nature, Manufacture and Uses;* Charles Griffin and Co., Ltd., London, 1924.

2. I. J. Mann, *Proc. Inst. Civil Engrs.* (London), 47:260 (1876–7).
3. D. A. Abrams, *Design of Concrete Mixtures*, Bull. 1; Structural Materials Research Lab., Lewis Inst., 1919.
4. A. T. Goldbeck, *Crushed Stone J.*, 17(5):3 (1942).
5. J. C. Witt, *Concrete*, 49:33 (1941).
6. A. N. Talbot and F. E. Richart, *The Strength of Concrete, Its Relation to the Cement, Aggregates, and Water*, Bull. 137; Univ. Ill. Eng. Exper. Sta., 1923.
7. J. C. Witt, The Proportioning of Materials in Process Engineering, *Trans. Am. Soc. Mech. Engrs.*, 60(6):493 (1938). *Concrete, Cement Mill Section*, 46:265 (1938).
8. Stanton Walker, *Estimating Quantities of Materials for Concrete*, Bull. 1; Natl. Sand and Gravel Assoc., 1927.
9. Stanton Walker, *Tables of Quantities of Materials for Concrete*, Bull. 4; Natl. Sand and Gravel Assoc., 1928.
10. A. T. Goldbeck, Bull. 1; Natl. Crushed Stone Assoc., 1927.
11. A. T. Goldbeck, *The Water-Ratio Specification for Concrete and Its Limitations*, Bull. 3; Natl. Crushed Stone Assoc., 1927.
12. A. T. Goldbeck, *Some Investigations in Proportioning of Concrete for Highways*, Bull. 7; Natl. Crushed Stone Assoc., 1931.
13. Fred Hubbard, *Materials Required per Cubic Yard of Slag, Stone and Gravel Concrete*, National Slag Assoc., 1939.
14. Stanton Walker, How to Design Concrete Mixtures, *J. Am. Concrete Inst.*, 10:211 (1939).
15. *Concrete Manual;* U.S. Bureau of Reclamation, Washington, D. C., 1938.
16. J. C. Witt, *Rock Products*, 44(12):75 (1941).
17. Fred Hubbard, *Materials Required per Cubic Yard of Concrete, Slag, Stone and Gravel Concrete, and Cement Mortar;* National Slag Assoc., 1940.
18. Stanton Walker and Delmar L. Bloem, *Estimating Proportions for Concrete*, Circular No. 68; National Sand and Gravel Association, June, 1962. Publication No. 69; National Ready Mixed Concrete Association, June, 1962.
19. A. T. Goldbeck and J. E. Gray, Bull. 11; National Crushed Stone Association, 1953.
20. *Concrete Tables*, National Slag Association, Lefax, Philadelphia (December 1958).

Chapter 21

CONCRETE TECHNOLOGY

This chapter deals primarily with the building material, concrete. After the mix has been designed and the concrete materials weighed, as explained in Chapter 20, there are three more steps before the concrete is ready for service. These are mixing, placing, and curing. The fundamentals of these will be given here. Some details will appear later in the chapter.

Mixing:

Concrete materials may be mixed by hand or by mechanical mixers. The hand operation seldom is employed except for small quantities. Mixing should continue until the mixture is as nearly homogeneous as practicable, but not until the initial set of the cement.

Placing:

This term is applied to the transfer of the mixture to the forms or other devices that support it during the plastic period.

Curing:

One of the most important factors in good concrete practice is curing. Reduced to simplest terms, curing is any arrangement or procedure by which cement and liquid water may be kept in contact for a time sufficient to permit hydration of the cement compounds. The time required is a function of several variables, one of the most important of which is temperature.*

---

* Much study is in progress on the methods of testing, and specifications for materials used in curing concrete. C. E. Proudley is Chairman of Subcommittee IIIg, ASTM Committee C-9.

In the industry, *concrete* has several meanings. It may refer to the concrete materials when they are in the mixer, or when they are removed from the mixer for placing, or to fully processed concrete. In this book, *concrete* refers to the fully processed material, with the exception of ready-mixed concrete. The customary nomenclature is followed.

Building Materials in General

For a given project, the value of any building material is based on its characteristics. Some of the most important characteristics may be grouped under the seven heads:

1. Resistance
2. Durability
3. Placement
4. Appearance
5. Economy
6. Density
7. Relation to other materials.

Resistance is the most important. The relative importance of the other groups is affected by a number of factors.

Resistance:

The first prerequisite of a building material is resistance to stresses, and to the flow of air, water, heat, light, electricity, and sound. In general, high resistance is preferable to low resistance, but there are many exceptions. For example, copper is chosen for some uses because it does permit the flow of air, water, or light; and for others because it has low resistance to the flow of heat and electricity or, as generally stated, high thermal and electrical conductivities.

Durability:

This is a general term for the resistance to weathering, high and low temperatures, corrosive substances carried by air and water, and animal and vegetable organisms. Some building materials must depend on the application of protective treatments, such as painting, creosoting, or glavanizing, for resistance to destructive agencies. Such treatments may be applied also for decorative purposes.

Placement:

Under this head, there are included the characteristics of a building material which determine the facility with which it may be employed in the erection of a structure. In what form does it reach the builder? What types of fasteners are required? Must it be cut and fit, or can it be molded?

Appearance:

A material may be decorative in itself, or it may be suitable for the application of decorative treatments. The decorative treatments may also increase the durability of the material as has been mentioned. A building material may be selected for utility regardless of its appearance; for both utility and appearance; or primarily for decorative purposes.

Economy:

Although economy must be taken into account in any project, its relative importance varies greatly. In some cases economy is most important, while in others it is only minor.

Density:

In some cases the weight per cubic foot is an important consideration. A relatively light weight building material requires lighter foundations, and it has insulation value for heat and sound.

Relation to other materials:

Many examples of this could be mentioned. Concrete protects steel from fire and corrosion. Steel reinforces foundations, concrete slabs, beams, columns, and other structural members.

Concrete as a Building Material

Concrete has a very high rank among building materials. This is demonstrated by large increases in quantity and in the number of uses. For some uses, concrete has displaced other materials almost completely; for others, its superiority is obvious; and for still others, it is in active competition. When the characteristics of concrete are compared with those of building materials in general, it should be remembered that some of the characteristics of concrete are not fixed, but can be varied by a number of procedures.

Resistance:

Concrete has high resistance to stresses, particularly in compression. Compressive strengths of 4,000 to 5,000 pounds per square inch at 28 days are common; and higher strengths may be obtained easily. Maney[1] has announced the development of a reinforced concrete column, which has supported a load as high as 83,000 pounds per square inch. Much ready-mixed concrete is sold on strength guarantees.[2] Concrete is opaque to light, except that thin sections prepared for microscopic examination are transluscent. When dry, its electrical conductivity is low. Its heat and sound conductivities are medium, but these are reduced greatly when light-weight aggregates are used, or when the unit weight is decreased by some other procedure.

Durability:

Without any protective treatments, the resistance to most destructive agents is high. Protection against some acids and some soluble sulfates is desirable.

Placement:

One of the greatest advantages of concrete over most other building materials is the facility of placement. It is cast into the shape desired, and requires no cutting, fitting, nor fastenings.

Appearance:

Without decoration, concrete has a pleasing appearance. There are many decorative processes that may be applied, such as the selection of special aggregates, exposure of aggregates, surface treatments of various types, and the application of pigments or other coloring materials, either integrally or on the surface. Also, concrete may be employed for decorating other materials.

Economy:

Concrete is one of the most economical materials of construction, when the low maintenance cost is taken into account. In general, all the components of concrete, with the exception of cement, are natural products, and cement is one of the lowest-priced manufactured commodities. This may be verified by comparing the price of a cubic yard of concrete with that of 1000 board feet of lumber, and with a ton of steel.

Density:

Ordinary concrete is a rather heavy material. As has been stated, however, the unit weight may be decreased by a number of procedures.

Relation to other materials:

Concrete excels most other materials. In connection with almost every construction project, concrete is used, or could be used, to advantage.

Concrete Manufacture

One of the most important factors in the growth of the concrete industry is the facility of placement and the diversity of manipulation which this makes possible. It seems sometimes that every possible modification in mixing and placing has already been employed, but something new appears frequently. The following outline indicates some of the principal procedures.

1. Handling cement and aggregates
    a) Premixing of cement and aggregates
    b) Conveying cement, aggregates, and water separately to the mixer
    c) Conveying cement and water to the aggregates in place (soil-cement)
    d) Pumping cement, water, and sand into coarse aggregates in place (pressure grouting of railroad ballast)
    e) Laying masonry units with mortar (mortar considered a type of concrete)
2. Mixing
    a) Hand
    b) Mechanical
3. Place of mixing
    a) Mixing on the job
    b) Central mixing (ready-mixed concrete)
    c) Transit mixing (ready-mixed concrete)
4. Place of casting
    a) Casting in place
    b) Precasting (concrete products)

5. Reinforcing
   a) Plain concrete
   b) Reinforced concrete
6. Forms
   a) Forms on bottom and sides
   b) Forms on one side, or bottom (Gunite)
7. Equipment for placing
   a) Buggies
   b) Chutes, or tremies
   c) Pumps (Pumpcrete)
   d) Nozzles (Gunite)
   e) Trowels (masonry units, or stucco)
   f) Combinations of two or more of these
8. Vibration
   a) Placing without vibration
   b) Placing with vibration
9. Prestressing of reinforcing steel
   a) No prestressing
   b) Prestressing.

Concrete may be divided roughly into three groups. The numbers and letters in parentheses refer to the preceding outline.

A. Concrete in general.
B. Ready-mixed concrete (3 b, c).
C. Concrete products (4 b).

The division lines are not well defined; in fact, there is much interrelation. Some of the concrete listed under A may be listed under B also.

The precasting of relatively large structural members is increasing surprisingly.

Ready-Mixed Concrete

Formerly it was necessary for the cement user, large or small, to obtain supplies of cement, water, and aggregates, and then mix them to produce his concrete. Now, if he prefers, he can buy ready-mixed concrete by the cubic yard, just as he buys any other manufactured commodity.

It is said that one commercial ready-mixed plant was in operation as early as 1913. In 1925, there were 25 plants; and in 1929, 100 plants. By 1941, the number of commercial plants had increased to 703, having an estimated total output in that year of 12 million cubic yards of concrete. It is estimated that during 1959 more than half the cement shipped went to ready-mixed concrete companies. An important factor in the rapid growth in the use of ready-mixed concrete is the work of the National Ready Mixed Concrete Association, organized in 1930.

The specifications for ready-mixed concrete, ASTM C 94–63, follow this chapter.

In *Recommended Practice for Measuring the Uniformity of Concrete Produced in Truck Mixers*, issued by the National Ready Mixed Concrete Association (Publication No. 105, April, 1962), we find

> The suggested procedures and criteria of evaluation are based on researches conducted in the Joint Research Laboratory of the National Ready Mixed Concrete Association at the University of Maryland.

Concrete Products

It has been stated that concrete may be either cast in place or precast. Precasting is the basis of the various branches of the concrete products industry. Although it is mentioned here as a division of the concrete industry, it really is a very large and important industry in its own right. Some types of concrete products have been manufactured for many years. Probably the first block was made 80 or 90 years ago.[2] A concrete pipe sewer was built at Mohawk, New York, in 1842.[3]

There are two types of concrete products. One type includes building units that are employed in groups, usually held together by mortar, for construction purposes. Examples are block, brick, roofing tile, siding shingles, floor tile, cast stone, and pipe. These are manufactured and sold in competition with clay products primarily, but they also compete with products made from steel, wood, and some other materials.

An interesting fact may be mentioned in comparing the manufacture of concrete products with corresponding clay products. In the case of concrete, the only high temperatures required are in connection with the manufacture of cement. Starting with the cement as delivered, no fuel is required, although relatively small quantities may be used to facilitate curing. Each clay unit requires individual heat treatment, which, in ordinary American manufacturing practice, requires more or less elaborate plant equipment. In some places,[4] clay products are manufactured with little heating, employing crude equipment. Such units could not be manufactured commercially in this country, because of their inferior quality, and the low production per man-hour.

The other type of concrete products includes items that are used individually, such as bath tubs, shower basins, septic tanks, sash weights, joists, posts, and piles. Concrete is very suitable for the manufacture of such items. The list of them is growing rapidly, and there is no reason to believe that such growth will not continue.

The following are some of the specifications of the American Society for Testing and Materials of interest in connection with concrete products:

C	55–64T	Concrete Building Brick
C	90–64T	Masonry Units, Concrete, Hollow Load-Bearing
C	129–64	Masonry Units, Concrete, Hollow Non-Load-Bearing
C	145–64T	Masonry Units, Concrete, Solid Load-Bearing
C	139–65	Masonry Units, Concrete, for Construction of Catch Basins and Manholes
C	14–65	Sewer Pipe, Concrete
C	76–65T	Culvert Pipe, Reinforced Concrete
C	118–65T	Irrigation Pipe, Concrete

Recent Developments

The activity and progress of the concrete products industry are indicated by the trade associations that have been organized. There are the National Concrete Masonry Association, the American Concrete Pipe Association, and many others.

During the last decade or two, the development of concrete has proceeded along two interrelated lines. The first is general improvement resulting from many items, such as developments in cements, more careful selection of concrete materials, better proportioning methods, more thorough curing, increasing technical data, and the like.

The second is a series of individual developments. Most of these are independent of one another, but dependent on the general improvement in concrete technique. Some are important now, and others seem destined to become important eventually. Some examples are the vibration of concrete, prestressing of reinforcing steel, ashlar block, architectural concrete, and soil-cement.

The desirability of agitating concrete during and just subsequent to placing has been known for years.[5] It was not, however, until studies revealed the desirable ranges of amplitude and frequency, convenient apparatus, and procedures for applying vibration either to the forms or to the concrete itself that vibration became commercially important. Vibration permits the use of less water for a given mixture of cement and aggregates. This means a lower water-cement ratio and, consequently, a concrete of higher strength for the same quantity of cement per cubic yard of concrete.

The prestressing of reinforcing steel is a procedure for increasing the strength of reinforced concrete. It has some commercial applications at present, and the application may be expected to increase greatly. There are some difficulties, but difficulties in the way of important procedures have been solved before.

Until a few years ago, it was the practice to use only one size of concrete block in a wall. Someone conceived the idea of making blocks in several lengths and thicknesses, thus permitting walls of ashlar design to be constructed, greatly increasing the attractiveness, with no increase in material and little increase in total cost.

Architectural concrete is used on the basic principle that concrete is in itself an architectural material. It can be used as a decorative material, and need not be used exclusively for load-bearing. Recognition of this principle has led to the construction of many beautiful buildings with a characteristic individual type of architecture not in any way imitating buildings constructed from other materials. A considerable advance in technique is the casting of decorative features, such as statuary, integrally. It was the previous practice to precast such items, and set them into the walls, with mortar. This is an advantage that architectural concrete has in competition with stone masonry.

Soil-cement is an important development in the concrete field, requiring large quantities of cement. It permits the use of soil as the aggregate, and it has been found to have great merit for secondary roads, and some other projects.

In the preface to the second edition, reference is made to the many changes that have come about since the publication of the first edition. In the technology of concrete the changes have resulted either from processes that originated since the first edition, or from those that have received much development and recognition since that time.

The improvements are so numerous that only a few of the processes can be mentioned here, and it is difficult to select them. It is not intended to suggest that these are the most important items.

A division into three groups on the basis of utility and appearance may be desirable.

1. Utility is the principal objective. Appearance receives much attention, and additional cost may be authorized in its behalf, but appearance is secondary.
2. Primarily, utility is the principal objective. Because of certain factors, however, appearance is almost equally important.
3. Appearance is the principal objective. If the construction is to continue sufficiently long to justify the cost, the other characteristics of concrete must be taken into account.

   Usually, but not always, an object is functional in addition to decorative, to ensure attention and appreciation.

Examples of structures illustrating each of these three groups are given in the next chapter.

Formerly, concrete blocks were manufactured primarily for their structural value. Decorative effects were considered separately—for example, the application of stucco and paint. Now one sees a great variety of units of open-work patterns, designed for decorative effects. At first the decorative units were marketed for general use, and a given design could be seen in many locations. The next step was to design large units for use in one structure exclusively. In fact, it is said that in one case the contract specifies that the mold be destroyed so that the same design cannot be employed for any other project.

The cement industry could not have reached its present status without basic developments in science and in engineering. But there are examples of cooperation. In many cases cement facilities science and engineering advancements. Some examples will be pointed out in Chapter 22.

Three of the important advances in concrete since the publication of the first edition are *prestressed concrete, shell concrete,* and *concrete products*—although none of these originated during the period.

Shell Design Seminar, sponsored by Illinois Institute of Technology and Portland Cement Association, was held at the Institute in June, 1962. The purpose was "to present to practicing architects and engineers some of the latest techniques of design and construction of thin shell structures in reinforced concrete."

In his paper "100,000,000 Years of Housing and City Planning," Greeley [6] explains that

> The oceans became so teeming with life — most of it predatory — that it wasn't safe to stay out nights.
>
> So the housing demand was so acute that things began to happen. Oysters and clams and periwinkles and whelks got themselves houses of synthetic limestone, with hinged doors—strong doors.

This goes back to really early times. It is interesting that today no better basic building substance than limestone (the principal raw material for cement manufacture) has been found.

REFERENCES

1. G. A. Maney, *Civil Eng.,* 14:496 (1944).
2. *Properties and Manufacture of Concrete Building Units,* Circ. 304; U.S. Bureau of Standards, Washington, D. C., 1926.
3. M. W. Loving, *Am. Concrete Pipe Assoc.,* Bull. 17, Chicago, 1938.
4. J. C. Witt, *Philippine J. Sci.,* 13, Sect. A(2):59 (1918).
5. D. A. Abrams, *Proc. Am. Concrete Inst.,* 15:63 (1919).
6. W. R. Greeley, *Journal American Institute of Architects,* 24(4): 147 (1955).

Standard Specifications for

READY-MIXED CONCRETE[1]

ASTM Designation: C 94 – 63

Adopted, 1958, Revised, 1961, 1962, 1963.[2]

This Standard of the American Society for Testing and Materials is issued under the fixed designation C 94; the final number indicates the year of original adoption as standard or, in the case of revision, the year of last revision.

Scope

1. (*a*) These specifications cover the requirements for ready-mixed concrete. Requirements for quality of materials and for proportions and quality of concrete shall be either as hereinafter specified or as specified by the purchaser by reference to applicable general specifications for concrete. In any case where the requirements of such general specifications are in conflict with these specifications, the requirements of the general specifications shall govern, unless otherwise specified by the purchaser.

(*b*) For the purpose of these specifications, ready-mixed concrete is portland-cement concrete manufactured for delivery to a purchaser in a plastic and unhardened state and delivered as hereinafter specified.

(*c*) These specifications do not cover the placement, consolidation, curing, or protection of the concrete after delivery to the purchaser.

Basis of Purchase

2. (*a*) The basis of purchase shall be the cubic yard of plastic and unhardened concrete as delivered to the purchaser.

(*b*) The volume of plastic and unhardened concrete in a given batch shall be determined from the total weight of the batch divided by the actual weight per cubic foot of the concrete. The total weight of the batch shall be calculated either as the sum of the weights of all materials, including water, entering the

---

[1] Under the standardization procedure of the Society, these specifications are under the jurisdiction of the ASTM Committee C-9 on Concrete and Concrete Aggregates.

[2] Prior to their present adoption as standard, these specifications were published as tentative from 1933 to 1935. They were adopted in 1935, and published as standard from 1935 to 1942, being revised in 1938, but were further revised and issued as tentative in 1941, being revised in 1942, 1943, and 1944. They were adopted in 1944 and published as standard from 1944 to 1947, being revised in 1947 and republished as tentative from 1947 to 1948. They were again adopted in 1948, and published as standard from 1948 to 1954. They were reverted to tentative and revised in 1954, and were published as tentative from 1954 to 1958, being revised in 1955 and 1958.

batch or as the net weight of the concrete in the batch as delivered. The weight per cubic foot shall be determined in accordance with the Method of Test for Weight per Cubic Foot, Yield, and Air Content (Gravimetric) of Concrete (ASTM Designation: C 138).[3]

Materials

3. In the absence of designated applicable specifications covering requirements for quality of materials, the following specifications shall govern:

(a) *Cement.*—Cement shall conform to the Specifications for Portland Cement (ASTM Designation: C 150),[3] the Specifications for Air-Entraining Portland Cement (ASTM Designation: C 175)[3] or the Specifications for Portland Blast-Furnace Slag Cement (ASTM Designation: C 205),[4] or the Specifications for Portland-Pozzolan Cement (ASTM Designation: C 340).[5] These different cements will produce concretes of different properties and should not be used interchangeably. The purchaser should specify the type or types required, but if no type is specified, the requirements of type I as prescribed in Specifications C 150 shall apply.

(b) *Aggregates.*—Aggregates shall conform to the Specifications for Concrete Aggregates (ASTM Designation: C 33).[3]

(c) *Water.*—Water shall be clear and free from injurious amounts of oil, acid, alkali, organic matter, or other deleterious substances.

(d) *Admixtures.*—Where it is desired to use admixtures, such use shall be provided for in the contract or in the designated applicable specifications. When air-entraining concrete is specified and an admixture is used to secure the desired air content, the admixture shall conform to the Specifications for Air-Entraining Admixtures for Concrete (ASTM Designation: C 260).[5]

Quality of Concrete

4. In the absence of designated applicable general specifications, the purchaser shall select one of the two following alternate bases for specifying the quality of the concrete:

Alternate No. 1:

(a) When the purchaser assumes responsibility for the design of the concrete mixture, he shall specify the following:

(1) Cement content in bags per cubic yard of concrete, or equivalent units.

(2) Designated size, or sizes, of coarse aggregate.

(3) Maximum allowable water content in gallons per bag of cement, or equivalent units, including surface moisture, but excluding water of absorption, of the aggregates (Note 1).

(4) Slump, or slumps, desired at the point of delivery (see Section 5 for acceptable tolerances).

(5) When air-entraining concrete is specified, the maximum and minimum limits for air content of samples taken from the transportation unit at the point of discharge (Note 2).

NOTE 1.—The purchaser, in selecting the requirements for water-cement ratio and compressive strength, should give consideration to requirements for durability, surface texture, and density, in addition to those for structural design. The purchaser is referred to the Recommended Practice for Selecting Proportions for Concrete of Committee 613-54, American Concrete Institute, for the selection of proportions that will result in concrete suitable for various types of structures and conditions of exposure.

NOTE 2.—Tests for air content, both preliminary to construction and routine tests for control purposes during construction, are required. For the range in aggregate sizes commonly used in ready-mixed concrete, the air content specified should be from 3 to 6 per cent. Any amount

---

[3] 1961 Book of ASTM Standards, Part 4.
[4] 1962 Supplement to Book of ASTM Standards, Part 4.
[5] Appears in this publication, see Contents in Numeric Sequence of ASTM Designations at front of book.

less than 3 per cent may not give the required resistance to weathering, which is the primary purpose of using air-entraining concrete, whereas an air content in excess of 6 per cent may reduce the strength without contributing additional protection.

(b) Prior to the actual delivery of the concrete, the manufacturer shall furnish a statement to the purchaser giving the properties of the materials and the proportions by weight (dry) of cement and of fine and coarse aggregates that are proposed to be used in the manufacture of each class of concrete ordered by the purchaser.

Alternate No. 2:

(a) When the purchaser requires the manufacturer to assume responsibility for the design of the concrete mixture, the purchaser shall specify the following:

(1) Minimum allowable compressive strength as determined on samples taken from the transportation unit at the point of discharge. The basis shall be the compressive strength at 28 days (Note 1). The purchaser shall specify the requirements for strength in terms of tests of standard specimens cured under standard laboratory conditions for moist curing (see Section 18).

(2) Designated size, or sizes, of coarse aggregate.

(3) Slump or slumps desired at the point of delivery (see Section 5 for acceptable tolerances).

(4) When air-entraining concrete is specified, the maximum and minimum limits for air content of samples taken from the transportation unit at the point of discharge (Note 2).

(b) Prior to the execution of the contract, the manufacturer shall furnish a statement to the purchaser, giving the proportions by weight (dry) of cement and of fine and coarse aggregates that will be used in the manufacture of each class of concrete ordered by the purchaser. He shall also furnish evidence satisfactory to the purchaser that the proportions selected will produce concrete of the quality specified.

Tolerances in Slump

5. When the specified slump is 3 in. or ess, the tolerance shall be plus or minus ½ in. When the specified slump is greater than 3 in., the tolerance shall be plus or minus 1 in.

Measuring Materials

6. (a) Cement shall be measured by weight or, if permitted by the purchaser, in bags of standard weight (Note 3). When cement is measured by weight, it shall be weighed on a scale separate from those used for other materials. When cement is measured in bags, no fraction of a bag shall be used unless weighed. The cement, as weighed, shall be within ±1 per cent of the required weight.

Note 3.—In the United States the standard weight is 94 lb.

(b) Aggregate shall be measured by weight. Batch weights shall be based on dry materials and shall be the required weights of dry materials plus the total weight of moisture (both absorbed and surface) contained in the aggregate. The individual aggregates, as weighed, shall be within ±2 per cent of the required weights and the total weight of the aggregates shall be within ±1 per cent of the required weight.

(c) Water shall be measured by volume or by weight. The device for the measurement of the water shall be readily adjustable and, under all operating conditions, shall have an accuracy within 1 per cent of the quantity of water required for the batch. The device shall be so arranged that the measurements will not be affected by variable pressures in the water supply line. Measuring tanks shall be equipped with outside taps and valves to provide for checking their calibration, unless other means are provided for readily and accurately determining the amount of water in the tank. In the case of truck mixers, if wash water is permitted to be used as a portion of the mixing water for succeeding batches, it shall be accurately measured in a separate tank, provided for the purpose, and taken into account in determining the amount of additional mixing water required. Water, as weighed or measured, shall be within ±1 per cent of the required amount.

(d) Powdered admixtures shall be measured by weight, and paste or liquid admixtures by weight or volume, within a limit of accuracy of 3 per cent. When admixtures are used in small quantities in proportion to the cement, as in the case of air-entraining admixtures, mechanical dispensing equipment is strongly recommended.

Batching Plant

7. (a) Bins with adequate separate compartments for fine aggregates and for each required size of coarse aggregate shall be provided in the batching plant. Each compartment shall be designed to discharge efficiently and freely into the weighing hopper. Means of control shall be provided so that, as the quantity desired in the weighing hopper is approached, the material may be added slowly and shut off with precision. A port or other opening for removing an overload and sampling materials from the hopper shall be provided. Weighing hoppers shall be constructed so as to eliminate accumulations of tare materials and to discharge fully.

(b) The scales for weighing aggregates and cement may be of either the beam type or the springless-dial type. Adequate standard test weights shall be available for checking accuracy. All exposed fulcrums, clevises, and similar working parts of scales shall be kept clean. When beam-

type scales are used, provision shall be made for indicating to the operator that the required load in the weighing hopper is being approached; the device shall indicate at least the last 200 lb of load. All weighing and indicating devices shall be in full view of the operator while charging the hopper, and he shall have convenient access to all controls.

Mixers and Agitators

8. (a) Mixers may be stationary mixers or truck mixers. Agitators may be truck mixers or truck agitators. Each mixer and agitator shall have attached thereto in a prominent place a metal plate or plates on which are plainly marked, for the various uses for which the equipment is designed, the capacity of the drum or container in terms of the volume of mixed concrete and the speed of rotation of the mixing drum, blades, or paddles. Stationary mixers shall be equipped with an acceptable timing device that will not permit the batch to be discharged until the specified mixing time has elapsed. Truck mixers shall be equipped with means by which the number of revolutions of the drum, blades, or paddles may be readily verified.

(b) The mixer, when loaded to capacity, shall be capable of combining the ingredients of the concrete within the specified time into a thoroughly mixed and uniform mass and of discharging the concrete with a satisfactory degree of uniformity as indicated in Paragraph (d).

(c) The agitator, when loaded to capacity, shall be capable of maintaining the mixed concrete in a thoroughly mixed and uniform mass and of discharging the concrete with a satisfactory degree of uniformity as indicated in Paragraph (d).

(d) The purchaser may, from time to time, make slump tests of individual samples taken at approximately the one-quarter and the three-quarter points of the load and if the slumps differ by more than 2 in., the mixer or agitator shall not be used unless the condition is corrected, except as provided in Paragraph (e).

(e) Use of the equipment may be permitted under conditions determined by the purchaser when operation with a longer mixing time or with a smaller load will permit the requirements of Paragraph (d) to be met.

(1) Mixed completely in a stationary mixer and the mixed concrete transported to the point of delivery in a truck agitator or in a truck mixer operating at agitating speed or in non-agitating equipment approved by the purchaser and meeting the requirements of Section 10(a). (Known as *central-mixed* concrete.)

(2) Mixed partially in a stationary mixer, and the mixing completed in a

TABLE I.—REQUIREMENTS ON UNIFORMITY OF CONCRETE TO ALLOW REDUCTION IN MIXING TIME FOR STATIONARY MIXERS.

Test	Requirement, Expressed as Maximum Permissible Difference in Results of Tests of Samples Taken from Two Locations in the Concrete Batch
Weight per cubic foot, calculated to an air-free basis, lb..........	1.0
Air content, per cent by volume of concrete.................	1.0
Slump:	
If average slump is 3 in. or less, in......................	3/4
If average slump is more than 3 in., per cent of average slump..	25
Coarse aggregate content, portion by weight of each sample retained on No. 4 (4760-mm) sieve, per cent.............	6.0
Cement content or unit weight of mortar:[a]	
Alternate No. 1.—Cement content[b] based on average for all comparative samples tested, per cent..................	7.0
Alternate No. 2.—Unit weight of air-free mortar[c] based on average for all comparative samples tested, per cent........	1.6
Average compressive strength at 7 days for each sample, based on average strength of all comparative test specimens, per cent.................................	7.5

[a] Alternate requirements.
[b] "Proposed Tentative Method of Test for Cement Content of Freshly Mixed Concrete," ASTM BULLETIN No. 239, July, 1959, p. 48.
[c] "Test for Variability of Constituents in Concrete," Designation 26, Bureau of Reclamation *Concrete Manual*, 6th edition, p. 447.

(f) Mixers and agitators shall be examined daily for changes in condition due to accumulations of hardened concrete or mortar or to wear of blades. When any such change of condition is found, the tests described in Paragraph (d) should be repeated.

Mixing and Delivery

9. (a) Ready-mixed concrete shall be mixed and delivered to the point designated by the purchaser by means of one of the following combinations of operations:

truck mixer. (Known as *shrink-mixed* concrete.)

(3) Mixed completely in a truck mixer. (Known as *transit-mixed* concrete.)

(b) Mixers and agitators shall be operated within the limits of capacity and speed of rotation designated by the manufacturer of the equipment.

(c) When a stationary mixer is used for the complete mixing of the concrete, mixing time shall be measured from the time all the solid materials are in the drum. The batch shall be so charged into the mixer that some water will enter in advance of the cement and aggregate, and all water shall be in the drum by the end of the first one fourth of the specified mixing time.

Where no mixer performance tests are made, the acceptable mixing time for mixers having capacities of 1 cu yd or less shall be not less than 1 min. For mixers of greater capacity, this minimum shall be increased 15 sec for each cubic yard or fraction thereof of additional capacity.

Where mixer performance tests have been made on given concrete mixtures in accordance with the testing program set forth in the following paragraphs, and the mixers have been charged to their rated capacity, the acceptable mixing time may be reduced for those particular circumstances to a point at which satisfactory mixing defined as follows shall have been accomplished:

(1) Samples of concrete for comparative purposes shall be obtained immediately after arbitrarily designated mixing times, in accordance with one of the following procedures:

Alternate Procedure 1.—The mixer shall be stopped, and the required samples removed by any suitable means from the concrete at equal distances from the front and back of the drum, or

Alternate Procedure 2.—As the mixer is being completely emptied, appropriate samples shall be obtained from the second and ninth-tenth portions of the batch. Any appropriate method may be used, provided the samples are representative of widely separated portions, but not the very ends of the batch.

(2) The samples of concrete shall be tested in accordance with methods listed in Section 18, and differences in test results for the two samples shall not exceed those given in Table I.

Mixer performance tests shall be repeated whenever the appearance of the concrete or the coarse aggregate content of samples selected as outlined in this section indicates that adequate mixing has not been accomplished.

(d) When a stationary mixer is used for partial mixing of the concrete (shrink-mixing), the mixing time in the stationary mixer may be reduced to the minimum required (about 30 sec) to intermingle the ingredients. Mixing shall be completed in a truck mixer by not less than 50 nor more than 100 revolutions of the drum or blades at the rate of rotation designated by the manufacturer of the equipment as mixing speed. Additional mixing, if any, shall be at the speed designated by the manufacturer of the equipment as agitating speed.

(e) When the concrete is mixed in a truck mixer loaded to its maximum capacity, the number of revolutions of the drum or blades at mixing speed shall be not less than 70 nor more than 100. If the batch is at least $\frac{1}{2}$ cu yd less than the maximum capacity, the number of revolutions at mixing speed may be reduced to not less than 50. All revolutions after 100 shall be at agitating speed.

(f) When a truck mixer or truck agitator is used for transporting concrete that has been completely mixed in a stationary mixer, mixing during transportation shall be at the speed designated by the manufacturer of the equipment as agitating speed.

(g) When a truck mixer or agitator is used for transporting concrete, the concrete shall be delivered to the site of the work and discharge shall be completed within $1\frac{1}{2}$ hr, or before the drum has been revolved 300 revolutions, whichever comes first, after the introduction of the mixing water to the cement and aggregates, or the introduction of the cement to the aggregates, unless a longer time is specifically authorized by the purchaser. In hot weather, or under conditions contributing to quick stiffening of the concrete, a time less than 1½ hr may be specified by the purchaser. When a truck mixer is used for the complete mixing of the concrete, the mixing operation shall begin within 30 min after the cement has been intermingled with the aggregates.

(h) Concrete delivered in cold weather shall have the applicable minimum temperature indicated in the following table. (The purchaser shall inform the producer as to the type of construction for which the concrete is intended.)

| Air Temperature deg Fahr | Minimum Concrete Temperature, deg Fahr | |
	Thin Sections and Unformed Slabs	Heavy Formed Sections and Mass Concrete
30 to 45..........	60	50
0 to 30..........	65	55
Below 0..........	70	60

The maximum temperature of concrete produced with heated aggregates, heated water, or both, shall at no time during its production or transportation exceed 90 F (Note 4).

NOTE 4.--When cement is at a low temperature and hot mixing water is used, flash set may occur if hot water is brought in direct contact with the cement.

(i) Every effort shall be made to maintain temperature of concrete produced during hot weather as low as possible. In some situations difficulty may be encountered with concrete at temperatures approaching 90 F.

Use of Nonagitating Equipment

10. Central-mixed concrete that is designed for the purpose may be transported in suitable nonagitating equipment approved by the purchaser. The proportions of the concrete shall be approved by the purchaser and the following limitations shall apply:

(a) Bodies of nonagitating equipment shall be smooth, watertight, metal containers equipped with gates that will permit control of the discharge of the concrete. Covers shall be provided for protection against the weather when required.

(b) The concrete shall be delivered to the site of the work in a thoroughly mixed and uniform mass and discharged with a satisfactory degree of uniformity as prescribed in Paragraph (c). Discharge shall be completed within 45 min after the introduction of the mixing water to the cement and aggregates.

(c) Slump tests of individual samples taken at approximately the one-quarter and the three-quarter points of the load during discharge shall not differ by more than 2 in. If the slumps differ by more than 2 in., the nonagitating equipment shall not be used unless the conditions are corrected as provided in Paragraph (d).

(d) If the requirements of Paragraph (c) are not met when the nonagitating equipment is operated at maximum capacity, for the maximum time of haul, and with the concrete mixed the minimum time, the equipment may still be used when operated using smaller loads, shorter hauls, or longer mixing times, or combinations thereof that will permit the requirements of Paragraph (c) to be met.

Inspection

11. Proper facilities shall be provided for the purchaser to inspect ingredients and processes used in the manufacture and delivery of the concrete. The manufacturer shall afford the inspector representing the purchaser all reasonable facilities, without charge, for securing samples to determine whether the concrete is being furnished in accordance with these specifications. All tests and inspections shall be so conducted as not to interfere unnecessarily with the manufacture and delivery of the concrete.

Sampling

12. Samples of concrete shall be ob-

tained in accordance with the Method of Sampling Fresh Concrete (ASTM Designation: C 172),[3] except in the case of individual samples secured to determine uniformity of consistency. In securing individual samples to determine uniformity of consistency, as provided in Sections 8(d) and 10(c), of these specifications, Method C 172 shall be followed, but the requirements shall be so modified to permit obtaining two samples, one at the one-quarter point and one at the three-quarter point of the load.

Slump and Air Content

13. Slump tests shall be made at the option of the purchaser. Determinations of air content shall be made at the option of the purchaser. If the measured slump or air content falls outside the limits specified, a check test shall be made immediately on another portion of the same sample. In the event of a second failure, the concrete shall be considered to have failed to meet the requirements of the specifications.

Certification

14. The manufacturer of the ready-mixed concrete shall furnish to the purchaser, for each type of concrete and when any change in composition is made, a statement showing the quantities of materials used in making the concrete.

Strength

15. (a) When strength is used as a basis for acceptance of concrete, standard specimens shall be made in accordance with the Method of Making and Curing Concrete Compression and Flexure Test Specimens in the Field (ASTM Designation: C 31).[4] The specimens shall be cured under standard moisture and temperature conditions in accordance with Section 7(a) and (b) of Method C 31 (see Section 18). Strength tests shall be made frequently by the purchaser and, in general, not less frequently than one strength test for each 50 loads of each class of concrete, except that in no case shall a given class of concrete be represented by less than three tests.

(b) For a strength test three standard test specimens shall be made from a composite sample secured as required in Section 12. The test result shall be the average of the strengths of the three specimens, except that, if one specimen in a test shows manifest evidence of improper sampling, molding, or testing, it shall be discarded and the remaining two strengths averaged. Should more than one specimen representing a given test show definite defects, due to improper sampling, molding, or testing, the entire test shall be discarded.

(c) The representative of the purchaser shall ascertain and record the delivery ticket number for the concrete and the exact location in the work at which each load represented by a strength test is deposited.

(d) To conform to the requirements of these specifications, the average of all of the strength tests (see Paragraph (b)) representing each class of concrete, as well as the average of any five consecutive strength tests representing each class of concrete, shall be equal to or greater than the specified strength, and not more than one test in ten shall have an average value less than 90 per cent of the specified strength.

Alternate Strength Basis

16. The provisions of Section 15 may be waived, at the option of the purchaser, if the manufacturer can furnish evidence satisfactory to the purchaser that concrete of the proportions, and made with the materials, that he proposes to use will have the specified strength.

Failure to Meet Strength Requirements

17. In the event that concrete tested in accordance with the requirements of Section 15 fails to meet the strength requirements of these specifications, the manufacturer of the ready-mixed concrete and the purchaser shall confer to determine whether agreement can be reached as to what adjustment, if any, shall be made. If an agreement on a mutually satisfactory adjustment cannot be reached by the manufacturer and the purchaser, a decision shall be made by a panel of three qualified engineers, one of whom shall be designated by the purchaser, one by the manufacturer, and the third chosen by these two members of the panel. The question of responsibility for the cost of such arbitration shall be determined by the panel. Its decision shall be binding, except as modified by a court decision.

Methods of Sampling and Testing

18. Methods of testing ready-mixed concrete shall be in accordance with the following methods of the American Society for Testing and Materials:

(a) *Compression Test Specimens.*—Method C 31, using standard moist curing in accordance with Section 7(a) and (b) of Method C 31.

(b) *Compression Tests.*—Method of Test for Compressive Strength of Molded Concrete Cylinders (ASTM Designation: C 39).[3]

(c) *Yield.*—Method C 138.

(d) *Air Content.*—Method C 138, Method of Test for Air Content of Freshly Mixed Concrete by the Volumetric Method (ASTM Designation: C 173),[3] or Method of Test for Air Content of Freshly Mixed Concrete by the Pressure Method (ASTM Designation: C 231).[3]

(e) *Slump.*—Method of Test for Slump of Portland Cement Concrete (ASTM Designation: C 143).[3]

(f) *Sampling Fresh Concrete.*—Method of Sampling Fresh Concrete (ASTM Designation: C 172).[3]

Chapter 22

SOME CONCRETE ACCOMPLISHMENTS

Some examples of concrete construction representing each of the three groups mentioned in Chapter 21 are listed here. The groups are not sharply separated. One group may tend to blend into another.

New records in concrete construction appear rather frequently. When it was completed, the thirty-nine story, 371-foot Executive House in Chicago (Fig. 22:1; photograph supplied by the Portland Cement Association.) was the highest reinforced concrete structure in the United States. The Chicago river in the foreground connects with the Chicago ship and drainage canal, one of the principal water ways of the world. In return for the millions of barrels of cement used in the construction of this waterway, and its environment, it serves the interests of the cement industry by transporting great tonnages of fuels, cement, and aggregates.

Figure 22:2 is taken from an artist's drawing of the east terminal of the Congress Street Expressway, Chicago, kindly supplied by the chief engineer of the Bureau of Engineering. The extensive use of concrete is evident. The Chicago river is in the background. The estimated cost of the project is one hundred million dollars, including right-of-way, construction, and engineering. Many records were set. The South Halsted Street interchange was planned for a capacity of 170,000 vehicles in a 24-hour period, to become "the busiest crossroads in the world."

Chicago is on the National System of Interstate and Defense Highways, which is to be 41,000 miles long and cost

41 billion dollars. The system is said to be the world's largest construction project.

The American Association of State Highway Officials has released a report on the road test near Ottawa, Illinois. The testing required two years, represented 17 million traffic miles, and cost 27 million dollars. The average serviceability index was very good for concrete, and fair for asphalt.

At the Dresden Nuclear Power Station at the junction of the Kankakee and DesPlaines rivers, near Chicago, the most spectacular external feature is the steel sphere. This is 190

Fig. 22:1. Executive House, Chicago
(*From* Portland Cement Association)

Fig. 22.2. CONGRESS STREET EXPRESSWAY, CHICAGO
(*From* Bureau of Engineering, Chicago)

feet in diameter, and consists of 3,500 tons of welded steel
plates 1.25 to 1.4 inches thick. The support and shielding
wall consists of 27,000 cubic yards of concrete, containing
heavy aggregates.

This group should include some reference to the immense
McCormick Place, one of the world's largest exposition cen-
ters, on Chicago's lake front. Its length of hundreds of feet,
its seating capacity for thousands, its many display facilities,
and multi-million-dollar cost have received so much world-
wide, well-deserved publicity that no additional figures need
be given here. Every visitor to Chicago should see it. In fact,
he could hardly avoid seeing it if he were anywhere in its
general locality.

The Baha'i Temple (Fig. 22:3) resulted from a blend of
chemistry, geology, engineering, architecture, and craftsmanship. John J. Early, a former president of the American Concrete Institute, frequently used "craftsmanship" in describing
his outstanding work with concrete. A full-sized model of the

dome was constructed in his studio before the concrete slabs were cast. His reputation for beautiful concrete had long been established.

This Temple is the only non-modern structure shown in both editions of *Portland Cement Technology*. Possibly this repetition is a matter of sentiment. During this construction, I never missed an opportunity to see and admire the progress of the work. White cement and white aggregates were used. The photograph was obtained from the Portland Cement Association.

The Mercer Museum, at Doylestown, Pennsylvania (Fig. 22:4), is an interesting reinforced concrete structure of distinctive architecture. Constructed in 1916, it is said to be "the world's first building literally built around and for a museum collection."

FIG. 22:3. BAHA'I TEMPLE, WILMETTE, ILLINOIS
(*From* Portland Cement Association)

FIG. 22:4. MERCER MUSEUM, DOYLESTOWN, PENNSYLVANIA
(*From* Bucks County Historical Society)

FIG. 22:5. OFFICE BUILDING, MIAMI, FLORIDA

My first glimpse of an office building under construction at 1111 Brickell Avenue on Miami's south shore attracted my interest, and this interest increased on my annual trips. By March, 1962, the building was completed and occupied, and inspection was permitted. The sun screen of specially-designed precast units extends nearly completely around the building. Decorations inside and out are beautiful. The architects, Polevitzky, Johnson, and Associates, Miami, gave permission to use the photograph shown in Figure 22:5.

FIG. 22:6. MARINA CITY, CHICAGO
(*From* ASHRAE *Journal*)

The record for height of the Executive House (Fig. 22:1) was not held for very long. The 60-story towers of Marina City, shown in Figure 22:6, makes this the world's highest reinforced concrete structure, and the highest apartment buildings. They are 588 feet high. Numerous techniques, innovations, and facilities could be mentioned. Two Linden climbing cranes were employed for lifting concrete and forms. Some light-weight aggregate, and some fiberglass forms were used. Over-all, it may be said that "Marine City represents not only an important advance in architecture and concrete engineering, but also a new concept in urban living."

Before reading about the American Cement Corporation building in *Pit and Quarry* (September, 1961) I happened to see it at a distance while in Los Angeles, without knowing its name or history. Quoting from the journal:

The new American Cement Corporation building in Los Angeles is an impressive example of a bold concept in modern architecture. It was designed to be a show place of functionally and esthetically applied concrete construction. Structural members are exposed—comprised of precast elements originating in sculptured forms that take full advantage of the plasticity of reinforced concrete.

The building consists of a nine-story tower (above the parking levels) with a central service core, and a load-bearing grid outer wall comprised of 2-ton precast "X" members, each one story tall, forming the striking exterior. Each floor of the tower is suspended between the elevator core and the exterior structural "X" members, providing a columnless interior from the fourth to the 13th floors.

In a departure from conventional construction and its single structural theme, there is a change of structure within the building to accommodate a functional change. The "structure within a structure" originates between the fourth and fifth floors. At the latter level, a heavily reinforced concrete "transition" box girder transfers the weight from the exterior walls to eight monolithic columns that carry the load to the foundations. Above the fifth floor are the offices, and below the fourth are parking levels for 400 cars.

The building was erected at a cost of $3½ million—considered 10 percent less than would be required for the average comparable office building, due to the use of reinforced concrete construction throughout.

Figure 22:7 shows some of the precast units employed in constructing a portion of the Maule Industries building, Miami. This building has numerous novel and decorative features.

Some years ago, a paper describing a local office building of Universal Atlas Cement Company was headed: "In

FIG. 22:7. MAULE INDUSTRIES BUILDING, MIAMI

which we practice what we preach." White cement, long publicized by the company, was used at the front of the building. Figure 22:8, the office building of the American Concrete Institute, called the incident to mind. The construction demonstrates the combination of utility and attractive appearance that can be obtained with concrete.

FIG. 22:8. AMERICAN CONCRETE ASSOCIATION BUILDING, DETROIT

Words that adequately describe the poetry and beauty of the Fountain of Time, at the west end of the Midway Plaisance, Chicago, are not to be expected in an engineering vocabulary. So, concrete will be the subject of these comments. A recent visit to the Lorado Taft laboratories at the University of Chicago served to remind me of some of the facts which I had forgotten.

When the construction of this masterpiece was discussed, bronze, granite, and marble were considered. Because of the length and massiveness the cost of each was far out of range of the funds available, and it seemed for a time that the project could not proceed. Then someone mentioned the utility and economy of concrete—and the statue has adorned the locality for a long time.

For many years Lincoln Road has been one of the principal merchandise streets of Miami Beach. A few years ago the store owners decided that a little glamor should not discourage business, and Lincoln Mall resulted. For several blocks motor traffic was discontinued, and from curb to curb so many decorative objects appeared that one could hardly place a pencil point without encountering one of them. There are attractive trees, shrubs, and flowers. Decorative concrete is employed for floors, walls, roofs, boundaries of flower-beds, pools, and fountains.

There are examples of somewhat similar decorative effects, with concrete predominating, at merchandise districts in the vicinity of Chicago. One of the most recent is Oakbrook Center. These merchandise districts differ from the mall type in that decorative figures are not confined between curbs. Within such a district there is a series of buildings for stores, offices, and restaurants, interspersed are various units of concrete decorations.

Advancement of the industry is accelerating to such an extent that record-breaking events are almost commonplace. The world's tallest concrete building is being constructed at 1000 Lake Shore Plaza, Chicago. The Verrazano-Narrows Bridge, in New York—the world's longest suspension bridge —required almost 400,000 cubic yards of concrete. A paper recently abstracted in *Chemical Abstracts* has 333 references on the technology of concrete used in the nuclear industry. The increase in the uses of concrete is unlimited.

Chapter 23

THE IRIS COLOR PROCESS

Concrete may be either concealed or exposed. Apparently the percentage of exposed concrete is increasing—particularly during the last decade or two. Color is an important asset for all concrete that remains in sight, and it seems that this asset has received less study and effort than some other assets that may be considered less important. This situation has one advantage. Future prospects are brighter for color development now than if such development has been active in the past.

From time to time there are indications that the importance of color in the industry is being increasingly recognized to some extent. According to *Modern Concrete*,[1]:

> As home owners become more conscious of the decorative value of colored concrete products, a new market is opening up for producers.

In the same paper, this statement appears:

> All exposed concrete masonry lends itself readily to color, and adding pigment is no different from adding other materials.

My experience and observation are not in accord with the second portion of this statement. Color technique has its own characteristics and this is one of the factors that have retarded the application of color in the industry.

To be effective, the color of concrete must be bright and retain its brightness, rather than become dim. If there is deterioration due to selection of color materials of questionable quality, the quantity or application of these materials, or careless workmanship, a decrease in public appreciation may be expected. There are two objectives for the use of

color—utility and decoration. The utility of color for concrete highways was first suggested some years ago. When
we take into account the fact that color is not applicable to
highways other than those of concrete, progress is far less
rapid than might be expected.

The following is from a paper on "Concrete of Color." [2]

Every one knows about the flow of color that is taking
place, with no limitation in sight. Painting, printing, electric
lighting, enameling, photography, moving pictures, tele-
vision have all been influenced. Practically all materials of
construction, including paper, wood, plastics, and metals,
have been affected directly or indirectly.

There have been many cyclic tendencies and interinflu-
ences. Color photography led to color printing, which in turn
became prominent in advertising. Colored pictures were
found to be much more effective than black and white on
magazine pages, and this realization accelerated the flow of
color.

For simplicity, we shall consider that portland cement
concrete includes any mixture which contains cement and
water—such as paste (cement plus water), mortar (cement
plus water plus fine aggregates), and concrete (cement plus
water plus fine and coarse aggregates).

In the application of color, and in decoration in general,
concrete holds a rather unique position, in comparison with
other materials. Some factors have encouraged the application of color, while other factors have retarded this application. We shall first mention the encouraging factors.

When the forms have been removed from an ordinary concrete wall, for example, neither the surface texture nor the
color is considered attractive by the average person. The
tendency is to modify either the texture or color, or both,
in some way.

Texture may be modified in advance by the use of special
form linings. Or the surface may be roughened by bush-
hammering, or by applying a coating of stucco. There are
many other procedures which might be mentioned. Color
treatments will be discussed in subsequent sections.

Recently I was in the Miami district for two months. During that period I did not see a single square foot of concrete in a residence or hotel structure that was not modified in some way—either by surface treatment or by color application. In many cases the influence of color is emphasized by multiple-tone effects.

Discouraging factors comprise the technical difficulties and limitations that have been encountered in coloring concrete. These have retarded the application of color to concrete in comparison with most other materials.

In the preparation of a colored concrete, one important factor is the color of the base cement. By this is meant the color of the cement resulting from the processing of the clinker, without any addition or special procedure to affect the color. The most desirable base is the "whitest" possible cement obtainable. Next, cements a shade darker, near-whites, blends of whites and grays, lightest gray cements, and so on, receive consideration. While in general the color of a gray cement is not an important characteristic, I have known of color comparisons made on five or ten cements, in which the buyer selects the cement of the lightest color, although the plant producing this cement was farthest away. Obviously, concrete of a pastel shade requires a base cement that is as close as possible to perfect white. However, it should be pointed out that much colored concrete is produced from gray cements. In some cases it is possible to prepare a color pigment in such a way that the color of the base cement is neutralized to some extent.

The search for methods of producing colored concrete has included the application of almost every conceivable technique, but comparatively little research has been applied directly to the problem.

It is believed that the following list contains most of the procedures that have been employed, but it may not be complete. Chronology has not been taken into account.

1. Selection of cement raw materials
2. Modification of clinkering procedures
3. Blending of finished cements
4. Mixing of pigment * with finished cements, manually or mechanically

* The term pigment denotes any black or colored substance that does not combine chemically with any other substance in the formation of clinker.

5. Intergrinding cement and pigment
6. Dusting concrete (before hardening) with mixture of cement and pigment
7. Painting finished concrete
8. Staining finished concrete
9. Use of colored aggregates, as in terrazzo

Each of these procedures has at least one disadvantage, such as cost, labor, or limited durability. These disadvantages may be outlined as follows:

1, 2, 3. The scope of available coloring is too limited.
  4. It is almost impossible to blend finely divided materials satisfactorily.
  5. Intergrinding introduces an additional unit process.
 6, 7. These are surface applications only.
  8. Staining is limited in a number of respects.
  9. Beautiful effects may be obtained, but cost and some other considerations limit the method.

Although concrete is one of the world's most important building materials, and the acceleration of the demand for color is known by all, the coloring of concrete (in comparison with other materials) has hardly been started.

At the beginning of the research leading to the nonaqueous process (see Chapter 16), color was not taken into account. Later many advantages for color application at this point of cement manufacture became apparent. As has been stated, most of the present procedures for preparing colored concrete by the cement user, involve the mixing of a dry cement and a dry pigment, each in the form supplied by the manufacturer. This is not a satisfactory method.

In the nonaqueous process, a pigment is processed with the clinker, employing the same liquid. Then a blend of cement and pigment slurries, in any desired proportion, is prepared, and after the separation of the liquid, the colored cement is ready for use.

Incidentally, this new process is advantageous for the preparation of special cements—such as white portland, high alumina, and masonry—and for introducing admixtures into standard portland cement.

Until cement plants are fully equipped to supply colored cements, there is an attractive field for some intermediate manufacturer. Or, manufacturers of concrete products may prepare their own colored cements. The method is simple:

1. Obtain from a paint manufacturer a selected pigment, ground to a paste in a specified liquid.
2. Stir together a small quantity of the paste, some of the same liquid, and some ordinary portland cement.
3. Evaporate the liquid, either with or without condensation and reuse.

All the major equipment units of an existing cement plant may be employed for the new process, and no additional major units would be required. Under ordinary conditions, the improved cement could be obtained over widely-separated locations in a relatively short time. When demand becomes sufficient, some manufacturer will produce the improved cement, and then competition will take over. Architects may look forward not only to improvements in present cementitious products, but also to the appearance of new products.

Duff A. Abrams, for many years director of concrete research for the Portland Cement Association, always prepared well in advance a detailed program for an investigation. His purpose was to provide as far as possible for any contingencies that might arise; and to avoid any after-thoughts that would require more materials or equipment, or additional tests. Every member of the staff taking part in the project received a copy and was asked for suggestions. Then the program was revised before the work was started. For one program there were seventeen revisions. These carefully planned programs were important factors in his work.

In my research, programs are prepared and are found most helpful. It often happens, however, that a solution of a problem, or the development of a process, does not result from following a program, but comes about rather informally from assembling facts that were apparently unrelated. The color process is an example. I did not plan any research in connection with its development.

During my first experience in a cement plant, white cement held no special interest for me, except curiosity. Visitors were not permitted at white plants. Then a white plant came within the range of my responsibility, and study of the raw materials, fuels, manufacturing process, and finished cement became necessary.

It was apparent that when cement and a pigment are mixed, the color of the cement is important. With a white cement a blend of the most distinct color is obtained, and the least percentage of pigment is required for a given result.

The modifications in technique necessary to produce a white cement—rather than a gray—suggested a reversal of these modifications to see what would happen. The result was the development of a black portland cement. The appearance was due to the cement itself. As no pigments were present there could be no deterioration of pigment. Some attractive effects were obtained by casting white cement mortars and black cement mortars in contact. My interest in color continued but no color research was planned.

Chapter 16 reports the development of the Nonaqueous Liquid Process. First, there was a grinding of raw materials in the presence of a liquid other than water, to produce a slurry. Then it was found that the grinding of clinker in the presence of a nonaqueous liquid to produce a cement slurry has some important advantages. Finally, a pigment can be ground in the presence of another portion of the same liquid in which the clinker has been ground, to produce a pigment slurry.

By mixing a cement slurry with one or more pigment slurries, in any proportions, a cement of color superior in color range and in uniformity, results. This liquid may be recovered and used repeatedly. All this has been done with the cooperation of individuals and organizations, including a large manufacturing paint company.

Some may wonder how the name of the process happened to be chosen. Iris is the goddess of the rainbow. "The rainbow is the bridge or road let down from heaven for her accommodation." [3] That was before the building of reinforced concrete superhighways.

In the cement industry, geographical names predominate. But occasionally one hears a name from mythology such as Atlas, Hercules, Medusa. (See list of manufacturers in Chapter 24.)

REFERENCES

1. *Modern Concrete*, July, 1960.
2. J. C. Witt, *Modern Concrete*, October, 1958.
3. Bulfinch's *Mythology*, p. 916; Thomas T. Crowell Co., New York (1947).

Chapter 24

MISCELLANY

In general each of the items in this chapter is applicable to one or more of the preceding chapters, or even to all of them. The material is assembled here to minimize repetition and cross-indexing. It would be well to glance over this chapter to find out the scope of its contents before studying *Portland Cement Technology* in detail.

LIBRARIES

Since my first occasion seriously to use library facilities, my appreciation of libraries and librarians has never ceased to increase. At first, a library meant to me just a place at which books and journals could be consulted. Then came the realization that assistance in locating information is always available. Then it became evident that a library does not just make publications available, but assists, in many ways, in the production of new publications. This is one of the most important functions of the library.

It has always been my good fortune to be near an excellent science library. First there was the Carnegie Library at Pittsburgh; later, the Bureau of Science Library in Manila; and still later, the John Crerar Library, in Chicago. My gratitude to each of them and to many other libraries visited from time to time will always remain. These include libraries maintained by universities, companies, and associations. I wish to mention especially the libraries of the Portland Cement Association, "a national organization to improve and extend the uses of concrete." The Public Relations Bureau of the Association has kindly supplied a number of photographs.

ABSTRACTS AND HANDBOOKS

Concentrated information may be considered the general term for this type of literature. Only a few of the many examples will be mentioned here. By using them, one can obtain a much wider scope of information than would be possible otherwise in the time available. Their importance increases with the rapid growth of scientific and technical literature.

"New problem: The Information Explosion" is the title of a paper by Karen G. Takle (*American Engineer*, May, 1962). The author states:

> excessive literature growth has increased the basic problem of getting necessary facts to the engineer. . . . the library as a research center in equal importance to the laboratory . . .

Chemical Abstracts

Published by the American Chemical Society, the scope of this journal is the abstracting of all papers involving chemistry, around the world. Information on other branches of science, and on engineering and technology is also to be found in these abstracts. The literature is covered in 74 sections (1965).

To the cement man, Section 22, *Cement and Concrete Products*, is the most important. Other sections that contain many items of interest include:

2. Analytical Chemistry
7. Thermodynamics, Thermochemistry, and Thermal Properties
16. Apparatus, Plant Equipment, and Unit Operations and Processes
21. Ceramics
26. Coal and Coal Derivatives
27. Petroleum and Petroleum Derivatives

As cement aids directly or indirectly in the manufacture of nearly every product, *Chemical Abstracts* contains much of general interest.

Engineering Index

This is an annual volume listing world-wide papers on engineering.

Journal of the American Concrete Institute

A section of this journal, entitled *Current Reviews*, presents abstracts on concrete construction.

Pit and Quarry Handbook

This annual publication contains much information on cement plant equipment and procedure, from raw materials to shipment of the finished products.

Mechanical Catalog

This is published annually by the American Society of Mechanical Engineers. It contains much engineering information not ordinarily found in catalogs. There is a list of the numerous ASME publications. The volume could be called a handbook rather than a catalog.

Consulting Engineer Specifications Data File

The first edition of this publication appeared in 1962. The scope is mechanical, electrical, and structural engineering.

General Sources of Information

These items may be mentioned:

> Papers presented before various organizations
> Discussions of these papers
> Scientific, engineering, and trade journals
> Publications of technical societies
> Federal, state, and municipal bulletins
> University bulletins
> Portland Cement Association publications
> Text and reference books

Related Reading

It is not good practice to limit one's reading to books, papers, and lectures, of which the titles contain the word *cement* or *concrete*. It should be remembered that much of the development of the cement industry has resulted from the application of information obtained from many branches of science, engineering, and industry in general. The following journals are received at my office:

Building Construction
Chemical Abstracts
Chemical and Engineering News
Civil Engineering
Chemical Engineering Progress
Concrete Products
Consulting Engineer
Industrial Research
Journal of the American Concrete Institute
Journal of the Power Division (ASCE)
Journal of the Waterways and Harbors Division (ASCE)
Materials Research and Standards
Mechanical Engineering
Pit and Quarry
Power Engineering
Rock Products
Science
Transactions, Illinois State Academy of Science

BIBLIOGRAPHIES

Bibliographies are particularly helpful in the study of any subject. The preparation of most of them represents much careful work. Of the many bibliographies that could be mentioned, these on the following subjects have been selected:

Constitution of Portland Cement Clinker: 1860-1925

Prepared by R. H. Bogue; published in *Concrete*, Vol. 29 (1–6) 1926; Vol. 30 (1, 2), 1927.

Setting and Hardening Processes in Portland Cement: 1780-1926

Prepared by R. H. Bogue; published in *Rock Products*, Vol. 31 (10, 12, 14, 16, 18, 20), 1928.

Cement and Concrete: 1917-1958

Published as Bulletin 100 of Research Department, Portland Cement Association, May 1959.

Chemistry of Cement

Proceedings of the Fourth International Symposium on the Chemistry of Cement, 1960; published by National Bureau of Standards, Washington, D.C.

Design and Control of Concrete Mixtures: 1892-1941

Prepared by National Ready-Mixed Concrete Association.

Design and Control of Concrete Mixtures: 1941-1961

Prepared by Stanton Walker, and Delmar L. Bloem; published as Circular 68, by National Sand and Gravel Association; Publication 69, by National Ready-Mixed Concrete Association (1962).

Add to these:

Cement and Concrete (1925-1947)

A comprehensive bibliography of cement and concrete by Floyd O. Slate, covering the world's scientific literature on cement and concrete, containing 40,000 references with subject, author, and chronological indexes. Published by the Joint Highways Research Project, Purdue University, Lafayette, Indiana.

Cement and Concrete (1905-1959)

The American Concrete Institute 55-year index (compiled by Robert G. Wieduck and Mary K. Hurd) is a combined author-and-subject index, with abstracts of papers published in the Journal.

Additional early bibliographic references may be found in two publications of the National Ready-Mixed Concrete Association:

Proportioning Concrete Mixtures (1941).
Control of Quality of Ready-Mixed Concrete (1945).
Both are by Stanton Walker.

Other publications of interest are:
Concrete: Properties and Manufacture, by T. N. W. Akroyd; published by Pergamon, London, 1962.

Concrete Technology (2 vols.): 1. Properties of Materials;
2. Practice, by D. F. Orchard; John Wiley and Sons,
New York, 1962.

*Analytical Studies of Relations Among Various Design
Criteria*, by Narbey Khachaturian, Iqbal Ali, and
L. T. Thorpe; Bulletin No. 463. University of Illinois
Engineering Experiment Station, 1962.

Application of Coal Petrography to Coal Preparation, by
John Harrison; Illinois Geological Survey, Urbana,
Illinois.

Formwork for Concrete, M. K. Hurd, Technical Editor;
American Concrete Institute, 1963. ACI Committee
622, Joseph R. Proctor, Chairman.

Strength of Reinforced Concrete Beams, by Sidney A.
Guralnik; awarded Collinwood Prize of ASCE for
1961; published in *Transactions* of ASCE, 1961.

ORGANIZATIONS

Most of the following organizations are related in some
way to portland cement technology.

AMERICAN ASSOCIATION OF ENGINEERS
AMERICAN ASSOCIATION OF STATE HIGHWAY OFFICIALS
AMERICAN CHEMICAL SOCIETY
AMERICAN CONCRETE INSTITUTE
AMERICAN CONCRETE PIPE ASSOCIATION
AMERICAN INSTITUTE OF CHEMICAL ENGINEERS
AMERICAN INSTITUTE OF CONSULTING ENGINEERS
AMERICAN INSTITUTE OF MINING AND METALLURGICAL AND
     PETROLEUM ENGINEERS
AMERICAN RAILWAY ENGINEERING ASSOCIATION
AMERICAN ROAD BUILDERS' ASSOCIATION
AMERICAN SOCIETY OF AGRICULTURAL ENGINEERS
AMERICAN SOCIETY OF CIVIL ENGINEERS
AMERICAN SOCIETY OF MECHANICAL ENGINEERS
AMERICAN SOCIETY FOR TESTING AND MATERIALS
AMERICAN STANDARDS ASSOCIATION
ASSOCIATED GENERAL CONTRACTORS OF AMERICA
ASSOCIATION OF CONSULTING CHEMISTS AND CHEMICAL
     ENGINEERS

BUILDING OFFICIALS CONFERENCE OF AMERICA
BUILDING RESEARCH ADVISORY BOARD, NATIONAL ACADEMY
   OF SCIENCES
CALCIUM CHLORIDE ASSOCIATION
CAST STONE INSTITUTE
CONCRETE REINFORCING STEEL INSTITUTE
CONCRETE VIBRATOR ASSOCIATION
CONSULTING ENGINEERS COUNCIL
ENGINEERS JOINT COUNCIL, INC.
HIGHWAY RESEARCH BOARD
INSTITUTE OF ELECTRICAL AND ELECTRONICS ENGINEERS
METAL LATH MANUFACTURERS' ASSOCIATION
MISSISSIPPI VALLEY CONFERENCE OF STATE HIGHWAY DE-
   PARTMENTS
NATIONAL ASSOCIATION OF SILO MANUFACTURERS
NATIONAL BOARD OF FIRE UNDERWRITERS
NATIONAL CONCRETE BURIAL VAULT ASSOCIATION
NATIONAL CONCRETE MASONRY ASSOCIATION
NATIONAL CRUSHED STONE ASSOCIATION
NATIONAL FIRE PROTECTION ASSOCIATION
NATIONAL HIGHWAY USERS CONFERENCE
NATIONAL LIME ASSOCIATION
NATIONAL READY MIXED CONCRETE ASSOCIATION
NATIONAL RIVERS AND HARBORS CONGRESS
NATIONAL SAFETY COUNCIL, CEMENT SECTION
NATIONAL SAND AND GRAVEL ASSOCIATION
NATIONAL SLAG ASSOCIATION
NATIONAL SOCIETY OF PROFESSIONAL ENGINEERS
NATIONAL TERRAZZO AND MOSAIC ASSOCIATION
NEW JERSEY STATE CONCRETE PRODUCTS ASSOCIATION
PORTLAND CEMENT ASSOCIATION
RAIL STEEL BAR ASSOCIATION
WIRE REINFORCEMENT INSTITUTE
WISCONSIN CONCRETE PRODUCTS ASSOCIATION

LOCATION OF CEMENT PLANTS
IN UNITED STATES AND PUERTO RICO

This listing is from a map showing the location of cement plants in the United States, Canada, and Mexico, prepared by *Pit and Quarry* Publications, Chicago, 1961. Process, capacity, and brand name are stated.

PRODUCING PLANTS

ALABAMA

Birmingham
Alpha Portland Cement Co.
  Wet process; 1,800,000 bbls.
  Brand: Alpha
Lehigh Portland Cement Co.
  Wet process; 1,840,000 bbls.
  Brand: Lehigh
Lone Star Cement Corp.
  Wet process; 1,900,000 bbls.
  Brand: Lone Star, Incor
Southern Cement Co.
  Div. Martin-Marietta Co.
  Dry process; 1,750,000 bbls.
  Brand: Magnolia
Demopolis
Lone Star Cement Corp.
  Dry process; 1,500,000 bbls.
  Brands: Lone Star, Incor
Leeds
Universal Atlas Cement Div.
U. S. Steel Corp.
  Wet process; 2,000,000 bbls.
  Brands: Universal, Atlas
Mobile
Ideal Cement Co.
  Wet process; 3,000,000 bbls.
  Brand: Ideal
Ragland
National Cement Co.
  Dry process; 1,500,000 bbls.
  Brand: Coosa
Roberta
Southern Cement Co.
  Div. Martin-Marietta Co.
  Dry process; 4,250,000 bbls.
  Brand: Magnolia

ARIZONA

Clarkdale
Phoenix Cement Co.
  Div. of American Cement
  Corp.
  Dry process; 1,800,000 bbls.
  2,600,000 bbls. in '62
  Brand: Phoenix
Rillito
Arizona Portland Cement Co.
  Dry process; 2,700,000 bbls.
  Brand: Arizona

ARKANSAS

Foreman
Arkansas Cement Corp.
  Wet process; 1,400,000 bbls.
  2,800,000 bbls. in '62
  Brand: Foreman
Okay
Ideal Cement Co.
  Wet process; 1,850,000 bbls.
  Brand: Ideal

CALIFORNIA

Colton
California Portland Cement Co.
  Dry process; 4,500,000 bbls.
  Brand: Colton
Davenport
Pacific Cement & Aggregates,
  Inc.
  Dry process; 2,800,000 bbls.
  Brand: Santa Cruz
Lucerne Valley
Permanente Cement Co.
  Cushenbury Plant
  Wet process; 2,700,000 bbls.
  Brand: Permanente
Mojave
California Portland Cement Co.
  Dry process; 6,000,000 bbls.
  Brand: Mojave
Monolith
Monolith Portland Cement Co.
  Wet process; 4,000,000 bbls.
  Brand: Monolith

Oro Grande
Riverside Cement Corp.
  Div. American Cement Corp.
  Dry process; 6,000,000 bbls.
  8,000,000 bbls. in '63
  Brands: Riverside, Bear
Permanente
Permanente Cement Co.
  Wet process; 8,500,000 bbls.
  Brands: Permanente, Yosem-
    ite, Kaiser, Santa Clara
Redding
Calaveras Cement Co.
  1,500,000 bbls. in '62
  Dry process
  Brand: Calaveras
Redwood City
Ideal Cement Co.
  Wet process; 2,550,000 bbls.
  Brand: Ideal
Riverside
Riverside Cement Co.
  Div. American Cement Corp.
  Dry process; 3,250,000 bbls.
  5,750,000 bbls. in '63
  Brand: Riverside
Riverside Cement Co.
  Div. American Cement Corp.
  Dry process; 300,000 bbls.
  500,000 bbls. in '63
  Brand: Riverside White
San Andreas
Calaveras Cement Co.
  Wet process; 4,500,000 bbls.
  Brand: Calaveras
San Juan Bautista
Ideal Cement Co.
  Wet process; 950,000 bbls.
  Brand: Ideal
Victorville
Southwestern Portland Cement
  Co.
  Wet process; 6,000,000 bbls.
  Brand: Victor

COLORADO

Florence
Ideal Cement Co.

  Wet and Dry process; 3,200,-
    000 bbls.
  Brand: Ideal
Fort Collins
Ideal Cement Co.
  Dry process; 2,750,000 bbls.
  Brand: Ideal

FLORIDA

Bunnell
Lehigh Portland Cement Co.
  Wet Process; 3,350,000 bbls.
  Brand: Lehigh
Miami
General Portland Cement Co.
  Florida Division
  Wet process; 2,500,000 bbls.
  Brand: Florida
Lehigh Portland Cement Co.
  Wet process; 2,500,000 bbls.
  Brand: Lehigh
Tampa
General Portland Cement Co.
  Florida Division
  Wet process; 7,000,000 bbls.
  Brands: Florida, Trinity

GEORGIA

Atlanta
Southern Cement Co.
  Div. Martin-Marietta Co.
  Dry process; 1,500,000 bbls.
    in '62
  Brand: Magnolia
Clinchfield
Penn-Dixie Cement Corp.
  Wet process; 1,222,000 bbls.
  2,372,000 bbls. in '62
  Brand: Penn-Dixie
Rockmart
Marquette Cement Mfg. Co.
  Dry process; 1,250,000 bbls.
  Brand: Marquette

HAWAII

Waianae, Honolulu
Permanente Cement Co.
  Wet process; 1,700,000 bbls.
  Brand: Permanente

Waipahu, Oahu
Hawaiian Cement Corp.
Dry process; 1,000,000 bbls.
Brand: Hawaiian

IDAHO

Inkom
Idaho Portland Cement Co.
Wet process; 800,000 bbls.
Brand: Eagle

ILLINOIS

Dixon
Medusa Portland Cement Co.
Dry process; 3,500,000 bbls.
Brand: Medusa

La Salle
Alpha Portland Cement Co.
Dry process; 1,500,000 bbls.
Brand: Alpha

Oglesby
Lehigh Portland Cement Co.
Dry process; 1,680,000 bbls.
Brand: Lehigh

Marquette Cement Manufacturing Co.
Dry process; 4,250,000 bbls.
Brand: Marquette

INDIANA

Buffington
Universal Atlas Cement Div.
U. S. Steel Corp.
Dry process; 9,700,000 bbls.
Brands: Universal, Atlas

Greencastle
Lone Star Cement Corp.
Wet process; 2,700,000 bbls.
Brands: Lone Star, Incor

Logansport
Louisville Cement Co.
1,200,000 bbls. in '62
Wet process
Brand: Speed Portland

Mitchell
Lehigh Portland Cement Co.
Dry process; 2,500,000 bbls.
Brand: Lehigh

Speed
Louisville Cement Co.
Dry process; 5,250,000 bbls.
Brand: Speed Portland

IOWA

Davenport (Linwood)
Dewey Portland Cement Co.
Div. Martin-Marietta Co.
Wet process; 3,600,000 bbls.
Brand: Dewey

Des Moines
Marquette Cement Manufacturing Co.
Wet process; 2,450,000 bbls.
Brand: Hawkeye-Marquette

Mason City
Lehigh Portland Cement Co.
Dry process; 3,340,000 bbls.
Brand: Lehigh

Northwestern States Portland Cement Co.
Dry process; 4,000,000 bbls.
Brand: Standard Portland

West Des Moines
Penn-Dixie Cement Corp.
Wet process; 2,340,000 bbls.
Brand: Penn-Dixie

KANSAS

Bonner Springs
Lone Star Cement Corp.
Wet process; 2,400,000 bbls.
Brands: Lone Star, Incor

Chanute
Ash Grove Lime & Portland Cement Co.
Wet process; 2,000,000 bbls.
Brand: Ash Grove Portland

Fredonia
General Portland Cement Co.
(Victor Division)
Wet process; 2,300,000 bbls.
Brands: Victor, Samson

Humboldt
Monarch Cement Co.
Dry process; 2,400,000 bbls.
Brand: Monarch

Independence
Universal Atlas Cement Division
U. S. Steel Corp.
Dry process; 2,200,000 bbls.
Brands: Universal, Atlas
Iola
Lehigh Portland Cement Co.
Wet process; 1,340,000 bbls.
Brand: Lehigh

KENTUCKY

Kosmosdale
Kosmos Portland Cement Co., Inc.
Dry process; 3,200,000 bbls.
4,000,000 bbls. in '63
Brand: Kosmos

LOUISIANA

Baton Rouge
Ideal Cement Co.
Wet process; 3,000,000 bbls.
Brand: Ideal
Lake Charles
Lone Star Cement Corp.
Wet process; 2,000,000 bbls.
Brand: Lone Star, Incor
New Orleans
Lone Star Cement Corp.
Wet process; 2,200,000 bbls.
Brands: Lone Star, Incor

MAINE

Thomaston
Dragon Cement Co.
Div. Martin-Marietta Co.
Wet process; 2,200,000 bbls.
Brand: Dragon

MARYLAND

Lime Kiln
Alpha Portland Cement Co.
Wet process; 2,250,000 bbls.
Brand: Alpha
Security
Marquette Cement Manufacturing Co.
Dry process; 2,200,000 bbls.
Brand: North American

Union Bridge
Lehigh Portland Cement Co.
Dry process; 3,500,000 bbls.
Brand: Lehigh

MICHIGAN

Alpena
Huron Portland Cement Co.
Sub. National Gypsum Co.
Dry process; 12,000,000 bbls.
13,800,000 bbls. in 1962
Brand: Huron
Bay City
Aetna Portland Cement Co.
Wet process; 3,200,000 bbls.
Brand: Aetna Portland
Cement City
General Portland Cement Co.
(Peninsular Div.)
Wet process; 1,200,000 bbls.
Brand: Peninsular
Detroit
Peerless Cement Co.
Div. American Cement Corp.
Wet process; 3,700,000 bbls.
Brand: Peerless
Peerless Cement Co.
Div. American Cement Corp.
Wet process; 1,800,000 bbls.
Brand: Peerless
Dundee
Dundee Cement Co.
Wet process; 6,000,000 bbls.
Brand: Dundee
Petoskey
Penn-Dixie Cement Corp.
Wet process; 2,772,000 bbls.
Brand: Penn-Dixie
Port Huron
Peerless Cement Co.
Div. American Cement Corp.
Wet process; 1,000,000 bbls.
Brand: Peerless
Wyandotte
Wyandotte Chemicals Corp.
Wet process; 2,000,000 bbls.
Brand: Wyandotte

MINNESOTA

Duluth
Universal Atlas Cement Div.
  U. S. Steel Corp.
  Dry process; 2,000,000 bbls.
  Brands: Universal, Atlas

MISSISSIPPI

Brandon
Marquette Cement Manufac-
  turing Co.
  Wet process; 1,300,000 bbls.
  Brand: Marquette
Redwood
Mississippi Valley Portland Ce-
  ment Co.
  Wet process; 700,000 bbls.
  1,700,000 bbls. in 1963
  Brand: Valley

MISSOURI

Cape Girardeau
Marquette Cement Manufac-
  turing Co.
  Wet process; 1,200,000 bbls.
  Brand: Marquette
Marquette Cement Manufac-
  turing Co.
  Dry process; 1,800,000 bbls.
  Brand: Marquette
Hannibal
Universal Atlas Cement Div.
  U. S. Steel Corp.
  Dry process; 2,200,000 bbls.
  Brands: Universal, Atlas
Lemay
Alpha Portland Cement Co.
  Wet process; 2,600,000 bbls.
St. Louis
Missouri Portland Cement Co.
  Wet process; 5,000,000 bbls.
  Brands: Red Ring, Red Ball,
  Red Diamond
Sugar Creek
Missouri Portland Cement Co.
  Dry process; 3,000,000 bbls.
  Brands: Red Ring, Red Ball,
  Red Diamond

MONTANA

Tridnet
Ideal Cement Co.
  Dry process; 1,500,000 bbls.
  Brand: Ideal

NEBRASKA

Louisville
Ash Grove Lime & Portland
  Cement Co.
  Wet process; 3,500,000 bbls.
  Brand: Ash Grove Portland
Superior
Ideal Cement Co.
  Wet process; 1,300,000 bbls.
  Brand: Ideal

NEW MEXICO

Tijeras
Ideal Cement Co.
  Dry process; 2,450,000 bbls.
  Brand: Ideal

NEW YORK

Buffalo
Lehigh Portland Cement Co.
  Wet process; 2,340,000 bbls.
  Brand: Lehigh
Penn-Dixie Cement Corp.
  Wet process; 2,016,000 bbls.
  Brand: Penn-Dixie
Catskill
Marquette Cement Mfg. Co.
  Wet process; 1,650,000 bbls.
  2,500,000 bbls. in '62
  Brand: North American Ce-
  ment
Cementon
Alpha Portland Cement Co.
  Dry process; 1,700,000 bbls.
  Brand: Alpha
Lehigh Portland Cement Co.
  Dry process; 2,650,000 bbls.
  Brand: Lehigh (Alsen)
Glens Falls
Glens Falls Portland Cement
  Co.
  Wet process; 1,850,000 bbls.
  Brand: Iron Clad

Howes Cave
Marquette Cement Mfg. Co.
  Dry process; 1,650,000 bbls.
  Brand: North American Ce-
  ment
Hudson
Lone Star Cement Corp.
  Wet process; 3,000,000 bbls.
  Brands: Lone Star, Incor
Universal Atlas Cement Div.
  U. S. Steel Corp.
  Dry process; 4,300,000 bbls.
  Brands: Universal, Atlas
Jamesville
Alpha Portland Cement Co.
  Wet process; 900,000 bbls.
  Brand: Alpha
Kingston
Hudson Cement Corp.
  2,000,000 bbls. estimated
  Wet process
Ravena
Atlantic Cement Co., Inc.
  10,000,000 bbls. in '62
  Wet process

OHIO

Barberton
Pittsburgh Plate Glass Co.
  Chemical Division
  Wet process; 1,500,000 bbls.
  Brand: Columbia Cement
Fairborn
Southwestern Portland Cement
  Co.
  Wet process; 3,300,000 bbls.
  Brand: Miami
Universal Atlas Cement Div.
  U. S. Steel Corp.
  Wet process; 2,500,000 bbls.
  Brands: Universal, Atlas
Ironton
Alpha Portland Cement Co.
  Dry process; 1,200,000 bbls.
  Brand: Alpha
Middle Branch
Diamond Portland Cement Co.
  Div. Flintkote Company

Dry process; 3,000,000 bbls.
  Brand: Diamond
Painesville
Standard Portland Cement
  Plant
  Cement-Coke Div.
  Diamond Alkali Co.
  Wet process; 2,700,000 bbls.
  Brand: Standard
Paulding
General Portland Cement Co.
  Wet process; 2,500,000 bbls.
  Brand: Peninsular
Silica
Medusa Portland Cement Co.
  Dry process; 1,450,000 bbls.
Superior
Marquette Cement Manufac-
  turing Co.
  Dry process; 1,250,000 bbls.
  Brand: Marquette
Zanesville
Columbia Cement Corp.
  Sub. Pittsburgh Plate Glass
  Co.
  Wet process; 3,000,000 bbls.
  Brand: Columbia

OKLAHOMA

Ada
Ideal Cement Co.
  Wet process; 5,900,000 bbls.
  Brand: Ideal
Dewey
Dewey Portland Cement Co.
  Div. Martin-Marietta Co.
  Dry process; 2,200,000 bbls.
  Brand: Dewey
Pryor
Oklahoma Cement Co.
  Dry process; 1,000,000 bbls.
  Brand: Oklahoma
Tulsa
Dewey Portland Cement Co.
  Div. Martin-Marietta Co.
  Dry process; 1,250,000 bbls.
  Brand: Dewey

OREGON
Gold Hill
Ideal Cement Co.
  Wet process; 700,000 bbls.
  Brand: Ideal
Lime
Oregon Portland Cement Co.
  Wet process; 1,200,000 bbls.
  Brand: Sun
Lake Oswego
Oregon Portland Cement Co.
  Wet process; 2,000,000 bbls.
  Brand: Oregon

PENNSYLVANIA
Bath
Keystone Portland Cement Co.
  Wet process; 3,300,000 bbls.
  Brand: Keystone
Penn-Dixie Cement Corp.
  Dry process; 2,052,000 bbls.
  Brand: Penn-Dixie
Bessemer
Bessemer Limestone & Cement
  Co.
  Wet process; 3,000,000 bbls.
  Brand: Bessemer
Bethlehem
National Portland Cement Co.
  Wet process; 2,000,000 bbls.
  Brand: Pioneer
Cementon
The Whitehall Cement Mfg. Co.
  Dry process; 3,000,000 bbls.
  Brand: Whitehall
Coplay
Coplay Cement Manufacturing
  Co.
  Dry process; 2,250,000 bbls.
  Brand: Saylors
Egypt
Giant Portland Cement Co.
  Dry process; 1,850,000 bbls.
  Brand: Giant
Evansville
Allentown Portland Cement Co.
  Dry process; 2,400,000 bbls.
  Brand: Allentown

Fogelsville
Lehigh Portland Cement Co.
  Dry process; 2,210,000 bbls.
  Brand: Lehigh
Martins Creek
Alpha Portland Cement Co.
  Dry process; 2,400,000 bbls.
Nazareth
Lone Star Cement Corp.
  Dry process; 3,600,000 bbls.
  Brands: Lone Star, Incor
Nazareth Cement Co.
  Dry process; 2,400,000 bbls.
  Brand: Nazareth
Penn-Dixie Cement Corp.
  Dry process; 1,836,000 bbls.
  Brand: Penn-Dixie
Neville Island
Green Bag Cement Co.
  Wet process; 920,000 bbls.
  1,750,000 bbls. in '62
  Brand: Green Bag
Northampton
Dragon Cement Co.
  Div. Martin-Marietta Co.
  Dry process; 2,600,000 bbls.
  Brand: Dragon
Universal Atlas Cement Div.
  U. S. Steel Corp.
  Wet process; 2,900,000 bbls.
  Brands: Universal, Atlas
Sandt's Eddy
Lehigh Portland Cement Co.
  Wet process; 1,770,000 bbls.
  Brand: Lehigh
Stockertown
Hercules Cement Co.
  Div. American Cement Corp.
  Dry process; 3,500,000 bbls.
  Brand: Hercules
Universal
Universal Atlas Cement Div.
  U. S. Steel Corp.
  Dry process; 2,600,000 bbls.
  Brands: Universal, Atlas

Wampum
Medusa Portland Cement Co.
   Dry process; 2,500,000 bbls.
   Brand: Medusa
West Conshohocken
Allentown Portland Cement Co.
   Wet process; 2,100,000 bbls.
   Brand: Allentown
West Winfield
Penn-Dixie Cement Corp.
   Wet process; 1,908,000 bbls.
   Brand: Penn-Dixie
York
Medusa Portland Cement Co.
   Wet process; 2,500,000 bbls.
   Brand: Medusa

SOUTH CAROLINA
Harleyville
Giant Portland Cement Co.
   Carolina Giant Division
   Wet process; 2,900,000 bbls.
   4,000,000 bbls. in 1962
   Brand: Giant

SOUTH DAKOTA
Rapid City
South Dakota Cement Plant
   Wet process; 3,300,000 bbls.
   Brand: Dacotah

TENNESSEE
Cowan
Marquette Cement Manufac-
   turing Co.
   Wet process; 1,020,000 bbls.
   Brand: Marquette
Kingsport
Penn-Dixie Cement Corp.
   Wet process; 1,620,000 bbls.
   Brand: Penn-Dixie
Knoxville
Volunteer Portland Cement Co.
   Div. of Ideal Cement Co.
   Wet process; 2,850,000 bbls.
   Brand: Volunteer
Nashville
Marquette Cement Manufac-
   turing Co.

Wet process; 1,200,000 bbls.
   Brand: Marquette
North Chattanooga
General Portland Cement Co.
   Signal Mountain Div.
   Wet process; 1,750,000 bbls.
   Brand: Signal Mountain
Richard City
Penn-Dixie Cement Corp.
   Wet process; 1,584,000 bbls.
   Brand: Penn-Dixie

TEXAS
Cementville
San Antonio Portland Cement
   Co.
   Wet process; 2,500,000 bbls.
   Brand: Alamo
Corpus Christi
Halliburton Portland Cement
   Co.
   Wet process; 1,400,000 bbls.
   Brand: Halliburton
Eagle Ford (P.O. Dallas)
General Portland Cement Co.
   Wet process; 3,650,000 bbls.
   Brand: Trinity
Dallas
Lone Star Cement Corp.
   Wet process; 4,000,000 bbls.
   Brands: Lone Star, Incor
El Paso
Southwestern Portland Cement
   Co.
   Dry process; 1,800,000 bbls.
   Brand: El Toro
Fort Worth
General Portland Cement Co.
   Trinity Division
   Wet process; 4,000,000 bbls.
   Brand: Trinity
Galena Park
Ideal Cement Co.
   Wet process; 4,780,000 bbls.
   Brand: Ideal

Houston
General Portland Cement Co.
  Trinity Division
  Wet process; 3,650,000 bbls.
  Brand: Trinity
Lone Star Cement Corp.
  Wet process; 3,300,000 bbls.
  Brands: Lone Star, Incor
Maryneal
Lone Star Cement Corp.
  Dry process; 2,900,000 bbls.
  Brands: Lone Star, Incor
Midlothian
Texas Industries, Inc.
  Wet process; 1,400,000 bbls.
  Brand: Texas Industries
Odessa
Southwestern Portland Cement
  Co.
  Dry process; 1,200,000 bbls.
  Brand: El Toro
Orange
Texas Portland Cement Co.
  Wet process; 660,000 bbls.
  Brands: Texas and Louisiana
San Antonio
Longhorn Portland Cement Co.
  Wet process; 2,700,000 bbls.
  Brand: Longhorn
Waco
Universal Atlas Cement Div.
  U. S. Steel Corp.
  Dry process; 1,100,000 bbls.
  Brands: Universal, Atlas

UTAH
Devil's Slide
Ideal Cement Co.
  Wet process; 1,900,000 bbls.
  Brand: Ideal
Salt Lake City
Portland Cement Co. of Utah
  Wet process; 1,000,000 bbls.
  Brand: Utah

VIRGINIA
Fordwick
Lehigh Portland Cement Co.
  Dry process; 1,690,000 bbls.
  Brand: Lehigh

Lone Star
Lone Star Cement Corp.
  Dry process; 3,400,000 bbls.
  Brands: Lone Star, Incor
Norfolk
Lone Star Cement Corp.
  Wet process; 2,300,000 bbls.
  Brands: Lone Star, Incor

WASHINGTON
Bellingham
Permanente Cement Co.
  Olympic Plant
  Wet process; 1,900,000 bbls.
  Brands: Olympic,
    Permanente
Concrete
Lone Star Cement Corp.
  Wet process; 1,700,000 bbls.
  Brands: Lone Star, Incor
Grotto
Ideal Cement Co.
  Wet process; 620,000 bbls.
  Brand: Ideal
Metaline Falls
Lehigh Portland Cement Co.
  Dry process; 1,290,000 bbls.
  Brand: Lehigh
Seattle
Lone Star Cement Corp.
  Wet process; 1,300,000 bbls.
  Brands: Lone Star, Incor
Spokane
Ideal Cement Co.
  Dry process; 700,000 bbls.
  Brand: Ideal

WEST VIRGINIA
Martinsburg
Standard Lime & Cement Co.
  Div. Martin-Marietta Co.
  Wet process; 4,200,000 bbls.
  Brand: Capitol

WISCONSIN
Manitowoc
Manitowoc Portland Cement
  Co.
  Sub. Medusa Portland Cement Co.

Wet process; 2,000,000 bbls.
Brand: Badger

Milwaukee
Marquette Cement Manufacturing Co.
Dry process; 1,250,000 bbls.
Brand: Marquette
Universal Atlas Cement Div.
U. S. Steel Corp.
Grinding only

Superior
Huron Portland Cement Co.
Grinding Only

WYOMING
Laramie
Monolith Portland Midwest Co.
Wet process; 1,100,000 bbls.
Brand: Monolith
PUERTO RICO
Ponce
Ponce Cement Corp.
Wet process; 4,500,000 bbls.
Brand: Ponce
San Juan
Puerto Rico Cement Corp.
Wet process; 3,000,000 bbls.
Brand: Puerto Rico

DISTRIBUTING PLANTS

ALASKA
1 **Anchorage**
Permanente Cement Co.
2 **Fairbanks**
Permanente Cement Co.
ARIZONA
1 **Phoenix**
Arizona Portland Cement Co.
CALIFORNIA
1 **Eureka**
Permanente Cement Co.
2 **Glendale**
Monolith Portland Cement Co.
3 **Inglewood**
Monolith Portland Cement Co.
4 **Laton**
Riverside Cement Corp.
5 **Long Beach**
Permanente Cement Co.
6 **Los Angeles**
Riverside Cement Co.
7 **Miramar**
Monolith Portland Cement Co.
8 **Norwalk**
California Portland Cement Co.
Monolith Portland Cement Co.

9 **Pacoima**
Monolith Portland Cement Co.
Redwood City
Permanente Cement Co.
10 **San Diego**
Riverside Cement Co.
11 **San Leandro**
Calaveras Cement Co.
12 **Sun Valley**
California Portland Cement Co.
13 **Torrance**
Monolith Portland Cement Co.
14 **West Sacramento**
Ideal Cement Co.
COLORADO
1 **Denver**
Monolith Portland Midwest Co.
CONNECTICUT
1 **Bridgeport**
Allentown Portland Cement Co.
DISTRICT OF COLUMBIA
1 **Washington**
Lone Star Cement Corp.
FLORIDA
1 **Ft. Lauderdale**
Ponce Products, Inc.

2 Jacksonville
Atlantic Cement Co., Inc.
Miami
Atlantic Cement Co., Inc.
Tampa
Atlantic Cement Co., Inc.
Ideal Cement Co.

GEORGIA
1 Savannah
Atlantic Cement Co., Inc.

HAWAII
1 Hilo
Permanente Cement Co.
2 Honolulu Harbor
Permanente Cement Co.
3 Kahului, Maui
Permanente Cement Co.

ILLINOIS
1 Chicago
Dundee Cement Co.
Marquette Cement Mfg. Co.
Medusa Portland Cement Co.
Penn-Dixie Cement Corp.
Universal Atlas Cement (2 plants)
2 Des Plaines
Dundee Cement Co.
Marquette Cement Mfg. Co.
3 Waukegan
Marquette Cement Mfg. Co. (1963)

INDIANA
1 Evansville
Kosmos Portland Cement Co.
(c/o Mead Johnson Terminal Corp.)

IOWA
1 Mason City
Northwestern States Portland Cement Co.

LOUISIANA
Baton Rouge
Lone Star Cement Corp.
Lake Charles
Ideal Cement Co.

New Orleans
Ideal Cement Co.

MARYLAND
1 Baltimore
Atlantic Cement Co., Inc.

MASSACHUSETTS
1 Boston
Atlantic Cement Co., Inc.
2 Framingham
North American Cement Corp.
3 Wilmington
Dragon Cement Co.

MICHIGAN
Detroit
Huron Portland Cement Co.
Penn-Dixie Cement Corp.
1 Grand Rapids
Peerless Cement Co.
2 Holland
Penn-Dixie Cement Corp.
3 Muskegon
Huron Portland Cement Co.
Universal Atlas Cement
4 Saginaw
Huron Portland Cement Co.
5 Schoolcraft
Peerless Cement Co.
6 St. Joseph
Huron Portland Cement Co.
Wyandotte
Wyandotte Chemicals Corp.

MINNESOTA
Duluth
Huron Portland Cement Co.
1 Minneapolis
Medusa Portland Cement Co.

MISSISSIPPI
1 Vicksburg
Marquette Cement Mfg. Co.

MISSOURI
St. Louis
Marquette Cement Mfg. Co.
Universal Atlas Cement

NEVADA
1 Las Vegas
Monolith Portland Cement Co.

NEW YORK
Buffalo
Huron Portland Cement Co.
1 College Point
North American Cement Corp.
2 New York City
Atlantic Cement Co., Inc.
3 Oswego
Huron Portland Cement Co.
4 Rochester
Rochester Portland Cement Corp.

NORTH CAROLINA
1 Wilmington
Ideal Cement Co.

OHIO
1 Bay Bridge
Medusa Portland Cement Co.
2 Cincinnati
Alpha Portland Cement Co.
Dundee Cement Co.
Kosmos Portland Cement Co.
3 Cleveland
Huron Portland Cement Co.
Medusa Portland Cement Co.
4 Independence
Dundee Cement Co.
5 Marietta
Green Bag Cement Co.
6 Toledo
Huron Portland Cement Co.

OREGON
1 Eugene
Ideal Cement Co.
2 Portland
Oregon Portland Cement Co.
Permanente Cement Co.
3 Springfield
Calaveras Cement Co.

RHODE ISLAND
1 East Providence
Glens Falls Portland Cement Co.

TENNESSEE
1 Memphis
Marquette Cement Mfg. Co.
Missouri Portland Cement Co.

TEXAS
Corpus Christi
Halliburton Portland Cement Co.

UTAH
Salt Lake City
Ideal Cement Co.
Portland Cement Co. of Utah

VIRGINIA
Norfolk
Atlantic Cement Co., Inc.

WASHINGTON
1 Pasco
Permanente Cement Co.
Seattle
Ideal Cement Co.
Permanente Cement Co.
2 Vancouver
Ideal Cement Co.

WEST VIRGINIA
1 Fairmont
Green Bag Cement Co.

WISCONSIN
1 Green Bay
Huron Portland Cement Co.
Universal Atlas Cement
Milwaukee
Huron Portland Cement Co.
Penn-Dixie Cement Corp.

TABLES

INTERNATIONAL ATOMIC WEIGHTS: 1961 *

	Symbol	Atomic Number	Atomic Weight1		Symbol	Atomic Number	Atomic Weight1
Actinium	Ac	89	227	Molybdenum	Mo	42	95.95
Aluminum	Al	13	26.98	Neodymium	Nd	60	144.27
Americium	Am	95	[243]	Neptunium	Np	93	[237]
Antimony	Sb	51	121.76	Neon	Ne	10	20.183
Argon	A	18	39.944	Nickel	Ni	28	58.69
Arsenic	As	33	74.91	Niobium			
Astatine	At	85	[210]	(Colubium)	Nb	41	92.91
Barium	Ba	56	137.36	Nitrogen	N	7	14.008
Berkelium	Bk	97	[245]	Nobelium	No	102	[...]
Beryllium	Be	4	9.013	Osmium	Os	76	190.2
Bismuth	Bi	83	209.00	Oxygen	O	8	16
Boron	B	5	10.82	Palladium	Pd	46	106.7
Bromine	Br	35	79.916	Phosphorus	P	15	30.975
Cadmium	Cd	48	112.41	Platinum	Pt	78	195.23
Calcium	Ca	20	40.08	Plutonium	Pu	94	[242]
Californium	Cf	98	[246]	Polonium	Po	84	210
Carbon	C	6	12.010	Potassium	K	19	39.100
Cerium	Ce	58	140.13	Praseodymium	Pr	59	140.92
Cesium	Cs	55	132.91	Promethium	Pm	61	[145]
Chlorine	Cl	17	35.457	Protactinium	Pa	91	231
Chromium	Cr	24	52.01	Radium	Ra	88	226.05
Cobalt	Co	27	58.94	Radon	Rn	86	222
Copper	Cu	29	63.54	Rhenium	Re	75	186.31
Curium	Cm	96	[243]	Rhodium	Rh	45	102.91
Dysprosium	Dy	66	162.46	Rubidium	Rb	37	85.48
Einsteinium	Es	99	[254]	Ruthenium	Ru	44	101.7
Erbium	Er	68	167.2	Samarium	Sm	62	150.43
Europium	Eu	63	152.0	Scandium	Sc	21	44.96
Fermium	Fm	100	[253]	Selenium	Se	34	78.96
Fluorine	F	9	19.00	Silicon	Si	14	28.09
Francium	Fr	87	[223]	Silver	Ag	47	107.880
Gadolinium	Gd	64	156.9	Sodium	Na	11	22.997
Gallium	Ga	31	69.72	Strontium	Sr	38	87.63
Germanium	Ge	32	72.60	Sulfur	S	16	32.0662
Gold	Au	79	197.2	Tantalum	Ta	73	180.88
Hafnium	Hf	72	178.6	Technetium	Tc	43	[99]
Helium	He	2	4.003	Tellurium	Te	52	127.61
Holmium	Ho	67	164.94	Terbium	Tb	65	159.2
Hydrogen	H	1	1.0080	Thallium	Tl	81	204.39
Indium	In	49	114.76	Thorium	Th	90	232.12
Iodine	I	53	126.91	Thulium	Tm	69	169.4
Iridium	Ir	77	193.1	Tin	Sn	50	118.70
Iron	Fe	26	55.85	Titanium	Ti	22	47.90
Krypton	Kr	36	83.80	Tungsten	W	74	183.92
Lanthanum	La	57	138.92	Uranium	U	92	238.07
Lead	Pb	82	207.21	Vanadium	V	23	50.95
Lithium	Li	3	6.940	Xenon	Xe	54	131.3
Lutetium	Lu	71	174.99	Ytterbium	Yb	79	173.04
Magnesium	Mg	12	24.32	Yttrium	Y	39	88.92
Manganese	Mn	25	54.93	Zinc	Zn	30	65.38
Mercury	Hg	80	200.61	Zirconium	Zr	40	91.22

* A value given in brackets denotes the mass number of the isotope of
longest known half-life.

CONVERSION FACTORS

(Unless otherwise indicated, mean values for caloric and B.t.u. are employed; U.S. units are used)

	From	*To*	*Multiply by*
LENGTH	Microns	Centimeters	0.0001
	Centimeters	Inches	0.3937
	Meters	Feet	3.28083
	Kilometers	Miles	0.62137
AREA	Square centimeters	Square inches	0.1550
	Acres	Square feet	43560
	Hectares	Acres	2.471044
	Hectares	Square feet	107639
VOLUME	Cubic meters	Cubic yards	1.3079
CAPACITY	Liters	Gallons (U.S.)	0.26418
	Gallons (U. S.)	Cubic feet	0.13368
	Cubic feet	Gallons (U. S.)	7.48052
	Gallons (British)	Gallons (U. S.)	1.20094
WEIGHT	Kilograms	Pounds	2.204622
	Metric tons	Tons (short)	1.10231
	Metric tons	Tons (long)	0.98421
	Metric tons	Barrels (376 pounds)	5.8633
	Tons (long)	Tons (short)	1.12
	Tons (long)	Barrels (376 pounds)	5.957
	Tons (short)	Barrels (376 pounds)	5.3191
	Tons (long) of coal per ton (long) of clinker	Pounds of coal per barrel of clinker	376
STRENGTH	Kilograms per square centimeter	Pounds per square inch	14.2233
TEMPERATURE	Degrees Centigrade	Degrees Fahrenheit	$(°C \times 1.8) + 32$
HEAT	Calories (kilogram)	B.t.u.	3.96753
	B.t.u. per ton (long) of clinker	B.t.u. per barrel of clinker	0.168
	B.t.u. per ton (short) of clinker	B.t.u. per barrel of clinker	0.188
ELECTRICITY	Kilowatt hours per ton (long) of clinker	Kilowatt hours per barrel of clinker	0.168
	Kilowatt hours per ton (short) of clinker	Kilowatt hours per barrel of clinker	0.188
POWER	Kilowatts	Horsepower	1.34102
	Horsepower	Kilowatts	0.746
	B.t.u. (per second)	Horsepower	1.41460
	Kilowatt hours	B.t.u.	3413.44

These equivalents were prepared in advance of the release of *Conversion Factors and Tables*, by D. T. Zimmerman and Irwin Devine, Industrial Research Service, Dover, N. H. (1944), but the values were checked with that publication.

SAID THE ENGINEER WHO DESIGNED
THE LEANING TOWER, "I SKIMPED A LITTLE
ON THE FOUNDATION... BUT NOBODY'LL
EVER KNOW THE DIFFERENCE!"

THE LEANING TOWER
(Permission of The Foundation Company, New York and Boston)

This table contains some factors for converting metric to English units, and for calculating data, originating in other countries, to equivalents in American cement practice.

Each of the terms *weight* and *mass* has more than one definition and the two terms are used interchangeably to some extent. In preparing the table, weight, rather than mass, was considered preferable.

The weight of a barrel of clinker is given as 376 pounds, the same as a barrel of cement. Actually, the weight is a few pounds less than this. The weight of clinker required for a barrel of cement is affected by a number of factors, the principal one of which is the quantity of gypsum added. It is the usual practice at a cement plant to select some value based on average practices, for example, 366 pounds.

CALCULATION TABLES *

These convenient tables have been used by me for a number of years. The values appear in three separate tables in some handbooks. When approximate square root and cube root values are satisfactory, the actual scope of such a table is much greater than the apparent scope. For example, 1060 is the maximum number appearing in the first column, for which the square root and the cube root are shown. However, in the squares and cubes columns, there are many numbers greater than 1060 for which the exact square root and the exact cube root may be found in the first column of the table. For any whole number less than 1,123,600, not shown in the squares column, the square root, correct to the nearest integer, may be found by referring to the first column of the table; for example, $\sqrt{997,000} = 998$, correct to the nearest integer. In the same way cube roots for numbers less than 1,191,016,000 may be found. By a similar procedure, approximately correct squares and cubes, of numbers not listed in the first column, may be obtained by reference to the columns of square roots and cube roots.

* Reproduced by permission of Lefax, Philadelphia, Pa.

CIRCUMFERENCE AND AREAS OF CIRCLES, SQUARES, CUBES, SQUARE ROOTS, CUBE ROOTS, AND RECIPROCALS *

No.	Circum.	Area	Square	Cube	Sq. Root	C. Root	Reciprocal
0.10	.314159	.007854	.0100	.00100	.31623	.46416	10.0000000
0.11	.345575	.009503	.0121	.00133	.33166	.47914	9.0909091
0.12	.376991	.011310	.0144	.00173	.34641	.49324	8.3333333
0.13	.408407	.013273	.0169	.00220	.36056	.50658	7.6923077
0.14	.439823	.015394	.0196	.00274	.37417	.51925	7.1428571
0.15	.471239	.017672	.0225	.00338	.38730	.53133	6.6666667
0.16	.502655	.020106	.0256	.00410	.40000	.54288	6.2500000
0.17	.534071	.022698	.0289	.00491	.41231	.55397	5.8823529
0.18	.565487	.025447	.0324	.00583	.42426	.56462	5.5555556
0.19	.596903	.028353	.0361	.00686	.43589	.57489	5.2631579
0.20	.628319	.031416	.0400	.00800	.44721	.58480	5.0000000
0.21	.659735	.034636	.0441	.00926	.45826	.59439	4.7619048
0.22	.691150	.038013	.0484	.01065	.46904	.60368	4.5454545
0.23	.722566	.041548	.0529	.01217	.47958	.61269	4.3478261
0.24	.753982	.045239	.0576	.01382	.48990	.62145	4.1666667
0.25	.785398	.049087	.0625	.01563	.50000	.62996	4.0000000
0.26	.816814	.053093	.0676	.01758	.50990	.63825	3.8461538
0.27	.848230	.057256	.0729	.01968	.51962	.64633	3.7037037
0.28	.879646	.061575	.0784	.02195	.52915	.65421	3.5714286
0.29	.911062	.066052	.0841	.02439	.53852	.66191	3.4482759
0.30	.942478	.070686	.0900	.02700	.54772	.66943	3.3333333
0.31	.973894	.075477	.0961	.02979	.55678	.67679	3.2258065
0.32	1.00531	.080425	.1024	.03277	.56569	.68399	3.1250000
0.33	1.03673	.085530	.1089	.03594	.57446	.69104	3.0303030
0.34	1.06814	.090792	.1156	.03930	.58310	.69795	2.9411765
0.35	1.09956	.096211	.1225	.04288	.59161	.70473	2.8571429
0.36	1.13097	.10179	.1296	.04666	.60000	.71138	2.7777778
0.37	1.16239	.10752	.1369	.05065	.60828	.71791	2.7027027
0.38	1.19381	.11341	.1444	.05487	.61644	.72432	2.6315789
0.39	1.22522	.11946	.1521	.05932	.62450	.73061	2.5641026
0.40	1.25664	.12566	.1600	.06400	.63246	.73681	2.5000000
0.41	1.28805	.13203	.1681	.06892	.64031	.74290	2.4390244
0.42	1.31947	.13854	.1764	.07409	.64807	.74889	2.3809524
0.43	1.35089	.14522	.1849	.07951	.65574	.75478	2.3255814
0.44	1.38230	.15205	.1936	.08518	.66332	.76059	2.2727273
0.45	1.41372	.15904	.2025	.09113	.67082	.76631	2.2222222
0.46	1.44513	.16619	.2116	.09734	.67823	.77194	2.1739130
0.47	1.47655	.17349	.2209	.10382	.68557	.77750	2.1276596
0.48	1.50796	.18096	.2304	.11059	.69282	.78297	2.0833333
0.49	1.52938	.18857	.2401	.11765	.70000	.78837	2.0408163
0.50	1.57080	.19635	.2500	.12500	.70711	.79370	2.0000000
0.51	1.60221	.20428	.2601	.132651	.71414	.79896	1.9607843
0.52	1.63363	.21237	.2704	.140608	.72111	.80415	1.9230769
0.53	1.66504	.22062	.2809	.148877	.72801	.80927	1.8867925
0.54	1.69646	.22902	.2916	.157464	.73485	.81433	1.8518519
0.55	1.72788	.23758	.3025	.166375	.74162	.81932	1.8181818
0.56	1.75929	.24630	.3136	.175616	.74833	.82426	1.7857143
0.57	1.79071	.25518	.3249	.185193	.75498	.82913	1.7543860
0.58	1.82212	.26421	.3364	.195112	.76158	.83396	1.7241379
0.59	1.85354	.27340	.3481	.205379	.76811	.83872	1.6949153
0.60	1.88496	.28274	.3600	.216000	.77460	.84343	1.6666667
0.61	1.91637	.29225	.3721	.226981	.78102	.84809	1.6393443
0.62	1.94779	.30191	.3844	.238328	.78740	.85270	1.6129032
0.63	1.97920	.31172	.3969	.250047	.79373	.85726	1.5873016
0.64	2.01062	.32170	.4096	.262144	.80000	.86177	1.5625000
0.65	2.04204	.33183	.4225	.274625	.80623	.86624	1.5384615
0.66	2.07345	.34212	.4356	.287496	.81240	.87066	1.5151515
0.67	2.10487	.35257	.4489	.300763	.81854	.87503	1.4925373
0.68	2.13628	.36317	.4624	.314432	.82462	.87937	1.4705882
0.69	2.16670	.37393	.4761	.328509	.83066	.88366	1.4492754
0.70	2.19912	.38485	.4900	.343000	.83666	.88790	1.4285714
0.71	2.23053	.39592	.5011	.357911	.84261	.89211	1.4084507
0.72	2.26195	.40715	.5184	.373248	.84853	.89628	1.3888889
0.73	2.29336	.41854	.5329	.389017	.85440	.90041	1.3698630
0.74	2.32478	.43008	.5476	.405224	.86023	.90450	1.3513514
0.75	2.35619	.44179	.5625	.421875	.86603	.90856	1.3333333
0.76	2.38761	.45365	.5776	.438976	.87178	.91258	1.3157895
0.77	2.41903	.46566	.5929	.456533	.87750	.91657	1.2987013

No.	Circum.	Area	Square	Cube	Sq. Root	C. Root	Reciprocal
0.78	2.45044	.47784	.6084	.474552	.88318	.92052	1.2820513
0.79	2.48186	.49017	.6241	.493039	.88882	.92443	1.2658228
0.80	2.51327	.50265	.6400	.512000	.89443	.92832	1.2500000
0.81	2.54469	.51530	.6561	.531441	.90000	.93217	1.2345679
0.82	2.57611	.52810	.6724	.551368	.90554	.93599	1.2195122
0.83	2.60752	.54106	.6889	.571787	.91104	.93978	1.2048193
0.84	2.63894	.55418	7056	.592704	.91652	.94354	1.1904762
0.85	2.67035	.56745	.7225	.614125	.92195	.94727	1.1764706
0.86	2.70177	.58088	.7396	.636056	.92736	.95097	1.1627907
0.87	2.73319	.59447	.7569	.658503	.93274	.95464	1.1494253
0.88	2.76460	.60821	.7744	.681472	.93808	.95828	1.1363636
0.89	2.79602	.62211	.7921	.704969	.94340	.96190	1.1235955
0.90	2.82743	.63617	.8100	.729000	.94868	.96549	1.1111111
0.91	2.85885	.65039	.8281	.753571	.95394	.96905	1.0989011
0.92	2.89027	.66476	.8464	.778688	.95917	.97259	1.0869565
0.93	2.92168	.67929	.8649	.804357	.96437	.97610	1.0752688
0.94	2.95310	.69395	.8836	.830584	.96954	.97959	1.0638298
0.95	2.98451	.70882	.9025	.857375	.97468	.98305	1.0526316
0.96	3.01593	.72382	.9216	.884736	.97980	.98648	1.0416667
0.97	3.04735	.73898	.9409	.912673	.98489	.98990	1.0309278
0.98	3.07876	.75430	.9604	.941192	.98995	.99329	1.0204082
0.99	3.11018	.76977	.9801	.970299	.99499	.99666	1.0101010
1.00	3.14159	.78540	1.00	1.000	1.00000	1.00000	1.000000
1.1	3.45575	.95033	1.21	1.331	1.049	1.032	.90909091
1.2	3.76991	1.1310	1.44	1.728	1.095	1.063	.83333333
1.3	4.08407	1.3273	1.69	2.197	1.140	1.091	.76923077
1.4	4.39823	1.5394	1.96	2.744	1.183	1.119	.71428571
1.5	4.71239	1.7672	2.25	3.375	1.225	1.145	.66666667
1.6	5.02655	2.0106	2.56	4.096	1.265	1.170	.62500000
1.7	5.34071	2.2698	2.89	4.913	1.304	1.193	.58823529
1.8	5.65487	2.5447	3.24	5.832	1.342	1.216	.55555556
1.9	5.96903	2.8353	3.61	6.859	1.378	1.239	.52631579
2.0	6.28319	3.1416	4.00	8.000	1.414	1.260	.50000000
2.2	6.91150	3.8013	4.84	10.648	1.483	1.301	.45454545
2.4	7.53932	4.5239	5.76	13.824	1.549	1.339	.41666667
2.6	8.16314	5.3093	6.76	17.576	1.612	1.375	.38461538
2.8	8.79646	6.1575	7.84	21.952	1.673	1 409	.35714286
3.0	9.42478	7.0686	9.00	27.000	1.732	1.442	.33333333
3.2	10.0531	8.0425	10.24	32.768	1.789	1.474	.31250000
3.4	10.6814	9.0792	11.56	39.304	1.844	1.504	.29411765
3.6	11.3097	10.179	12.96	46.656	1.897	1.533	.27777778
3.8	11.9381	11.341	14.44	54.872	1.949	1.560	.26315789
4.0	12.5661	12.566	16.00	64.000	2.000	1.587	.25000000
4.2	13.1047	13.854	17.64	74.088	2.049	1.613	.23809524
4.4	13.8230	15.205	19.36	85.184	2.098	1.639	.22727270
4.6	14.4513	16.619	21.16	97.336	2.145	1.663	.21739130
4.8	15.0796	18.096	23.04	110.592	2 191	1.687	.20833333
5.0	15 7080	19.635	25 00	125 000	2.236	1.710	.20000000
5.2	16.3363	21.237	27 04	140.608	2.280	1.732	.19230769
5.4	16.9646	22.902	29.16	157.464	2.324	1.754	.18518519
5.6	17.5929	24.630	31.36	175.610	2.366	1.776	.17857143
5.8	18 2212	26.421	33.64	195.112	2.408	1.797	.17241379
6.0	18.8196	28.274	36.00	216.000	2.449	1.817	.16666667
6.2	19.4779	30.191	38.44	238.328	2.490	1.837	.16129032
6.4	20.1062	32.170	40.96	262.144	2.530	1.857	.15625000
6.6	20.7345	34.212	43.56	287.496	2.569	1.876	.15151515
6.8	21.3628	36.317	46.24	314.432	2.608	1.895	.14705882
7.0	21.9911	38.485	49.00	343.000	2.646	1.913	.14285714
7.2	22.6195	40.715	51.84	373.248	2.683	1.931	.13888889
7.4	23.2478	43 008	54.76	405.224	2.720	1.949	.13513514
7.6	23.8761	45.365	57.76	438.976	2.757	1.966	.13157895
7.8	24.5044	47.784	60.84	474.552	2.793	1.983	.12820513
8.0	25.1327	50.265	64.00	512.000	2.828	2.000	.12500000
8.2	25.7611	52.810	67.24	551.368	2.864	2.017	12195122
8.4	26.3894	55.418	70.56	592.704	2.898	2.033	.11904762
8.6	27 0177	58.088	73.96	636.056	2.933	2.049	.11627907
8.8	27.6460	30.821	77.44	681.472	2.966	2.065	.11363636
9.0	28.2743	63.617	81.00	729.000	3.000	2.080	.11111111
9.2	28.9027	66.476	84.64	778.688	3.033	2.095	.10869565
9.4	29.5310	69.398	88.36	830.584	3.066	2.110	.10638298
9.6	30.1593	72.382	92.16	884.736	3.098	2.125	.10416667
9.8	30.7876	75.430	•96.04	941.192	3.130	2.140	.10204082
10	31.4159	78.540	100	1000	3.1622777	2.1544347	.10000000
11	34.5575	95.033	121	1331	3.3166248	2.2239801	.09090909

No.	Circum.	Area	Square	Cube	Sq. Root	C. Root	Reciprocal
12	37.6991	113.10	144	1728	3.4641016	2.2894286	.083333333
13	40.8407	132.73	169	2197	3.6055513	2.3513347	.076923077
14	43.9823	153.94	196	2744	3.7416574	2.4101422	.071428571
15	47.1239	176.72	225	3375	3.8729833	2.4662121	.066666667
16	50.2655	201.06	256	4096	4.0000000	2.5198421	.062500000
17	53.4071	226.98	289	4913	4.1231056	2.5712816	.058823529
18	56.5487	254.47	324	5832	4.2426407	2.6207414	.055555556
19	59.6903	283.53	361	6859	4.3588989	2.6684016	.052631579
20	62.8319	314.16	400	8000	4.4721360	2.7144177	.050000000
21	65.9734	346.36	441	9261	4.5825757	2.7589243	.047619048
22	69.1150	380.13	484	10648	4.6904158	2.8020393	.045454545
23	72.2566	415.48	529	12167	4.7958315	2.8438670	.043478261
24	75.3982	452.39	576	13824	4.8989795	2.8844991	.041666667
25	78.5398	490.87	625	15625	5.0000000	2.9240177	.040000000
26	81.6814	530.93	676	17576	5.0990195	2.9624960	.038461538
27	84.8230	572.56	729	19683	5.1961524	3.0000000	.037037037
28	87.9646	615.75	784	21952	5.2915026	3.0365889	.035714286
29	91.1062	660.52	841	24389	5.3851648	3.0723168	.034482759
30	94.2478	706.86	900	27000	5.4772256	3.1072325	.033333333
31	97.3894	754.77	961	29791	5.5677644	3.1413806	.032258065
32	100.531	804.25	1024	32768	5.6568542	3.1748021	.031250000
33	103.673	855.30	1089	35937	5.7445626	3.2075343	.030303030
34	106.814	907.92	1156	39304	5.8309519	3.2396118	.029411765
35	109.956	962.11	1225	42875	5.9160798	3.2710663	.028571429
36	113.097	1017.9	1296	46656	6.0000000	3.3019272	.027777778
37	116.239	1075.2	1369	50653	6.0827625	3.3322215	.027027027
38	119.381	1134.1	1444	54872	6.1644140	3.3619754	.026315789
39	122.522	1194.6	1521	59319	6.2449980	3.3912114	.025641026
40	125.664	1256.6	1600	64000	6.3245553	3.4199519	.025000000
41	128.805	1320.3	1681	68921	6.4031242	3.4482172	.024390244
42	131.947	1385.4	1764	74088	6.4807407	3.4760266	.023809524
43	135.088	1452.2	1849	79507	6.5574385	3.5033981	.023255814
44	138.230	1520.5	1936	85184	6.6332496	3.5303483	.022727273
45	141.372	1590.4	2025	91125	6.7082039	3.5568933	.022222222
46	144.513	1661.9	2116	97336	6.7823300	3.5830479	.021739130
47	147.655	1734.9	2209	103823	6.8556546	3.6088261	.021276596
48	150.796	1809.6	2304	110592	6.9282032	3.6342411	.020833333
49	153.938	1885.7	2401	117649	7.0000000	3.6593057	.020408163
50	157.080	1963.5	2500	125000	7.0710678	3.6840314	.020000000
51	160.221	2042.8	2601	132651	7.1414284	3.7084295	.019607843
52	163.363	2123.7	2704	140608	7.2111026	3.7325111	.019230769
53	166.504	2206.2	2809	148877	7.2801099	3.7562858	.018867925
54	169.646	2290.2	2916	157464	7.3484692	3.7797631	.018518519
55	172.788	2375.8	3025	166375	7.4161985	3.8029525	.018181818
56	175.929	2463.0	3136	175616	7.4833148	3.8258624	.017857143
57	179.071	2551.8	3249	185193	7.5498344	3.8485011	.017543860
58	182.212	2642.1	3364	195112	7.6157731	3.8708766	.017241379
59	185.354	2734.0	3481	205379	7.6811457	3.8929965	.016949153
60	188.496	2827.4	3600	216000	7.7459667	3.9148676	.016666667
61	191.637	2922.5	3721	226981	7.8102497	3.9364972	.016393443
62	194.779	3019.1	3844	238328	7.8740079	3.9578915	.016129032
63	197.920	3117.2	3969	250047	7.9372539	3.9790571	.015873016
64	201.062	3217.0	4096	262144	8.0000000	4.0000000	.015625000
65	204.204	3318.3	4225	274625	8.0622577	4.0207256	.015384615
66	207.345	3421.2	4356	287496	8.1240384	4.0412401	.015151515
67	210.487	3525.7	4489	300763	8.1853528	4.0615480	.014925373
68	213.628	3631.7	4624	314432	8.2462113	4.0816551	.014705882
69	216.770	3739.3	4761	328509	8.3066239	4.1015661	.014492754
70	219.911	3848.5	4900	343000	8.3666003	4.1212853	.014285714
71	223.053	3959.2	5041	357911	8.4261498	4.1408178	.014084507
72	226.195	4071.5	5184	373248	8.4852814	4.1601676	.013888889
73	229.336	4185.4	5329	389017	8.5440037	4.1793390	.013698630
74	232.478	4300.8	5476	405224	8.6023253	4.1983364	.013513514
75	235.619	4417.9	5625	421875	8.6602540	4.2171633	.013333333
76	238.761	4536.5	5776	438976	8.7177979	4.2358236	.013157895
77	241.903	4656.6	5929	456533	8.7749644	4.2543210	.012987013
78	245.044	4778.4	6084	474552	8.8317609	4.2726586	.012820513
79	248.186	4901.7	6241	493039	8.8881944	4.2908404	.012658228
80	251.327	5026.5	6400	512000	8.9442719	4.3088695	.012500000
81	254.469	5153.0	6561	531441	9.0000000	4.3267487	.012345679
82	257.611	5281.0	6724	551368	9.0553851	4.3444815	.012195122
83	260.752	5410.6	6889	571787	9.1104336	4.3620707	.012048193
84	263.894	5541.8	7056	592704	9.1651514	4.3795191	.011904762
85	267.035	5674.5	7225	614125	9.2195445	4.3968296	.011764706

No.	Circum.	Area	Square	Cube	Sq. Root	Cube Root	Reciprocal
86	270.177	5808.8	7396	636056	9.2736185	4.4140049	.011627907
87	273.319	5944.7	7569	658503	9.3273791	4.4310476	.011494253
88	276.460	6082.1	7744	681472	9.3808315	4.4479602	.011363636
89	279.602	6221.1	7921	704969	9.4339811	4.4647451	.011235955
90	282.743	6361.7	8100	729000	9.4868330	4.4814047	.011111111
91	285.885	6503.9	8281	753571	9.5393920	4.4979414	.010989011
92	289.027	6647.6	8464	778688	9.5916630	4.5143574	.010869565
93	292.168	6792.9	8649	804357	9.6436508	4.5306549	.010752688
94	295.310	6939.8	8836	830584	9.6953597	4.5468359	.010638298
95	298.451	7088.2	9025	857375	9.7467943	4.5629026	.010526316
96	301.593	7238.2	9216	884736	9.7979590	4.5788570	.010416667
97	304.734	7389.8	9409	912673	9.8488578	4.5947009	.010309278
98	307.876	7543.0	9604	941192	9.8994949	4.6104363	.010204082
99	311.018	7697.7	9801	970299	9.9498744	4.6260650	.010101010
100	314.159	7854.0	10000	1000000	10.0000000	4.6415888	.010000000
101	317.301	8011.8	10201	1030301	10.0498756	4.6570095	.009900990
102	320.442	8171.3	10404	1061208	10.0995049	4.6723287	.009803622
103	323.584	8332.3	10609	1092727	10.1488916	4.6875482	.009708738
104	326.726	8494.9	10816	1124864	10.1980390	4.7026694	.009615385
105	329.867	8659.0	11025	1157625	10.2469508	4.7176940	.009523810
106	333.009	8824.7	11236	1191016	10.2956301	4.7326235	.009433962
107	336.150	8992.0	11449	1225043	10.3440804	4.7474594	.009345794
108	339.292	9160.9	11664	1259712	10.3923048	4.7622032	.009259259
109	342.434	9331.3	11881	1295029	10.4403065	4.7768562	.009174312
110	345.575	9503.3	12100	1331000	10.4880885	4.7914199	.009090909
111	348.717	9676.9	12321	1367631	10.5356538	4.8058955	.009009009
112	351.859	9852.0	12544	1404928	10.5830052	4.8202845	.008928571
113	355.000	10028.8	12769	1442897	10.6301458	4.8345881	.008849558
114	358.112	10207.0	12996	1481544	10.6770783	4.8488076	.008771930
115	361.283	10386.9	13225	1520875	10.7238053	4.8629442	.008695652
116	364.425	10568.3	13456	1560896	10.7703296	4.8769990	.008620690
117	367.566	10751.3	13689	1601613	10.8166538	4.8909732	.008547009
118	370.709	10935.9	13924	1643032	10.8627805	4.9048681	.008474576
119	373.859	11122.0	14161	1685159	10.9087121	4.9186847	.008403361
120	376.991	11309.7	14400	1728000	10.9544512	4.9324242	.008333333
121	380.133	11499.0	14641	1771561	11.0000000	4.9460874	.008264463
122	383.274	11689.9	14884	1815848	11.0453610	4.9596757	.008196721
123	386.416	11882.3	15129	1860867	11.0905365	4.9731898	.008130081
124	389.557	12076.3	15376	1906624	11.1355287	4.9866310	.008064516
125	392.699	12271.9	15625	1953125	11.1803399	5.0000000	.008000000
126	395.841	12469.0	15876	2000376	11.2249722	5.0132979	.007936508
127	398.982	12667.7	16129	2048383	11.2694277	5.0265257	.007874016
128	402.124	12868.0	16384	2097152	11.3137085	5.0396842	.007812500
129	405.265	13069.8	16641	2146689	11.3578167	5.0527743	.007751938
130	408.407	13273.2	16900	2197000	11.4017543	5.0657970	.007692308
131	411.549	13478.2	17161	2248091	11.4455231	5.0787531	.007633588
132	414.690	13684.8	17424	2299968	11.4891253	5.0916434	.007575758
133	417.832	13892.9	17689	2352637	11.5325626	5.1044687	.007518797
134	420.973	14102.6	17956	2406104	11.5758369	5.1172299	.007462687
135	424.115	14313.9	18225	2460375	11.6189500	5.1299278	.007407407
136	427.257	14526.7	18496	2515456	11.6619038	5.1425632	.007352941
137	430.398	14741.1	18769	2571353	11.7046999	5.1551367	.007299270
138	433.540	14957.1	19044	2628072	11.7473401	5.1676493	.007246377
139	436.681	15174.7	19321	2685619	11.7898261	5.1801015	.007194245
140	439.823	15393.8	19600	2744000	11.8321596	5.1924941	.007142857
141	442.965	15614.5	19881	2803221	11.8743421	5.2048279	.007092199
142	446.106	15836.8	20164	2863288	11.9163753	5.2171034	.007042254
143	449.248	16060.6	20449	2924207	11.9582607	5.2293215	.006993007
144	452.389	16286.0	20736	2985984	12.0000000	5.2414828	.006944444
145	455.531	16513.0	21025	3048625	12.0415946	5.2535879	.006896552
146	458.673	16741.6	21316	3112136	12.0830460	5.2656374	.006849315
147	461.814	16971.7	21609	3176523	12.1243557	5.2776321	.006802721
148	464.956	17203.4	21904	3241792	12.1655251	5.2895725	.006756757
149	468.097	17436.6	22201	3307949	12.2065556	5.3014592	.006711409
150	471.239	17671.5	22500	3375000	12.2474487	5.3132928	.006666667
151	474.330	17907.9	22801	3442951	12.2882057	5.3250740	.006622517
152	477.522	18145.8	23104	3511808	12.3288280	5.3368033	.006578947
153	430.664	18385.4	23409	3581577	12.3693169	5.3484812	.006535948
154	483.805	18626.5	23716	3652264	12.4096736	5.3601084	.006493506
155	486.947	18869.2	24025	3723875	12.4498996	5.3716854	.006451613
156	490.088	19113.5	24336	3796416	12.4899960	5.3832126	.006410256
157	493.230	19359.3	24649	3869893	12.5299641	5.3946907	.006369427
158	496.372	19606.7	24964	3944312	12.5698051	5.4061202	.006329114
159	499.513	19855.7	25281	4019679	12.6095202	5.4175015	.006289308
160	502.655	20106.2	25600	4096000	12.6491106	5.4288352	.006250000

No.	Circum.	Area	Square	Cube	Sq. Root	Cube Root	Reciprocal
161	505.796	20358.3	25921	4173281	12.6885775	5.4401218	.006211180
162	508.938	20612.0	26244	4251528	12.7279221	5.4513618	.006172840
163	512.080	20867.2	26569	4330747	12.7671453	5.4625550	.006134969
164	515.221	21124.1	26896	4410944	12.8062485	5.4737037	.006097561
165	518.363	21382.5	27225	4492125	12.8452326	5.4848066	.006060606
166	521.504	21642.4	27556	4574296	12.8840987	5.4958647	.006024096
167	524.646	21904.0	27889	4657463	12.9228480	5.5068784	.005988024
168	527.788	22167.1	28224	4741632	12.9614814	5.5178484	.005952381
169	530.929	22431.8	28561	4826809	13.0000000	5.5287748	.005917160
170	534.071	22698.0	28900	4913000	13.0384048	5.5396583	.005882353
171	537.212	22965.8	29241	5000211	13.0766968	5.5504991	.005847953
172	540.354	23235.2	29584	5088448	13.1148770	5.5612978	.005813953
173	543.496	23506.2	29929	5177717	13.1529464	5.5720546	.005780347
174	546.637	23778.7	30276	5268024	13.1909060	5.5827702	.005747126
175	549.779	24052.8	30625	5359375	13.2287566	5.5934447	.005714286
176	552.920	24328.5	30976	5451776	13.2664992	5.6040787	.005681818
177	556.062	24605.7	31329	5545233	13.3041347	5.6146724	.005649718
178	559.203	24884.6	31684	5639752	13.3416641	5.6252263	.005617978
179	562.345	25164.9	32041	5735339	13.3790882	5.6357408	.005586592
180	565.487	25446.9	32400	5832000	13.4164079	5.6462162	.005555556
181	568.623	25730.4	32761	5929741	13.4536240	5.6566528	.005524862
182	571.770	26015.5	33124	6028568	13.4907376	5.6670511	.005494505
183	574.911	26302.2	33489	6128487	13.5277493	5.6774114	.005464481
184	578.053	26590.4	33856	6229504	13.5646600	5.6877340	.005434783
185	581.195	26889.3	34225	6331625	13.6014705	5.6980192	.005405405
186	584.336	27171.6	34596	6434856	13.6381817	5.7082675	.005376344
187	587.478	27464.6	34969	6539203	13.6747943	5.7184791	.005347594
188	590.619	27759.1	35344	6644672	13.7113092	5.7286543	.005319149
189	593.761	28055.2	35721	6751269	13.7477271	5.7387936	.005291005
190	596.903	28352.9	36100	6859000	13.7840488	5.7488971	.005263158
191	600.044	28652.1	36481	6967871	13.8202750	5.7589652	.005235602
192	603.186	28952.9	36864	7077888	13.8564065	5.7689982	.005208333
193	606.327	29255.3	37249	7189057	13.8924440	5.7789966	.005181347
194	609.469	29559.3	37636	7301384	13.9283883	5.7889604	.005154639
195	612.611	29864.8	38025	7414875	13.9642400	5.7988900	.005128205
196	615.752	30171.9	38416	7529536	14.0000000	5.8087857	.005102041
197	618.894	30480.5	38809	7645373	14.0356688	5.8186479	.005076142
198	622.035	30790.8	39204	7762392	14.0712473	5.8284767	.005050505
199	625.177	31102.6	39601	7880599	14.1067360	5.8382725	.005025126
200	628.319	31415.9	40000	8000000	14.1421356	5.8480355	.005000000
201	631.460	31730.9	40401	8120601	14.1774469	5.8577660	.004975124
202	634.602	32047.4	40804	8242408	14.2126704	5.8674643	.004950495
203	637.743	32365.5	41209	8365427	14.2478068	5.8771307	.004926108
204	640.585	32685.1	41616	8489664	14.2828569	5.8867653	.004901961
205	644.026	33006.4	42025	8615125	14.3178211	5.8963685	.004878049
206	647.168	33329.2	42436	8741816	14.3527001	5.9059406	.004854369
207	650.310	33653.5	42849	8869743	14.3874946	5.9154817	.004830918
208	653.451	33979.5	43264	8998912	14.4222051	5.9249921	.004807692
209	656.593	34307.0	43681	9129329	14.4568323	5.9344721	.004784689
210	659.734	34636.1	44100	9261000	14.4913767	5.9439220	.004761905
211	662.876	34966.7	44521	9393931	14.5258390	5.9533418	.004739336
212	666.018	35298.9	44944	9528128	14.5602198	5.9627320	.004716981
213	669.159	35632.7	45369	9663597	14.5945195	5.9720926	.004694836
214	672.301	35968.1	45796	9800344	14.6287388	5.9814240	.004672897
215	675.442	36305.0	46225	9938375	14.6628783	5.9907264	.004651163
216	678.584	36643.5	46656	10077696	14.6969385	6.0000000	.004629630
217	681.726	36983.6	47089	10218313	14.7309199	6.0092450	.004608295
218	684.867	37325.3	47524	10360232	14.7648231	6.0184617	.004587156
219	688.009	37668.5	47961	10503459	14.7986486	6.0276502	.004566210
220	691.150	38013.3	48400	10648000	14.8323970	6.0368107	.004545455
221	694.292	38359.6	48841	10793861	14.8660687	6.0459435	.004524887
222	697.434	38707.6	49284	10941048	14.8996644	6.0550489	.004504505
223	700.575	39057.1	49729	11089567	14.9331845	6.0641270	.004484305
224	703.717	39408.1	50176	11239424	14.9666295	6.0731779	.004464286
225	706.858	39760.8	50625	11390625	15.0000000	6.0822020	.004444444
226	710.000	40115.0	51076	11543176	15.0332964	6.0911994	.004424779
227	713.142	40470.8	51529	11697083	15.0665192	6.1001702	.004405286
228	716.283	40828.1	51984	11852352	15.0996689	6.1091147	.004385965
229	719.425	41187.1	52441	12008989	15.1327460	6.1180332	.004366812
230	722.566	41547.6	52900	12167000	15.1657509	6.1269257	.004347826
231	725.708	41909.6	53361	12326391	15.1986842	6.1357924	.004329004
232	728.849	42273.3	53824	12487168	15.2315462	6.1446337	.004310345
233	731.991	42638.5	54289	12649337	15.2643375	6.1534495	.004291845
234	735.133	43005.3	54756	12812904	15.2970585	6.1622401	.004273504
235	738.274	43373.6	55225	12977875	15.3297097	6.1710058	.004255319

No.	Circum.	Area	Square	Cube	Sq. Root	Cube Root	Reciprocal
236	741.416	43743.5	55696	13144256	15.3622915	6.1797466	.004237288
237	744.557	44115.0	56169	13312053	15.3948043	6.1884628	.004219409
238	747.699	44486.1	56644	13481272	15.4272486	6.1971544	.004201681
239	750.841	44862.7	57121	13651919	15.4596248	6.2058218	.004184100
240	753.982	45238.9	57600	13824000	15.4919334	6.2144650	.004166667
241	757.124	45616.7	58081	13997521	15.5241747	6.2230843	.004149378
242	760.265	45996.1	58564	14172488	15.5563492	6.2316797	.004132231
243	763.407	46377.0	59049	14348907	15.5884573	6.2402515	.004115226
244	766.549	46759.5	59536	14526784	15.6204994	6.2487998	.004098361
245	769.690	47143.5	60025	14706125	15.6524758	6.2573248	.004081633
246	772.832	47529.2	60516	14886936	15.6843871	6.2658266	.004065041
247	775.973	47916.4	61009	15069223	15.7162336	6.2743054	.004048583
248	779.115	48305.1	61504	15252992	15.7480157	6.2827613	.004032258
249	782.257	48695.5	62001	15438249	15.7797338	6.2911946	.004016064
250	785.398	49087.4	62500	15625000	15.8113883	6.2996053	.004000000
251	788.540	49480.9	63001	15813251	15.8429795	6.3079935	.003984064
252	791.681	49875.9	63504	16003008	15.8745079	6.3163596	.003968254
253	794.823	50272.6	64009	16194277	15.9059737	6.3247035	.003952569
254	797.965	50670.8	64516	16387064	15.9373775	6.3330256	.003937003
255	801.106	51070.5	65025	16581375	15.9687194	6.3413257	.003921569
256	804.248	51471.9	65536	16777216	16.0000000	6.3496042	.003906250
257	807.389	51874.8	66049	16974593	16.0312195	6.3578611	.003891051
258	810.531	52279.2	66564	17173512	16.0623784	6.3660968	.003875969
259	813.672	52685.3	67081	17373979	16.0934769	6.3743111	.003861004
260	816.814	53092.9	67600	17576000	16.1245155	6.3825043	.003846154
261	819.956	53502.1	68121	17779531	16.1554944	6.3906765	.003831418
262	823.097	53912.9	68644	17984728	16.1864141	6.3988279	.003816794
263	826.239	54325.2	69169	18191447	16.2172747	6.4069585	.003802281
264	829.380	54739.1	69696	18399744	16.2480768	6.4150687	.003787879
265	832.522	55154.6	70225	18609625	16.2788206	6.4231583	.003773585
266	835.664	55571.6	70756	18821096	16.3095064	6.4312276	.003759398
267	838.805	55990.3	71289	19034163	16.3401346	6.4392767	.003745318
268	841.947	56410.4	71824	19248832	16.3707055	6.4473057	.003731343
269	845.088	56832.2	72361	19465109	16.4012195	6.4553148	.003717472
270	848.230	57255.5	72900	19683000	16.4316767	6.4633041	.003703704
271	851.372	57680.4	73441	19902511	16.4620776	6.4712736	.003690037
272	854.513	58106.9	73984	20123648	16.4924225	6.4792236	.003676471
273	857.655	58534.9	74529	20346417	16.5227116	6.4871541	.003663004
274	860.796	58964.6	75076	20570824	16.5529454	6.4950653	.003649635
275	863.938	59395.7	75625	20796875	16.5831240	6.5029572	.003636364
276	867.080	59828.5	76176	21024576	16.6132477	6.5108300	.003623188
277	870.221	60262.8	76729	21253933	16.6433170	6.5186839	.003610108
278	873.363	60698.7	77284	21484952	16.6733320	6.5265189	.003597122
279	876.504	61136.2	77841	21717639	16.7032931	6.5343351	.003584229
280	879.646	61575.2	78400	21952000	16.7332005	6.5421326	.003571429
281	882.788	62015.8	78961	22188041	16.7630546	6.5499116	.003558719
282	885.929	62458.0	79524	22425768	16.7928556	6.5576722	.003546099
283	889.071	62901.8	80089	22665187	16.8226038	6.5654144	.003533569
284	892.212	63347.1	80656	22906304	16.8522995	6.5731385	.003521127
285	895.354	63794.0	81225	23149125	16.8819430	6.5808443	.003508772
286	898.495	64242.4	81796	23393656	16.9115345	6.5885323	.003496503
287	901.637	64692.5	82369	23639903	16.9410743	6.5962023	.003484321
288	904.779	65144.1	82944	23887872	16.9705627	6.6038545	.003472222
289	907.920	65597.2	83521	24137569	17.0000000	6.6114890	.003460208
290	911.062	66052.0	84100	24389000	17.0293864	6.6191060	.003448276
291	914.203	66508.3	84681	24642171	17.0587221	6.6267054	.003436426
292	917.345	66966.2	85264	24897088	17.0880075	6.6342874	.003424658
293	920.487	67425.7	85849	25153757	17.1172428	6.6418522	.003412969
294	923.628	67886.7	86436	25412184	17.1464282	6.6493998	.003401361
295	926.770	68349.3	87025	25672375	17.1755640	6.6569302	.003389831
296	929.911	68813.5	87616	25934336	17.2046505	6.6644437	.003378378
297	933.053	69279.2	88209	26198073	17.2336879	6.6719403	.003367003
298	936.195	69746.5	88804	26463592	17.2626765	6.6794200	.003355705
299	939.336	70215.4	89401	26730899	17.2916165	6.6868831	.003344482
300	942.478	70685.8	90000	27000000	17.3205081	6.6943295	.003333333
301	945.619	71157.9	90601	27270901	17.3493516	6.7017593	.003322259
302	948.761	71631.5	91204	27543608	17.3781472	6.7091729	.003311258
303	951.903	72106.6	91809	27818127	17.4068952	6.7165700	.003300330
304	955.044	72583.4	92416	28094464	17.4355958	6.7239508	.003289474
305	958.186	73061.7	93025	28372625	17.4642492	6.7313155	.003278689
306	961.327	73541.5	93636	28652616	17.4928557	6.7386641	.003267974
307	964.469	74023.0	94249	28934443	17.5214155	6.7459967	.003257329
308	967.611	74506.0	94864	29218112	17.5499288	6.7533134	.003246753
309	970.752	74990.6	95481	29503629	17.5783958	6.7606143	.003236246
310	973.894	75476.8	96100	29791000	17.6068169	6.7678995	.003225806

No.	Circum.	Area	Square	Cube	Sq. Root	Cube Root	Reciprocal
311	977.035	75964.5	96721	30080231	17.6351921	6.7751690	.003215434
312	980.177	76453.8	97344	30371328	17.6635217	6.7824229	.003205128
313	983.318	76944.7	97969	30664297	17.6918060	6.7896613	.003194888
314	986.460	77437.1	98596	30959144	17.7200451	6.7968844	.003184713
315	989.602	77931.1	99225	31255875	17.7482393	6.8040921	.003174603
316	992.743	78426.7	99856	31554496	17.7763888	6.8112847	.003164557
317	995.885	78923.9	100489	31855013	17.8044938	6.8184620	.003154574
318	999.026	79422.6	101124	32157432	17.8325545	6.8256242	.003144654
319	1002.17	79922.9	101761	32461759	17.8605711	6.8327714	.003134796
320	1005.31	80424.8	102400	32768000	17.8885438	6.8399037	.003125000
321	1008.45	80928.2	103041	33076161	17.9164729	6.8470213	.003115265
322	1011.59	81433.2	103684	33386248	17.9443584	6.8541240	.003105590
323	1014.73	81939.8	104329	33698267	17.9722008	6.8612120	.003095975
324	1017.88	82448.0	104976	34012224	18.0000000	6.8682855	.003086420
325	1021.02	82957.7	105625	34328125	18.0277564	6.8753443	.003076923
326	1024.16	83469.0	106276	34645976	18.0554701	6.8823888	.003067485
327	1027.30	83981.8	106929	34965783	18.0831413	6.8894188	.003058104
328	1030.44	84496.3	107584	35287552	18.1107703	6.8964345	.003048780
329	1033.58	85012.3	108241	35611289	18.1383571	6.9034359	.003039514
330	1036.73	85529.9	108900	35937000	18.1659021	6.9104232	.003030303
331	1039.87	86049.0	109561	36264691	18.1934054	6.9173964	.003021148
332	1043.01	86569.7	110224	36594368	18.2208672	6.9243556	.003012048
333	1046.15	87092.0	110889	36926037	18.2482876	6.9313008	.003003003
334	1049.29	87615.9	111556	37259704	18.2756669	6.9382321	.002994012
335	1052.43	88141.3	112225	37595375	18.3030052	6.9451496	.002985075
336	1055.58	88668.3	112896	37933056	18.3303028	6.9520533	.002976190
337	1058.72	89196.9	113569	38272753	18.3575598	6.9589434	.002967359
338	1061.86	89727.0	114244	38614472	18.3847763	6.9658198	.002958580
339	1065.00	90258.7	114921	38958219	18.4119526	6.9726826	.002949853
340	1068.14	90792.0	115600	39304000	18.4390889	6.9795321	.002941176
341	1071.28	91326.9	116281	39651821	18.4661853	6.9863681	.002932551
342	1074.42	91863.3	116964	40001688	18.4932420	6.9931906	.002923977
343	1077.57	92401.3	117649	40353607	18.5202592	7.0000000	.002915452
344	1080.71	92940.9	118336	40707584	18.5472370	7.0067962	.002906977
345	1083.85	93482.0	119025	41063625	18.5741756	7.0135791	.002898551
346	1086.99	94024.7	119716	41421736	18.6010752	7.0203490	.002890173
347	1090.13	94569.0	120409	41781923	18.6279360	7.0271058	.002881844
348	1093.27	95114.9	121104	42144192	18.6547581	7.0338497	.002873563
349	1096.42	95662.3	121801	42508549	18.6815417	7.0405806	.002865330
350	1099.56	96211.3	122500	42875000	18.7082869	7.0472987	.002857143
351	1102.70	96761.8	123201	43243551	18.7349940	7.0540041	.002849003
352	1105.84	97314.0	123904	43614208	18.7616630	7.0606967	.002840909
353	1108.98	97867.7	124609	43986977	18.7882942	7.0673767	.002832861
354	1112.12	98423.0	125316	44361864	18.8148877	7.0740440	.002824859
355	1115.27	98979.8	126025	44738875	18.8414437	7.0806988	.002816901
356	1118.41	99538.2	126736	45118016	18.8679623	7.0873411	.002808989
357	1121.55	100098.2	127449	45499293	18.8944436	7.0939709	.002801120
358	1124.69	100659.8	128164	45882712	18.9208879	7.1005885	.002793296
359	1127.83	101222.9	128881	46268279	18.9472953	7.1071937	.002785515
360	1130.97	101787.6	129600	46656000	18.9736660	7.1137866	.002777778
361	1134.11	102353.9	130321	47045881	19.0000000	7.1203674	.002770083
362	1137.26	102921.7	131044	47437928	19.0262976	7.1269360	.002762431
363	1140.40	103491.1	131769	47832147	19.0525589	7.1334925	.002754821
364	1143.54	104062.1	132496	48228544	19.0787840	7.1400370	.002747253
365	1146.68	104634.7	133225	48627125	19.1049732	7.1465695	.002739726
366	1149.82	105208.8	133956	49027896	19.1311265	7.1530901	.002732240
367	1152.96	105784.5	134689	49430863	19.1572441	7.1595988	.002724796
368	1156.11	106361.8	135424	49836032	19.1833261	7.1660957	.002717391
369	1159.25	106940.6	136161	50243409	19.2093727	7.1725809	.002710027
370	1162.39	107521.0	136900	50653000	19.2353841	7.1790544	.002702703
371	1165.53	108103.0	137641	51064811	19.2613603	7.1855162	.002695418
372	1168.67	108686.5	138384	51478848	19.2873015	7.1919663	.002688172
373	1171.81	109271.7	139129	51895117	19.3132079	7.1984050	.002680965
374	1174.96	109858.4	139876	52313624	19.3390796	7.2048322	.002673797
375	1178.10	110446.6	140625	52734375	19.3649167	7.2112479	.002666667
376	1181.24	111036.5	141376	53157376	19.3907194	7.2176522	.002659574
377	1184.38	111627.9	142129	53582633	19.4164878	7.2240450	.002652520
378	1187.52	112220.8	142884	54010152	19.4422221	7.2304268	.002645503
379	1190.66	112815.4	143641	54439939	19.4679223	7.2367972	.002638522
380	1193.81	113411.5	144400	54872000	19.4935887	7.2431565	.002631579
381	1196.95	114009.2	145161	55306341	19.5192213	7.2495045	.002624672
382	1200.09	114608.4	145924	55742968	19.5448203	7.2558415	.002617801
383	1203.23	115209.3	146689	56181887	19.5703858	7.2621675	.002610966
384	1206.37	115811.7	147456	56623104	19.5959177	7.2684824	.002604167
385	1209.51	116415.6	148225	57066625	19.6214169	7.2747864	.002597403

No.	Circum.	Area	Square	Cube	Sq. Root	Cube Root	Reciprocal
386	1212.65	117021.2	148996	57512456	19.6468827	7.2810794	.002590674
387	1215.80	117628.3	149769	57960603	19.6723156	7.2873617	.002583979
388	1218.94	118237.0	150544	58411072	19.6977156	7.2936330	.002577320
389	1222.08	118847.2	151321	58863869	19.7230829	7.2998936	.002570694
390	1225.22	119459.1	152100	59319000	19.7484177	7.3061436	.002564103
391	1228.36	120072.5	152881	59776471	19.7737199	7.3123828	.002557545
392	1231.50	120687.4	153664	60236288	19.7989899	7.3186114	.002551020
393	1234.65	121304.0	154449	60698457	19.8242276	7.3248295	.002544529
394	1237.79	121922.1	155236	61162984	19.8494332	7.3310369	.002538071
395	1240.93	122541.8	156025	61629875	19.8746069	7.3372339	.002531646
396	1244.07	123163.0	156816	62099136	19.8997487	7.3434205	.002525253
397	1247.21	123785.8	157609	62570773	19.9248588	7.3495966	.002518892
398	1250.35	124410.2	158404	63044792	19.9499373	7.3557624	.002512563
399	1253.50	125036.2	159201	63521199	19.9749844	7.3619178	.002506266
400	1256.64	125663.7	160000	64000000	20.0000000	7.3680630	.002500000
401	1259.78	126292.8	160801	64481201	20.0249844	7.3741979	.002493766
402	1262.92	126923.5	161604	64964808	20.0499377	7.3803227	.002487562
403	1266.06	127555.7	162409	65450827	20.0748599	7.3864373	.002481390
404	1269.20	128189.6	163216	65939264	20.0997512	7.3925418	.002475248
405	1272.35	128824.9	164025	66430125	20.1246118	7.3986363	.002469136
406	1275.49	129461.9	164836	66923416	20.1494417	7.4047206	.002463054
407	1278.63	130100.4	165649	67419143	20.1742410	7.4107950	.002457002
408	1281.77	130740.5	166464	67917312	20.1990099	7.4168595	.002450980
409	1284.91	131382.2	167281	68417929	20.2237484	7.4229142	.002444988
410	1288.05	132025.4	168100	68921000	20.2484567	7.4289589	.002439024
411	1291.19	132670.2	168921	69426531	20.2731349	7.4349938	.002433090
412	1294.34	133316.6	169744	69934528	20.2977831	7.4410189	.002427184
413	1297.48	133964.6	170569	70444997	20.3224014	7.4470342	.002421308
414	1300.62	134614.1	171396	70957944	20.3469899	7.4530399	.002415459
415	1393.76	135265.2	172225	71473375	20.3715488	7.4590359	.002409639
416	1306.90	135917.9	173056	71991296	20.3960781	7.4650223	.002403846
417	1310.04	136572.1	173889	72511713	20.4205779	7.4709991	.002398082
418	1313.19	137227.9	174724	73034632	20.4450483	7.4769664	.002392344
419	1316.33	137885.3	175561	73560059	20.4694895	7.4829242	.002386635
420	1319.47	138544.2	176400	74088000	20.4939015	7.4888724	.002380952
421	1322.61	139204.8	177241	74618461	20.5182845	7.4948113	.002375297
422	1325.75	139866.9	178084	75151448	20.5426386	7.5007406	.002369668
423	1328.99	140530.5	178929	75686967	20.5669638	7.5066607	.002364066
424	1332.04	141195.7	179776	76225024	20.5912603	7.5125715	.002358491
425	1335.18	141862.5	180625	76765625	20.6155281	7.5184730	.002352941
426	1338.32	142530.9	181476	77308776	20.6397674	7.5243652	.002347418
427	1341.46	143200.9	182329	77854483	20.6639783	7.5302482	.002341920
428	1344.60	143872.4	183184	78402752	20.6881609	7.5361221	.002336449
429	1347.74	144545.5	184041	78953589	20.7123152	7.5419867	.002331002
430	1350.88	145220.1	184900	79507000	20.7364414	7.5478423	.002325581
431	1354.03	145896.4	185761	80062991	20.7605395	7.5536888	.002320186
432	1357.17	146574.2	186624	80621568	20.7846097	7.5595263	.002314815
433	1360.31	147253.5	187489	81182737	20.8086520	7.5653548	.002309469
434	1363.45	147934.5	188356	81746504	20.8326667	7.5711743	.002304147
435	1366.59	148617.0	189225	82312875	20.8566536	7.5769849	.002298851
436	1369.73	149301.1	190096	82881856	20.8806130	7.5827865	.002293578
437	1372.88	149986.7	190969	83453453	20.9045450	7.5885793	.002288330
438	1376.02	150673.9	191844	84027672	20.9284495	7.5943633	.002283105
439	1379.16	151362.7	192721	84604519	20.9523268	7.6001385	.002277904
440	1382.30	152053.1	193600	85184000	20.9761770	7.6059049	.002272727
441	1385.44	152745.0	194481	85766121	21.0000000	7.6116626	.002267574
442	1388.58	153438.5	195364	86350888	21.0237960	7.6174116	.002262443
443	1391.73	154133.6	196249	86938307	21.0475652	7.6231519	.002257336
444	1394.87	154830.3	197136	87528384	21.0713075	7.6288837	.002252252
445	1398.01	155528.5	198025	88121125	21.0950231	7.6346067	.002247191
446	1401.15	156228.3	198916	88716536	21.1187121	7.6403213	.002242152
447	1404.29	156929.6	199809	89314623	21.1423745	7.6460272	.002237136
448	1407.43	157632.6	200704	89915392	21.1660105	7.6517247	.002232143
449	1410.58	158337.1	201601	90518849	22.1896201	7.6574138	.002227171
450	1413.72	159043.1	202500	91125000	21.2132034	7.6630943	.002222222
451	1416.86	159750.8	203401	91733851	21.2367606	7.6687665	.002217295
452	1420.00	160460.0	204304	92345408	21.2602916	7.6744303	.002212389
453	1423.14	161170.8	205209	92959677	21.2837967	7.6800857	.002207506
454	1426.28	161883.1	206116	93576664	21.3072758	7.6857328	.002202643
455	1129.42	162597.1	207025	94196375	21.3307290	7.6913717	.002197802
456	1432.57	163312.6	207936	94818816	21.3541565	7.6970023	.002192982
457	1435.71	164029.6	208849	95443993	21.3775583	7.7026246	.002188184
458	1438.85	164748.3	209764	96071912	21.4009346	7.7082388	.002183406
459	1441.99	165468.5	210681	96702579	21.4242853	7.7138448	.002178649
460	1445.13	166190.3	211600	97336000	21.4476106	7.7194426	.002173913

No.	Circum.	Area	Square	Cube	Sq. Root	Cube Root	Reciprocal
461	1448.27	166913.6	212521	97972181	21.4709106	7.7250325	.002169197
462	1451.42	167638.5	213444	98611128	21.4941853	7.7306141	.002164502
463	1454.56	168365.0	214369	99252847	21.5174348	7.7361877	.002159827
464	1457.70	169093.1	215296	99897344	21.5406592	7.7417532	.002155172
465	1460.84	169822.7	216225	100544625	21.5638587	7.7473109	.002150538
466	1463.98	170553.9	217156	101194696	21.5870331	7.7528606	.002145923
467	1467.12	171286.7	218089	101847563	21.6101828	7.7584023	.002141328
468	1470.27	172021.1	219024	102503232	21.6333077	7.7639361	.002136752
469	1473.41	172757.0	219961	103161709	21.6564078	7.7694620	.002132196
470	1476.55	173494.5	220900	103823000	21.6794834	7.7749801	.002127660
471	1479.69	174233.5	221841	104487111	21.7025344	7.7804904	.002123142
472	1482.83	174974.1	222784	105154048	21.7255610	7.7859928	.002118644
473	1485.97	175716.4	223729	105823817	21.7485632	7.7914875	.002114165
474	1489.11	176460.1	224676	106496424	21.7715411	7.7969745	.002109705
475	1492.26	177205.5	225625	107171875	21.7944947	7.8024538	.002105263
476	1495.40	177952.4	226576	107850176	21.8174242	7.8079254	.002100840
477	1498.54	178700.9	227529	108531333	21.8403297	7.8133892	.002096436
478	1501.68	179450.9	228484	109215352	21.8632111	7.8188456	.002092050
479	1504.82	180202.5	229441	109902239	21.8860686	7.8242942	.002087683
480	1507.96	180955.7	230400	110592000	21.9089023	7.8297353	.002083333
481	1511.11	181710.5	231361	111284641	21.9317122	7.8351688	.002079002
482	1514.25	182466.8	232324	111980168	21.9544984	7.8405949	.002074689
483	1517.39	183224.8	233289	112678587	21.9772610	7.8460134	.002070393
484	1520.53	183984.2	234256	113379904	22.0000000	7.8514244	.002066116
485	1523.67	184745.3	235225	114084125	22.0227155	7.8568281	.002061586
486	1526.81	185507.9	236196	114791256	22.0454077	7.8622242	.002057613
487	1529.96	186272.1	237169	115501303	22.0680765	7.8676130	.002053388
488	1533.10	187037.9	238144	116214272	22.0907220	7.8729944	.002049180
489	1536.24	187805.2	239121	116930169	22.1133444	7.8783684	.002044990
490	1539.38	188574.1	240100	117649000	22.1359436	7.8837352	.002040816
491	1542.52	189344.6	241081	118370771	22.1585198	7.8890946	.002036660
492	1545.66	190116.6	242064	119095488	22.1810730	7.8944468	.002032520
493	1548.81	190890.2	243049	119823157	22.2036033	7.8997917	.002028398
494	1551.95	191665.4	244036	120553784	22.2261108	7.9051294	.002024291
495	1555.09	192442.2	245025	121287375	22.2485955	7.9104599	.002020202
496	1558.23	193220.5	246016	122023936	22.2710575	7.9157832	.002016129
497	1561.37	194000.4	247009	122763473	22.2934968	7.9210994	.002012072
498	1564.51	194781.9	248004	123505992	22.3159136	7.9264085	.002008032
499	1567.65	195564.9	249001	124251499	22.3383079	7.9317104	.002004008
500	1570.80	196349.5	250000	125000000	22.3606798	7.9370053	.002000000
501	1573.94	197135.7	251001	125751501	22.3830293	7.9422931	.001996008
502	1577.08	197923.5	252004	126506008	22.4053565	7.9475739	.001992032
503	1580.22	198712.8	253009	127263527	22.4276615	7.9528477	.001988072
504	1583.36	199503.7	254016	128024064	22.4499443	7.9581144	.001984127
505	1586.50	200296.2	255025	128787625	22.4722051	7.9633743	.001980198
506	1589.65	201090.2	256036	129554216	22.4944438	7.9686271	.001976285
507	1592.79	201885.8	257049	130323843	22.5166605	7.9738731	.001972387
508	1595.93	202683.0	258064	131096512	22.5388553	7.9791122	.001968504
509	1599.07	203481.7	259081	131872229	22.5610283	7.9843444	.001964637
510	1602.21	204282.1	260100	132651000	22.5831796	7.9895697	.001960784
511	1605.35	205084.0	261121	133432831	22.6053091	7.9947883	.001956947
512	1608.50	205887.4	262144	134217728	22.6274170	8.0000000	.001953125
513	1611.64	206692.5	263169	135005697	22.6495033	8.0052049	.001949318
514	1614.78	207499.1	264196	135796744	22.6715681	8.0104032	.001945525
515	1617.92	208307.2	265225	136590875	22.6936114	8.0155946	.001941748
516	1621.06	209117.0	266256	137388096	22.7156334	8.0207794	.001937984
517	1624.20	209928.3	267289	138188413	22.7376340	8.0259574	.001934236
518	1627.35	210741.2	268324	138991832	22.7596134	8.0311287	.001930502
519	1630.49	211555.6	269361	139798359	22.7815715	8.0362935	.001926782
520	1633.63	212371.7	270400	140608000	22.8035085	8.0414515	.001923077
521	1636.77	213189.3	271441	141420761	22.8254244	8.0466030	.001919386
522	1639.91	214008.4	272484	142236648	22.8473193	8.0517479	.001915709
523	1643.05	214829.2	273529	143055667	22.8691933	8.0568862	.001912046
524	1646.20	215651.5	274576	143877824	22.8910463	8.0620180	.001908397
525	1649.34	216475.4	275625	144703125	22.9128785	8.0671432	.001904762
526	1652.48	217300.8	276676	145531576	22.9346899	8.0722620	.001901141
527	1655.62	218127.9	277729	146363183	22.9564806	8.0773743	.001897533
528	1658.76	218956.4	278784	147197952	22.9782506	8.0824800	.001893939
529	1661.90	219786.6	279841	148035889	23.0000000	8.0875794	.001890359
530	1665.04	220618.3	280900	148877000	23.0217289	8.0926723	.001886792
531	1668.19	221451.7	281961	149721291	23.0434372	8.0977589	.001883239
532	1671.33	222286.5	283024	150568768	23.0651252	8.1028390	.001879699
533	1674.47	223123.0	284089	151419437	23.0867928	8.1079128	.001876173
534	1677.61	223961.0	285156	152273304	23.1084400	8.1129803	.001872659
535	1680.75	224800.6	286225	153130375	23.1300670	8.1180414	.001869159

No.	Circum.	Area	Square	Cube	Sq. Root	Cube Root	Reciprocal
536	1683.89	225641.8	287296	153990656	23.1516738	8.1230962	.001865672
537	1687.04	226484.5	288369	154854153	23.1732605	8.1281447	.001862197
538	1690.18	227328.8	289444	155720872	23.1948270	8.1331870	.001858736
539	1693.32	228174.7	290521	156590819	23.2163735	8.1382230	.001855288
540	1696.46	229022.1	291600	157464000	23.2379001	8.1432529	.001851852
541	1699.60	229871.1	292681	158340421	23.2594067	8.1482765	.001848429
542	1702.74	230721.7	293764	159220088	23.2808935	8.1532939	.001845018
543	1705.88	231573.9	294849	160103007	23.3023604	8.1583051	.001841621
544	1709.03	232427.6	295936	160989184	23.3238076	8.1633102	.001838255
545	1712.17	233282.9	297025	161878625	23.3452351	8.1683092	.001834862
546	1715.31	234139.8	298116	162771336	23.3666429	8.1733020	.001831502
547	1718.45	234998.2	299209	163667323	23.3880311	8.1782888	.001828154
548	1721.59	235858.2	300304	164566592	23.4093998	8.1832695	.001824818
549	1724.73	236719.8	301401	165469149	23.4307490	8.1882441	.001821494
550	1727.88	237582.9	302500	166375000	23.4520788	8.1932127	.001818182
551	1731.02	238447.7	303601	167284151	23.4733892	8.1981753	.001814882
552	1734.16	239314.0	304704	168196608	23.4946802	8.2031319	.001811594
553	1737.30	240181.8	305809	169112377	23.5159520	8.2080825	.001808318
554	1740.44	241051.3	306916	170031464	23.5372046	8.2130271	.001805054
555	1743.58	241922.3	308025	170953875	23.5584380	8.2179657	.001801802
556	1746.73	242794.9	309136	171879616	23.5796522	8.2228985	.001798561
557	1749.87	243669.0	310249	172808693	23.6008474	8.2278254	.001795332
558	1753.01	244544.7	311364	173741112	23.6220236	8.2327463	.001792115
559	1756.15	245422.0	312481	174676879	23.6431808	8.2376614	.001788909
560	1759.29	246300.9	313600	175616000	23.6643191	8.2425706	.001785714
561	1762.43	247181.3	314721	176558481	23.6854386	8.2474740	.001782531
562	1765.58	248063.3	315844	177504328	23.7065392	8.2523715	.001779359
563	1768.72	248946.9	316969	178453547	23.7276210	8.2572633	.001776199
564	1771.86	249832.0	318096	179406144	23.7486842	8.2621492	.001773050
565	1775.00	250718.7	319225	180362125	23.7697286	8.2670294	.001769912
566	1778.14	251607.0	320356	181321496	23.7907545	8.2719039	.001766784
567	1781.28	252496.9	321489	182284263	23.8117618	8.2767726	.001763668
568	1784.42	253388.3	322624	183250432	23.8327506	8.2816355	.001760563
569	1787.57	254281.3	323761	184220009	23.8537209	8.2864928	.001757469
570	1790.71	255175.9	324900	185193000	23.8746728	8.2913444	.001754386
571	1793.85	256072.0	326041	186169411	23.8956063	8.2961903	.001751313
572	1796.99	256969.7	327184	187149248	23.9165215	8.3010304	.001748252
573	1800.13	257869.0	328329	188132517	23.9374184	8.3058651	.001745201
574	1803.27	258769.9	329476	189119224	23.9582971	8.3106941	.001742160
575	1806.42	259672.3	330625	190109375	23.9791576	8.3155175	.001739130
576	1809.56	260576.3	331776	191102976	24.0000000	8.3203353	.001736111
577	1812.70	261481.8	332929	192100033	24.0208243	8.3251475	.001733102
578	1815.84	262389.0	334084	193100552	24.0416306	8.3299542	.001730114
579	1818.98	263297.7	335241	194104539	24.0624188	8.3347553	.001727116
580	1822.12	264207.9	336400	195112000	24.0831891	8.3395509	.001724138
581	1825.27	265119.8	337561	196122941	24.1039416	8.3443410	.001721170
582	1828.41	266033.2	338724	197137368	24.1246762	8.3491256	.001718213
583	1831.55	266948.2	339889	198155287	24.1453929	8.3539047	.001715266
584	1834.69	267864.8	341056	199176704	24.1660919	8.3586784	.001712329
585	1837.83	268782.9	342225	200201625	24.1867732	8.3634466	.001709402
586	1840.97	269702.6	343396	201230056	24.2074369	8.3682095	.001706485
587	1844.11	270623.9	344569	202262003	24.2280829	8.3729668	.001703578
588	1847.26	271546.7	345744	203297472	24.2487113	8.3777188	.001700680
589	1850.40	272471.1	346921	204336469	24.2693222	8.3824653	.001697793
590	1853.54	273397.1	348100	205379000	24.2899156	8.3872065	.001694915
591	1856.68	274324.7	349281	206425071	24.3104916	8.3919423	.001692047
592	1859.82	275253.8	350464	207474688	24.3310501	8.3966729	.001689189
593	1862.96	276184.5	351649	208527857	24.3515913	8.4013981	.001686341
594	1866.11	277116.8	352836	209584584	24.3721152	8.4061180	.001683502
595	1869.25	278050.6	354025	210644875	24.3926218	8.4108326	.001680672
596	1872.39	278986.0	355216	211708736	24.4131112	8.4155419	.001677852
597	1875.53	279923.0	356409	212776173	24.4335834	8.4202460	.001675042
598	1878.67	280861.5	357604	213847192	24.4540385	8.4249448	.001672241
599	1881.81	281801.7	358801	214921799	24.4744765	8.4296383	.001669449
600	1884.96	282743.3	360000	216000000	24.4948974	8.4343267	.001666667
601	1888.10	283686.6	361201	217081801	24.5153013	8.4390098	.001663894
602	1891.24	284631.4	362404	218167208	24.5356883	8.4436877	.001661130
603	1894.38	285577.8	363609	219256227	24.5560583	8.4483605	.001658375
604	1897.52	286525.8	364816	220348864	24.5764115	8.4530281	.001655629
605	1900.66	287475.4	366025	221445125	24.5967478	8.4576906	.001652893
606	1903.81	288426.5	367236	222545016	24.6170673	8.4623479	.001650165
607	1906.95	289379.2	368449	223648543	24.6373700	8.4670001	.001647446
608	1910.09	290333.4	369664	224755712	24.6576560	8.4716471	.001644737
609	1913.23	291289.3	370881	225866529	24.6779254	8.4762892	.001642036
610	1916.37	292246.7	372100	226981000	24.6981781	8.4809261	.001639344

No.	Circum.	Area	Square	Cube	Sq. Root	Cube Root	Reciprocal
611	1919.51	293205.6	373321	228099131	24.7184142	8.4855579	.001636661
612	1922.65	294166.2	374544	229220928	24.7386338	8.4901848	.001633987
613	1925.80	295128.3	375769	230346397	24.7588368	8.4948065	.001631321
614	1928.94	296092.0	376996	231475544	24.7790234	8.4994233	.001628664
615	1932.08	297057.2	378225	232608375	24.7991935	8.5040350	.001626016
616	1935.22	298024.1	379456	233744896	24.8193473	8.5086417	.001623377
617	1938.36	298992.4	380689	234885113	24.8394847	8.5132435	.001620746
618	1941.50	299962.4	381924	236029032	24.8596058	8.5178403	.001618123
619	1944.65	300934.0	383161	237176659	24.8797106	8.5224321	.001615509
620	1947.79	301907.1	384400	238328000	24.8997992	8.5270189	.001612903
621	1950.93	302881.7	385641	239483061	24.9198716	8.5316009	.001610306
622	1954.07	303858.0	386884	240641848	24.9399278	8.5361780	.001607717
623	1957.21	304835.8	388129	241804367	24.9599679	8.5407501	.001605136
624	1960.35	305815.2	389376	242970624	24.9799920	8.5453173	.001602564
625	1963.50	306796.2	390625	244140625	25.0000000	8 5498797	.001600000
626	1966.64	307778.7	391876	245314376	25.0199920	8.5544372	.001597444
627	1969.78	308762.8	393129	246491883	25.0399681	8.5589899	.001594896
628	1972.92	309748.5	394384	247673152	25.0599282	8.5635377	.001592357
629	1976.06	310735.7	395641	248858189	25.0798724	8.5680807	.001589825
630	1979.20	311724.5	396900	250047000	25.0998008	8 5726189	.001587302
631	1982.35	312714.9	398161	251239591	25.1197134	8.5771523	.001584786
632	1985.49	313706.9	399424	252435968	25.1396102	8.5816809	.001582278
633	1988.63	314700.4	400689	253636137	25.1594913	8.5862047	.001579779
634	1991.77	315695.5	401956	254840104	25.1793566	8.5907238	.001577287
635	1994.91	316692.2	403225	256047875	25.1992063	8.5952380	.001574803
636	1998.05	317690.4	404496	257259456	25.2190404	8.5997476	.001572327
637	2001.19	318690.2	405769	258474853	25.2388589	8.6042525	.001569859
638	2004.34	319691.6	407044	259694072	25.2586619	8.6087526	.001567398
639	2007.48	320694.6	408321	260917119	25.2784493	8.6132480	.001564945
640	2010.62	321699.1	409600	262144000	25.2982213	8.6177388	.001562500
641	2013.76	322705.2	410881	263374721	25.3179778	8.6222248	.001560062
642	2016.90	323712.9	412164	264600288	25.3377189	8.6267063	.001557632
643	2020.04	324722.1	413449	265847707	25.3574447	8.6311830	.001555210
644	2023.19	325732.9	414736	267089984	25.3771551	8.6356551	.001552795
645	2026.33	326745.3	416025	268336125	25.3968502	8 6401226	.001550388
646	2029.47	327759.2	417316	269586136	25.4165301	8.6445855	.001547988
647	2032.61	328774.7	418609	270840023	25.4361947	8.6490437	.001545595
648	2035.75	329791.8	419904	272097702	25.4558441	8.6534974	.001543210
649	2038.89	330810.5	421201	273359449	25.4754784	8.6579465	.001540832
650	2042.04	331830.7	422500	274625000	25.4950976	8.6623911	.001538462
651	2045.18	332852.5	423801	275894451	25.5147016	8.6668310	.001536098
652	2048.32	333875.9	425104	277167808	25.5342907	8.6712665	.001533742
653	2051.46	334900.9	426409	278445077	25.5538647	8.6756974	.001531394
654	2054.60	335927.4	427716	279726264	25.5734237	8.6801237	.001529052
655	2057.74	336955.5	429025	281011375	25.5929678	8.6845456	.001526718
656	2060.88	337985.1	430336	282300416	25.6124969	8.6889630	.001524390
657	2064.03	339016.3	431649	283593393	25.6320112	8.6933759	.001522070
658	2067.17	340049.1	432964	284890312	25.6515107	8.6977843	.001519757
659	2070.31	341083.5	434281	286191179	25.6709953	8.7021882	.001517451
660	2073.45	342119.4	435600	287496000	25.6904652	8.7065877	.001515152
661	2076.59	343157.0	436921	288804781	25.7099203	8.7109827	.001512859
662	2079.73	344196.0	438244	290117528	25.7293607	8.7153734	.001510574
663	2082.88	345236.7	439569	291434247	25.7487864	8.7197596	.001508296
664	2086.02	346278.9	440896	292754944	25.7681975	8.7241414	.001506024
665	2089.16	347322.7	442225	294079625	25.7875939	8.7285187	.001503759
666	2092.30	348368.1	443556	295408296	25.8069758	8.7328918	.001501502
667	2095.44	349415.0	444889	296740963	25.8263431	8.7372604	.001499250
668	2098.58	350463.5	446224	298077632	25.8456960	8.7416246	.001497006
669	2101.73	351513.6	447561	299418309	25.8650343	8.7459846	.001494768
670	2104.87	352565.2	448900	300763000	25.8843582	8.7503401	.001492537
671	2108.01	353618.5	450241	302111711	25.9036677	8.7546913	.001490313
672	2111.15	354673.2	451584	303464448	25.9229628	8.7590383	.001488095
673	2114.29	355729.6	452929	304821217	25.9422435	8.7633809	.001485884
674	2117.43	356787.5	454276	306182024	25.9615100	8.7677192	.001483680
675	2120.58	357847.0	455625	307546875	25.9807621	8.7720532	.001481481
676	2123.72	358908.1	456976	308915776	26.0000000	8.7763830	.001479290
677	2126.86	359970.8	458329	310288733	26.0192237	8.7807084	.001477105
678	2130.00	361035.0	459684	311665752	26.0384331	8.7850296	.001474926
679	2133.14	362100.8	461041	313046839	26.0576284	8.7893466	.001472754
680	2136.28	363168.1	462400	314432000	26.0768096	8.7936593	.001470588
681	2139.42	364237.0	463761	315821241	26.0959767	8.7979679	.001468429
682	2142.57	365307.5	465124	317214568	26.1151297	8.8022721	.001466276
683	2145.71	366379.6	466489	318611987	26.1342687	8.8065722	.001464129
684	2148.85	367453.2	467856	320013504	26.1533937	8.8108681	.001461988
685	2151.99	368528.5	469225	321419125	26.1725047	8.8151598	.001459854

No.	Circum.	Area	Square	Cube	Sq. Root	Cube Root	Reciprocal
686	2155.13	369605.2	470596	322828856	26.1916017	8.8194474	.001457726
687	2158.27	370683.6	471969	324242703	26.2106848	8.8237307	.001455604
688	2161.42	371763.5	473344	325660672	26.2297541	8.8280099	.001453488
689	2164.56	372845.0	474721	327082769	26.2488095	8.8322850	.001451379
690	2167.70	373928.1	476100	328509000	26.2678511	8.8365559	.001449275
691	2170.84	375012.7	477481	329939371	26.2868789	8.8408227	.001447178
692	2173.98	376098.9	478864	331373888	26.3058929	8.8450854	.001445087
693	2177.12	377186.7	480249	332812557	26.3248932	8.8493440	.001443001
694	2180.27	378276.0	481636	334255384	26.3438797	8.8535985	.001440922
695	2183.41	379367.0	483025	335702375	26.3628527	8.8578489	.001438849
696	2186.55	380459.4	484416	337153536	26.3818119	8.8620952	.001436782
697	2189.69	381553.5	485809	338608873	26.4007576	8.8663375	.001434720
698	2192.83	382649.1	487204	340068392	26.4196896	8.8705757	.001432665
699	2195.97	383746.3	488601	341532099	26.4386081	8.8748090	.001430615
700	2199.11	384845.1	490000	343000000	26.4575131	8.8790400	.001428571
701	2202.26	385945.4	491401	344472101	26.4764046	8.8832661	.001426534
702	2205.40	387047.4	492804	345948408	26.4952826	8.8874882	.001424501
703	2208.54	388150.8	494209	347428927	26.5141472	8.8917063	.001422475
704	2211.68	389255.9	495616	348913664	26.5329983	8.8959204	.001420455
705	2214.82	390362.5	497025	350402625	26.5518361	8.9001304	.001418440
706	2217.96	391470.7	498436	351895816	26.5706605	8.9043366	.001416431
707	2221.11	392580.5	499849	353393243	26.5894716	8.9085387	.001414427
708	2224.25	393691.8	501264	354894912	26.6082694	8.9127369	.001412429
709	2227.39	394804.7	502681	356400829	26.6270539	8.9169311	.001410437
710	2230.53	395919.2	504100	357911000	26.6458252	8.9211214	.001408451
711	2233.67	397035.3	505521	359425431	26.6645833	8.9253078	.001406470
712	2236.81	398152.9	506944	360944128	26.6833281	8.9294902	.001404494
713	2239.96	399272.1	508369	362467097	26.7020598	8.9336687	.001402525
714	2243.10	400392.8	509796	363994344	26.7207784	8.9378433	.001400560
715	2246.24	401515.2	511225	365525875	26.7394839	8.9420140	.001398601
716	2249.38	402639.1	512656	367061696	26.7581763	8.9461809	.001396648
717	2252.52	403764.6	514089	368601813	26.7768557	8.9503438	.001394700
718	2255.66	404891.6	515524	370146232	26.7955220	8.9545029	.001392758
719	2258.81	406020.2	516961	371694959	26.8141754	8.9586581	.001390821
720	2261.95	407150.4	518400	373248000	26.8328157	8.9628095	.001388889
721	2265.09	408282.2	519841	374805361	26.8514432	8.9669570	.001386963
722	2268.23	409415.5	521284	376367048	26.8700577	8.9711007	.001385042
723	2271.37	410550.4	522729	377933067	26.8886593	8.9752400	.001383126
724	2274.51	411686.9	524176	379503424	26.9072481	8.9793766	.001381215
725	2277.65	412824.9	525625	381078125	26.9258240	8.9835089	.001379310
726	2280.80	413964.5	527076	382657176	26.9443872	8.9876373	.001377410
727	2283.94	415105.7	528529	384240583	26.9629375	8.9917620	.001375516
728	2287.08	416248.5	529984	385828352	26.9814751	8.9958829	.001373626
729	2290.22	417392.8	531441	387420489	27.0000000	9.0000000	.001371742
730	2293.36	418538.7	532900	389017000	27.0185122	9.0041134	.001369863
731	2296.50	419686.2	534361	390617891	27.0370117	9.0082229	.001367989
732	2299.65	420835.2	535824	392223168	27.0554985	9.0123288	.001366120
733	2302.79	421985.8	537289	393832837	27.0739727	9.0164309	.001364256
734	2305.93	423138.0	538756	395446904	27.0924344	9.0205293	.001362398
735	2309.07	424291.7	540225	397065375	27.1108834	9.0246239	.001360544
736	2312.21	425447.0	541696	398688256	27.1293199	9.0287149	.001358696
737	2315.35	426603.9	543169	400315553	27.1477439	9.0328021	.001356852
738	2318.50	427762.4	544644	401947272	27.1661554	9.0368857	.001355014
739	2321.64	428922.4	546121	403583419	27.1845544	9.0409655	.001353180
740	2324.78	430084.0	547600	405224000	27.2029410	9.0450419	.001351351
741	2327.92	431247.2	549081	406869021	27.2213152	9.0491142	.001349528
742	2331.06	432412.0	550564	408518488	27.2396769	9.0531831	.001347709
743	2334.20	433578.3	552049	410172407	27.2580263	9.0572482	.001345895
744	2337.34	434746.2	553536	411830784	27.2763634	9.0613098	.001344086
745	2340.49	435915.6	555025	413493625	27.2946881	9.0653677	.001342252
746	2343.63	437086.6	556516	415160936	27.3130006	9.0694220	.001340483
747	2346.77	438259.2	558009	416832723	27.3313007	9.0734726	.001338668
748	2349.91	439433.4	559504	418508992	27.3495887	9.0775197	.001336898
749	2353.05	440609.2	561001	420189749	27.3678644	9.0815631	.001335113
750	2356.19	441786.5	562500	421875000	27.3861279	9.0856030	.001333333
751	2359.34	442965.4	564001	423564751	27.4043792	9.0896392	.001331558
752	2362.48	444145.8	565504	425259008	27.4226184	9.0936719	.001329787
753	2365.62	445327.8	567009	426957777	27.4408455	9.0977010	.001328021
754	2368.76	446511.4	568516	428661064	27.4590604	9.1017265	.001326260
755	2371.90	447696.6	570025	430368875	27.4772633	9.1057485	.001324503
756	2375.04	448883.3	571536	432081216	27.4954542	9.1097669	.001322751
757	2378.19	450071.6	573049	433798093	27.5136330	9.1137818	.001321004
758	2381.33	451261.5	574564	435519512	27.5317998	9.1177931	.001319261
759	2384.47	452453.0	576081	437245479	27.5499546	9.1218010	.001317523
760	2387.61	453646.0	577600	438976000	27.5680975	9.1258053	.001315789

PORTLAND CEMENT TECHNOLOGY

No.	Circum.	Area	Square	Cube	Sq. Root	Cube Root	Reciprocal
761	2390.75	454840.6	579121	440711081	27.5862284	9.1295061	.001314060
762	2393.89	456036.7	580644	442450728	27.6043475	9.1338034	.001312336
763	2397.04	457234.5	582169	444194947	27.6224546	9.1377971	.001310616
764	2400.18	458433.8	583696	445943744	27.6405499	9.1417874	.001308901
765	2403.32	459634.6	585225	447697125	27.6586334	9.1457742	.001307190
766	2406.46	460837.1	586756	449455096	27.6767050	9.1497576	.001305483
767	2409.60	462041.1	588289	451217663	27.6947648	9.1537375	.001303781
768	2412.74	463246.7	589824	452984832	27.7128129	9.1577139	.001302083
769	2415.88	464453.8	591361	454756609	27.7308492	9.1616869	.001300390
770	2419.03	465662.6	592900	456533000	27.7488739	9.1656565	.001298701
771	2422.17	466872.9	594441	458314011	27.7668868	9.1696225	.001297017
772	2425.31	468084.7	595984	460099648	27.7848880	9.1735852	.001295337
773	2428.45	469298.2	597529	461889917	27.8028775	9.1775445	.001293661
774	2431.59	470513.2	599076	463684824	27.8208555	9.1815003	.001291990
775	2434.73	471729.8	600625	465484375	27.8388218	9.1854527	.001290323
776	2437.88	472947.9	602176	467288576	27.8567766	9.1894018	.001288660
777	2241.02	474167.7	603729	469097433	27.8747197	9.1933474	.001287001
778	2444.16	475388.9	605284	470910952	27.8926514	9.1972897	.001285347
779	2447.30	476611.8	606841	472729139	27.9105715	9.2012286	.001283697
780	2450.44	477836.2	608400	474552000	27.9284801	9.2051641	.001282051
781	2453.58	479062.3	609961	476379541	27.9463772	9.2090962	.001280410
782	2456.73	480289.8	611524	478211768	27.9642629	9.2130250	.001278772
783	2459.87	481519.0	613089	480048687	27.9821372	9.2169505	.001277139
784	2463.01	482749.7	614656	481890304	28.0000000	9.2208726	.001275510
785	2466.15	483982.0	616225	483736625	28.0178515	9.2247914	.001273885
786	2469.29	485215.8	617796	485587656	28.0356915	9.2287068	.001272265
787	2472.43	486451.3	619369	487443403	28.0535203	9.2326189	.001270648
788	2475.58	487688.3	620944	489303872	28.0713377	9.2365277	.001269036
789	2478.72	488926.9	622521	491169069	28.0891438	9.2404333	.001267427
790	2481.86	490167.0	624100	493039000	28.1069386	9.2443355	.001265823
791	2485.00	491408.7	625681	494913671	28.1247222	9.2482344	.001264223
792	2488.14	492652.0	627264	496793088	28.1424946	9.2521300	.001262626
793	2491.28	493896.9	628849	498677257	28.1602557	9.2560224	.001261034
794	2494.42	495143.3	630436	500566184	28.1780056	9.2599114	.001259446
795	2497.57	496391.3	632025	502459875	28.1957444	9.2637973	.001257862
796	2500.71	497640.8	633616	504358336	28.2134720	9.2676798	.001256281
797	2503.85	498892.0	635209	506261573	28.2311884	9.2715592	.001254705
798	2506.99	500144.7	636804	508169592	28.2488938	9.2754352	.001253133
799	2510.13	501399.0	638401	510082399	28.2665881	9.2793081	.001251564
800	2513.27	502654.8	640000	512000000	28.2842712	9.2831777	.001250000
801	2516.42	503912.3	641601	513922401	28.3019434	9.2870440	.001248439
802	2519.56	505171.2	643204	515849608	28.3196045	9.2909072	.001246883
803	2522.70	506431.8	644809	517781627	28.3372546	9.2947671	.001245330
804	2525.84	507693.9	646416	519718464	28.3548938	9.2986239	.001243781
805	2528.98	508957.6	648025	521660125	28.3725219	9.3024775	.001242236
806	2532.12	510222.9	649636	523606616	28.3901391	9.3063278	.001240695
807	2535.27	511489.8	651249	525557943	28.4077454	9.3101750	.001239157
808	2538.41	512758.2	652864	527514112	28.4253408	9.3140190	.001237624
809	2541.55	514028.2	654481	529475129	28.4420253	9.3178599	.001236094
810	2544.69	515299.7	656100	531441000	28.4604989	9.3216975	.001234568
811	2547.83	516572.9	657721	533411731	28.4780617	9.3255320	.001233046
812	2550.97	517847.6	659344	535387328	28.4956137	9.3293634	.001231527
813	2554.11	519123.8	660969	537367797	28.5131549	9.3331916	.001230012
814	2557.26	520401.7	662596	539353144	28.5306852	9.3370167	.001228501
815	2560.40	521681.1	664225	541343375	28.5482048	9.3408386	.001226994
816	2563.54	522962.1	665856	543338496	28.5657137	9.3446575	.001225490
817	2566.68	524244.6	667489	545338513	28.5832119	9.3484731	.001223990
818	2569.82	525528.8	669124	547343432	28.6006993	9.3522857	.001222494
819	2572.96	526814.5	670761	549353259	28.6181760	9.3560952	.001221001
820	2576.11	528101.7	672400	551368000	28.6356421	9.3599016	.001219512
821	2579.25	529390.6	674041	553387661	28.6530976	9.3637049	.001218027
822	2582.39	530681.0	675684	555412248	28.6705424	9.3675051	.001216545
823	2585.53	531973.0	677329	557441767	28.6879766	9.3713022	.001215067
824	2588.67	533266.5	678976	559476224	28.7054002	9.3750963	.001213592
825	2591.81	534561.6	680625	561515625	28.7228132	9.3788873	.001212121
826	2594.96	535858.3	682276	563559976	28.7402157	9.3826752	.001210654
827	2598.10	537156.6	683929	565609283	28.7576077	9.3864600	.001209190
828	2601.24	538456.4	685584	567663552	28.7749891	9.3902419	.001207729
829	2604.38	539757.8	687241	569722789	28.7923601	9.3940206	.001206273
830	2607.52	541060.8	688900	571787000	28.8097206	9.3977964	.001204819
831	2610.66	542365.3	690561	573856191	28.8270706	9.4015691	.001203369
832	2613.81	543671.5	692224	575930368	28.8444102	9.4053387	.001201923
833	2616.95	544979.2	693889	578009537	28.8617394	9.4091054	.001200480
834	2620.09	546288.4	695556	580093704	28.8790582	9.4128690	.001199041
835	2623.23	547599.2	697225	582182875	28.8963666	9.4166297	.001197605

No.	Circum.	Area	Square	Cube	Sq. Root	Cube Root	Reciprocal
836	2626 37	548911 6	698896	584277056	28 9136646	9 4203873	001196172
837	2629 51	550225.6	700569	586376253	28.9309523	9 4241420	001194743
838	2632 65	551541.2	702244	588480472	28.9482297	9 4278936	.001193317
839	2635 80	552858.3	703921	590589719	28 9654967	9 4316423	.001191895
840	2638 94	554176 9	705600	592704000	28.9827535	9 4353880	.001190476
841	2642 08	555497.2	707281	594823321	29 0000000	9 4391307	.001189061
842	2645.22	556819.0	708964	596947688	29.0172363	9.4428704	.001187648
843	2648.36	558142.4	710649	599077107	29 0344623	9 4466072	.001186240
844	2651 50	559467.4	712336	601211584	29 0516781	9.4503410	.001184834
845	2654 65	560793 9	714025	603351125	29.0688837	9 4540719	.001183432
846	2657.79	562122 0	715716	605495736	29.0860791	9 4577909	.001182033
847	2660 93	563451 7	717409	607645423	29 1032644	9 4615249	.001180638
848	2664 07	564783 0	719104	609800192	29 1204396	9 4652470	.001179245
849	2667.21	566115 8	720801	611960049	29 1376046	9 4689661	.001177856
850	2670.35	567450 2	722500	614125000	29 1547595	9 4726824	.001176471
851	2673.50	568786 1	724201	616295051	29 1719043	9.4763957	.001175088
852	2676 64	570123 7	725904	618470208	29.1890390	9 4801061	.001173709
853	2679 78	571462 8	727609	620650477	29 2061637	9.4838136	.001172333
854	2682.92	572803 5	729316	622835864	29 2232784	9 4875182	.001170960
855	2686 06	574145 7	731025	625026375	29.2403830	9 4912200	.001169591
856	2689.20	575489 5	732786	627222016	29.2574777	9 4949188	.001168224
857	2692.34	576834.9	734449	629422793	29.2745623	9.4986147	.001166861
858	2695 49	578181 9	736164	631628712	29.2916370	9.5023078	.001165501
859	2698 63	579530 4	737881	633839779	29.3087018	9 5059980	.001164144
860	2701 77	580880.5	739600	636056000	29 3257566	9.5096854	.001162791
861	2704.91	582232.2	741321	638277381	29.3428015	9.5133699	.001161440
862	2708.05	583585 4	743044	640503928	29.3598365	9.5170515	.001160093
863	2711.19	584940.2	744769	642735647	29.3768616	9.5207303	.001158749
864	2714.34	586296.6	746496	644972544	29.3938769	9 5244063	.001157407
865	2717.48	587654.5	748225	647214625	29.4108823	9.5280794	.001156069
866	2720.62	589014.1	749956	649461896	29.4278779	9.5317497	.001154734
867	2723.76	590375.2	751689	651714363	29.4448637	9.5354172	.001153403
868	2726 90	591737.8	753424	653972032	29.4618397	9.5390818	.001152074
869	2730.04	593102.1	755161	656234909	29.4788059	9.5427437	.001150748
870	2733.19	594467 9	756900	658503000	29.4957624	9.5464027	.001149425
871	2736.33	595835.3	758641	660776311	29.5127091	9.5500589	.001148106
872	2739.47	597204.2	760384	663054848	29.5296461	9.5537123	.001146789
873	2742.61	598574.7	762129	665338617	29.5465734	9.5573630	.001145475
874	2745.75	599946.8	763876	667627624	29.5634910	9.5610108	.001144165
875	2748.89	601320.5	765625	669921875	29.5803989	9.5646559	.001142857
876	2752.04	602695.7	767376	672221376	29.5972972	9.5682982	.001141553
877	2755.18	604072.5	769129	674526133	29.6141858	9.5719377	.001140251
878	2758.32	605450.9	770884	676836152	29.6310648	9.5755745	.001138952
879	2761.46	606830 8	772641	679151439	29.6479342	9.5792085	.001137656
880	2764.60	608212.3	774400	681472000	29.6647939	9.5828397	.001136364
881	2767 74	609595.4	776161	683797841	29.6816442	9.5864682	.001135074
882	2770.88	610980.1	777924	686128968	29.6984848	9.5900939	.001133787
883	2774.03	612366.3	779689	688465387	29.7153159	9.5937169	.001132503
884	2777.17	613754.1	781456	690807104	29.7321375	9.5973373	.001131222
885	2780.31	615143.5	783225	693154125	29.7489496	9.6009548	.001129944
886	2783.45	616534.4	784996	695506456	29.7657521	9.6045696	.001128668
887	2786.59	617926.9	786769	697864103	29.7825452	9.6081817	.001127396
888	2789.73	619321.0	788544	700227072	29.7993289	9.6117911	.001126126
889	2792.88	620716.7	790321	702595369	29.8161030	9.6153977	.001124859
890	2796.02	622113.9	792100	704969000	29.8328678	9.6190017	.001123596
891	2799.16	623512.7	793881	707347971	29.8496231	9.6226030	.001122334
892	2802.30	624913.0	795664	709732288	29 8663690	9.6262016	.001121076
893	2805.44	626315 0	797449	712121957	29.8831056	9.6297975	.001119821
894	2808.58	627718.5	799236	714516984	29.8998328	9.6333907	.001118568
895	2811.73	629123.6	801025	716917375	29.9165506	9.6369812	.001117318
896	2814.87	630530.2	802816	719323136	29.9332591	9.6405690	.001116071
897	2818.01	631938.4	804609	721734273	29.9499583	9.6441542	.001114827
898	2821.15	633348.2	806404	724150792	29.9666481	9.6477367	.001113586
899	2824.29	634759.6	808201	726572699	29.9833287	9.6513166	.001112347
900	2827 43	636172.5	810000	729000000	30.0000000	9.6548938	.001111111
901	2830.58	637587 0	811801	731432701	30.0166620	9.6584684	.001109878
902	2833 72	639003 1	813604	733870808	30.0333148	9.6620403	001108647
903	2836.86	640420.7	815409	736314327	30.0499584	9.6656096	.001107420
904	2840.00	641840.0	817216	738763264	30.0665928	9.6691762	.001106195
905	2843.14	643260.7	819025	741217625	30.0832179	9.6727403	.001104972
906	2846.28	644693.1	820836	743677416	30.0998339	9.6763017	.001103753
907	2849.42	646107.0	822649	746142643	30.1164407	9.6798604	.001102536
908	2852.57	647532.5	824464	748613312	30.1330383	9.6834166	.001101322
909	2855.71	648959.6	826281	751089429	30.1496269	9.6869701	.001100110
910	2858.85	650388.2	828100	753571000	30.1662063	9.6905211	.001098901

No.	Circum.	Area	Square	Cube	Sq. Root	Cube Root	Reciprocal
911	2861.99	651818 4	829921	756058031	30.1827765	9.6940694	.001097695
912	2865.13	653250.2	831744	758550528	30.1993377	9.6976151	001096491
913	2868.27	654683.6	833569	761048497	30 2158899	9.7011583	.001095290
914	2871.42	656118.5	835396	763551944	30.2324329	9.7046989	.001094092
915	2874.56	657555.0	837225	766060875	30.2489669	9.7082369	.001092896
916	2877.70	658993 0	839056	768575296	30.2654919	9.7117723	.001091703
917	2880.84	660432.7	840889	771095213	30 2820079	9.7153051	001090513
918	2883.98	661873.9	842724	773620632	30.2985148	9.7188354	.001089325
919	2887 12	663316.7	844561	776151559	30 3150128	9.7223631	.001088139
920	2890 27	664761.0	846400	778688000	30.3315018	9 7258883	.001086957
921	2893 41	666206.9	848241	781229961	30.3479818	9.7294109	001085776
922	2896.55	667654.4	850084	783777448	30.3644529	9.7329309	001084599
923	2899.69	669103 5	851929	786330467	30.3809151	9 7364484	.001083423
924	2902.83	670554.1	853776	788889024	30.3973683	9 7399634	.001082251
925	2905.97	672006.3	855625	791453125	30.4138127	9.7434758	.001081081
926	2909.11	673460.1	857476	794022776	30 4302481	9.7469857	.001079914
927	2912.26	674915.4	859329	796597983	30.4466747	9.7504930	.001078749
928	2915.40	676372.3	861184	799178752	30.4630924	9.7539979	.001077586
929	2918.54	677830.8	863041	801765089	30.4795013	9.7575002	.001076426
930	2921.68	679290.9	864900	804357000	30.4959014	9.7610001	.001075269
931	2924.82	680752.5	866761	806954491	30.5122926	9.7644974	.001074114
932	2927.96	682215.7	868624	809557568	30 5286750	9.7679922	.001072961
933	2931.11	683680.5	870489	812166237	30.5450487	9.7714845	.001071811
934	2934.25	685146.8	872356	814780504	30.5614136	9.7749743	.001070664
935	2937.39	686614.7	874225	817400375	30.5777697	9.7784616	.001069519
936	2940.53	688084.2	876096	820025856	30.5941171	9.7819466	.001068376
937	2943.67	689555.2	877969	822656953	30.6104557	9.7854288	.001067236
938	2946.81	691027.9	879844	825293672	30.6267857	9.7889087	.001066098
939	2949.96	692502.1	881721	827936019	30.6431069	9.7923861	.001064963
940	2953.10	693977.8	883600	830584000	30.6594194	9.7958611	.001063830
941	2956.24	695455.2	885481	833237621	30.6757233	9.7993336	.001062699
942	2959.38	696934.1	887364	835896888	30.6920185	9.8028036	.001061571
943	2962.52	698414.5	889249	838561807	30.7083051	9.8062711	.001060445
944	2965.66	699896.6	891136	841232384	30.7245830	9.8097362	.001059322
945	2968.81	701380.2	893025	843908625	30.7408523	9.8131989	.001058201
946	2971.95	702865.4	894916	846590536	30.7571130	9.8166591	.001057082
947	2975.09	704352.1	896809	849278123	30.7733651	9.8201169	.001055966
948	2978.23	705840.5	898704	851971392	30.7896086	9.8235723	.001054852
949	2981.37	707330.4	900601	854670349	30.8058436	9.8270252	.001053741
950	2984.51	708821.8	902500	857375000	30 8220700	9 8304757	.001052632
951	2987.65	710314.9	904401	860085351	30.8382879	9.8339238	.001051525
952	2990.80	711809.5	906304	862801408	30.8544972	9.8373695	.001050420
953	2993.94	713305.7	908209	865523177	30.8706981	9.8408127	.001049318
954	2997.08	714803.4	910116	868250664	30.8868904	9.8442536	.001048218
955	3000.22	716302.8	912025	870983875	30.9030743	9.8476920	.001047120
956	3003.36	717803.7	913936	873722816	30.9192497	9.8511280	.001046025
957	3006.50	719306.1	915849	876467493	30.9354166	9.8545617	.001044932
958	3009.65	720810.2	917764	879217912	30.9515751	9.8579929	.001043841
959	3012.79	722315.8	919681	881974079	30.9677251	9.8614218	.001042753
960	3015.93	723823.0	921600	884736000	30.9838668	9.8648483	.001041667
961	3019.07	725331.7	923521	887503681	31.0000000	9.8682724	.001040583
962	3022.21	726842.0	925444	890277128	31.0161248	9.8716941	.001039501
963	3025.35	728353.9	927369	893056347	31.0322413	9.8751135	.001038422
964	3028.50	729867.4	929296	895841344	31.0483494	9.8785305	.001037344
965	3031.64	731382.4	931225	898632125	31.0644491	9.8819451	.001036269
966	3034.78	732899.0	933156	901428696	31.0805405	9.8853574	.001035197
967	3037.92	734417.2	935089	904231063	31.0966236	9.8887673	.001034126
968	3041.06	735936.9	937024	907039232	31.1126984	9.8921749	.001033058
969	3044.20	737458.2	938961	909853209	31.1287648	9.8955801	.001031992
970	3047.34	738981.1	940900	912673000	31.1448230	9.8989830	.001030928
971	3050.49	740505.6	942841	915498611	31.1608729	9.9023835	.001029866
972	3053.63	742031.6	944784	918330048	31.1769145	9.9057817	.001028807
973	3056.77	743559.2	946729	921167317	31.1929479	9.9091776	.001027749
974	3059.91	745088.4	948676	924010424	31.2089731	9.9125712	.001026694
975	3063.05	746619.1	950625	926859375	31.2249900	9.9159624	.001025641
976	3066.19	748151.4	952576	929714176	31.2409987	9.9193513	.001024590
977	3069.34	749685.3	954529	932574833	31.2569992	9.9227379	.001023541
978	3072.48	751220.8	956484	935441352	31.2729915	9.9261222	.001022495
979	3075.62	752757.8	958441	938313739	31.2889757	9.9295042	.001021450
980	3078.76	754296.4	960400	941192000	31.3049517	9.9328839	.001020408
981	3081.90	755836.6	962361	944076141	31.3209195	9.9362613	.001019368
982	3085.04	757378.3	964324	946966168	31.3368792	9.9396363	.001018330
983	3088.19	758921.6	966289	949862087	31.3528308	9.9430092	.001017294
984	3091.33	760466.5	968256	952763904	31.3687743	9.9463797	.001016260
985	3094.47	762012.9	970225	955671625	31.3847097	9 9497479	001015228

No.	Circum.	Area	Square	Cube	Sq. Root	Cube Root	Reciprocal
986	3097.61	763561.0	972196	958585256	31.4006369	9.9531138	.001014199
987	3100.75	765110.5	974169	961504803	31.4165561	9.9564775	.001013171
988	3103.89	766661.7	976144	964430272	31.4324673	9.9598389	.001012146
989	3107.04	768214.4	978121	967361669	31.4483704	9.9631981	.001011122
990	3110.18	769768.7	980100	970299000	31.4642654	9.9665549	.001010101
991	3113.32	771324.6	982081	973242271	31.4801525	9.9699095	.001009082
992	3116.46	772882.1	984064	976191488	31.4960315	9.9732619	.001008065
993	3119.60	774441.1	986049	979146657	31.5119025	9.9766120	.001007049
994	3122.74	776001.7	988036	982107784	31.5277655	9.9799599	.001006036
995	3125.88	777563.8	990025	985074875	31.5436206	9.9833055	.001005025
996	3129.03	779127.5	992016	988047936	31.5594677	9.9866488	.001004016
997	3132.17	780692.8	994009	991026973	31.5753068	9.9899900	.001003009
998	3135.31	782259.7	996004	994011992	31.5911380	9.9933289	.001002004
999	3138.45	783828.2	998001	997002999	31.6069613	9.9966656	.001001001
1000	3141.59	785398.2	1000000	1000000000	31.6227766	10.0000000	.001000000
1001	3144.73	786968.7	1002001	1003003001	31.6385840	10.0033322	.0009990010
1002	3147.88	788543.9	1004004	1006012008	31.6543836	10.0066622	.0009980040
1003	3151.02	790118.3	1006009	1009027027	31.6701752	10.0099899	.0009970090
1004	3154.16	791694.2	1008016	1012048064	31.6859590	10.0133155	.0009960159
1005	3157.30	793271.6	1010025	1015075125	31.7017349	10.0166389	.0009950249
1006	3160.44	794850.7	1012036	1018108216	31.7175030	10.0199601	.0009940358
1007	3163.57	796428.8	1014049	1021147343	31.7332633	10.0232791	.0009930487
1008	3166.73	798016.0	1016064	1024192512	31.7490157	10.0265958	.0009920635
1009	3169.87	799584.7	1018081	1027243729	31.7647603	10.0299104	.0009910803
1010	3173.01	801185.0	1020100	1030301000	31.7804972	10.0332228	.0009900990
1011	3176.15	802771.9	1022121	1033304331	31.7962262	10.0365330	.0009891197
1012	3179.29	804360.4	1024144	1036433728	31.8119474	10.0398410	.0009881423
1013	3182.43	805950.4	1026169	1039509197	31.8276609	10.0431469	.0009871668
1014	3185.57	807542.0	1028196	1042590744	31.8433666	10.0464506	.0009861933
1015	3188.72	809137.7	1030225	1045678375	31.8590646	10.0497521	.0009852217
1016	3191.86	810732.4	1032256	1048772096	31.8747549	10.0530514	.0009842520
1017	3195.00	812328.8	1034289	1051871913	31.8904374	10.0563485	.0009832842
1018	3198.14	813929.6	1036324	1054977832	31.9061123	10.0596435	.0009823183
1019	3201.29	815526.1	1038361	1058089859	31.9217704	10.0629364	.0009813543
1020	3204.42	817128.2	1040400	1061208000	31.9374388	10.0662271	.0009803922
1021	3207.56	818731.2	1042441	1064332261	31.9530906	10.0695156	.0009794319
1022	3210.70	820335.8	1044484	1067462648	31.9687347	10.0728020	.0009784736
1023	3213.85	821942.0	1046529	1070509167	31.9843712	10.0760863	.0009775171
1024	3216.99	823549.7	1048576	1073741824	32.0000000	10.0793684	.0009765625
1025	3220.13	825158.9	1050625	1076890625	32.0156212	10.0826484	.0009756098
1026	3223.27	826769.8	1052676	1080045576	32.0312348	10.0859262	.0009746589
1027	3226.41	828382.2	1054729	1083206683	32.0468407	10.0892019	.0009737098
1028	3229.55	829996.2	1056784	1086373952	32.0624391	10.0924755	.0009727626
1029	3232.70	831611.8	1058841	1089547389	32.0780298	10.0957469	.0009718173
1030	3235.84	833228.9	1060900	1092727000	32.0936131	10.0990163	.0009708738
1031	3238.98	834847.6	1062961	1095912791	32.1091887	10.1022835	.0009699321
1032	3242.12	836467.9	1065024	1099104768	32.1247568	10.1055487	.0009689922
1033	3245.26	838089.7	1067089	1102302937	32.1403173	10.1088117	.0009680542
1034	3248.40	839713.2	1069156	1105507304	32.1558704	10.1120726	.0009671180
1035	3251.55	841338.1	1071225	1108717875	32.1714159	10.1153314	.0009661836
1036	3254.69	842964.7	1073296	1111934656	32.1869539	10.1185882	.0009652510
1037	3257.83	844592.8	1075369	1115157653	32.2024844	10.1218425	.0009643202
1038	3260.97	846222.5	1077444	1118386872	32.2180074	10.1250953	.0009633911
1039	3264.11	847853.8	1079521	1121622319	32.2335229	10.1283457	.0009624639
1040	3267.25	849486.7	1081600	1124864000	32.2490310	10.1315941	.0009615385
1041	3270.40	851121.1	1083681	1128111921	32.2645316	10.1348403	.0009606148
1042	3273.54	852757.1	1085764	1131366088	32.2800248	10.1380845	.0009596929
1043	3276.68	854394.6	1087849	1134626507	32.2955105	10.1413266	.0009587728
1044	3279.82	856033.7	1089936	1137893184	32.3109888	10.1445667	.0009578544
1045	3282.96	857674.4	1092025	1141166125	32.3264598	10.1478047	.0009569378
1046	3286.10	859316.7	1094116	1144445336	32.3419233	10.1510406	.0009560229
1047	3289.25	860960.5	1096209	1147730823	32.3573794	10.1542744	.0009551098
1048	3292.39	862605.9	1098304	1151022592	32.3728281	10.1575062	.0009541985
1049	3295.53	864252.9	1100401	1154320649	32.3882695	10.1607359	.0009532888
1050	3298.67	865901.5	1102500	1157625000	32.4037035	10.1639636	.0009523810
1051	3301.81	867551.6	1104601	1160935651	32.4191301	10.1671893	.0009514748
1052	3304.95	869203.3	1106704	1164252608	32.4345495	10.1704129	.0009505703
1053	3308.10	870856.6	1108809	1167575877	32.4499615	10.1736344	.0009496676
1054	3311.24	872411.4	1110916	1170905464	32.4653662	10.1768539	.0009487666
1055	3314.38	874167.8	1113025	1174241375	32.4807635	10.1800714	.0009478673
1056	3317.52	875825.8	1115136	1177583616	32.4961536	10.1832868	.0009469697
1057	3320.66	877485.3	1117249	1180932193	32.5115364	10.1865002	.0009460738
1058	3323.80	879146.4	1119364	1184287112	32.5269119	10.1897116	.0009451796
1059	3326.94	880809.1	1121481	1187648379	32.5422802	10.1929209	.0009442871
1060	3330.09	882473.4	1123600	1191016000	32.5576412	10.1961283	.0009433962

Index*

---

* Nearly every topic in portland cement technology has several elements. The most important of these elements are material, process, and equipment. For example, limestone is crushed in gyratory crushers; coal is ground in unit mills; and so on. If each topic were listed under each element, the index, covering many pages, would be unnecessarily long. In this index, preference is given to material, rather than to process or equipment.

www.ingramcontent.com/pod-product-compliance
Lightning Source LLC
Chambersburg PA
CBHW052011230326
41598CB00078B/2507